Differential Models, Numerical Simulations and Applications

Differential Models, Numerical Simulations and Applications

Editor

Gabriella Bretti

MDPI • Basel • Beijing • Wuhan • Barcelona • Belgrade • Manchester • Tokyo • Cluj • Tianjin

Editor
Gabriella Bretti
Istituto per le Applicazioni del
Calcolo "M. Picone" Consiglio
Nazionale delle Ricerche
Italy

Editorial Office
MDPI
St. Alban-Anlage 66
4052 Basel, Switzerland

This is a reprint of articles from the Special Issue published online in the open access journal *Axioms* (ISSN 2075-1680) (available at: https://www.mdpi.com/journal/axioms/special_issues/differential_models_numerical_simulations).

For citation purposes, cite each article independently as indicated on the article page online and as indicated below:

LastName, A.A.; LastName, B.B.; LastName, C.C. Article Title. *Journal Name* **Year**, *Volume Number*, Page Range.

ISBN 978-3-0365-2299-9 (Hbk)
ISBN 978-3-0365-2300-2 (PDF)

© 2021 by the authors. Articles in this book are Open Access and distributed under the Creative Commons Attribution (CC BY) license, which allows users to download, copy and build upon published articles, as long as the author and publisher are properly credited, which ensures maximum dissemination and a wider impact of our publications.

The book as a whole is distributed by MDPI under the terms and conditions of the Creative Commons license CC BY-NC-ND.

Contents

About the Editor vii

Gabriella Bretti
Differential Models, Numerical Simulations and Applications
Reprinted from: *Axioms* **2021**, *10*, 260, doi:10.3390/axioms10040260 1

Gabriella Bretti, Adele De Ninno, Roberto Natalini, Daniele Peri and Nicole Roselli
Estimation Algorithm for a Hybrid PDE–ODE Model Inspired by Immunocompetent
Cancer-on-Chip Experiment
Reprinted from: *Axioms* **2021**, *10*, 243, doi:10.3390/axioms10040243 5

Benoît Fabrèges, Frédéric Lagoutière, Sébastien Tran Tien and Nicolas Vauchelet
Relaxation Limit of the Aggregation Equation with Pointy Potential
Reprinted from: *Axioms* **2021**, *10*, 108, doi:10.3390/axioms10020108 35

Maya Briani, Emiliano Cristiani and Paolo Ranut
Macroscopic and Multi-Scale Models for Multi-Class Vehicular Dynamics with Uneven Space
Occupancy: A Case Study
Reprinted from: *Axioms* **2021**, *10*, 102, doi:10.3390/axioms10020102 57

Ankush Aggarwal, Damiano Lombardi and Sanjay Pant
An Information-Theoretic Framework for Optimal Design: Analysis of Protocols for Estimating
Soft Tissue Parameters in Biaxial Experiments
Reprinted from: *Axioms* **2021**, *10*, 79, doi:10.3390/axioms10020079 79

Maria Francesca Carfora and Isabella Torcicollo
A Fractional-in-Time Prey–Predator Model with Hunting Cooperation: Qualitative Analysis,
Stability and Numerical Approximations
Reprinted from: *Axioms* **2021**, *10*, 78, doi:10.3390/axioms10020078 95

Gabriella Bretti
A Quadratic Mean Field Games Model for the Langevin Equation
Reprinted from: *Axioms* **2021**, *10*, 68, doi:10.3390/axioms10020068 109

Fasma Diele, Carmela Marangi and Angela Martiradonna
Non-Standard Discrete RothC Models for Soil Carbon Dynamics
Reprinted from: *Axioms* **2021**, *10*, 56, doi:10.3390/axioms10020056 119

Elisabetta Vallarino, Michele Piana, Alberto Sorrentino and Sara Sommariva
The Role of Spectral Complexity in Connectivity Estimation
Reprinted from: *Axioms* **2020**, *10*, 35, doi:10.3390/axioms10010035 141

Marco Scianna and Luigi Preziosi
A Cellular Potts Model for Analyzing Cell Migration across Constraining Pillar Arrays
Reprinted from: *Axioms* **2021**, *10*, 32, doi:10.3390/axioms10010032 157

Eleonora Messina and Antonia Vecchio
Analysis of the Transient Behaviour in the Numerical Solution of Volterra Integral Equations
Reprinted from: *Axioms* **2021**, *10*, 23, doi:10.3390/axioms10010023 179

Maria Laura Delle Monache, Karen Chi, Yong Chen, Paola Goatin, Ke Han, Jing-mei Qiu and Benedetto Piccoli
A Three-Phase Fundamental Diagram from Three-Dimensional Traffic Data
Reprinted from: *Axioms* **2021**, *10*, 17, doi:10.3390/axioms10010017 . . . . **191**

Simone Göttlich, Michael Herty and Gediyon Weldegiyorgis
Input-to-State Stability of a Scalar Conservation Law with Nonlocal Velocity
Reprinted from: *Axioms* **2020**, *10*, 12, doi:10.3390/axioms10010012 . . . . **211**

About the Editor

Gabriella Bretti She graduated in Mathematics and received a Ph.D. in Applied Mathematics from University of Rome "La Sapienza", Italy. She is a permanent researcher at IAC-CNR in Rome. Her research interests are focused on modeling and numerical methods for nonlinear PDEs, flows in heterogeneous media, inverse problems, differential models in mathematical biology, medicine and cultural heritage. In an interdisciplinary framework, her main research interests consist of the study of the available data coming from laboratory experiments or clinical measures in order to extract the most significant underlying features of the observed p henomena. Her expertise consists of the development of mathematical models and related simulation and optimization tools in order to describe and forecast the evolution of a complex systems. Research and Industrial Projects: She has been involved in many national projects and in technology transfer to Italian companies (Autovie Venete, OCTO Telematics) for the development of a predictive software for managing vehicular traffic on the Italian highways. She has been involved in Italian regional projects for the development of mathematical algorithms for the conservation of Cultural Heritage (CH), ADAMO and SISMI DTC Project Latium. Currently, she is involved in the ESA project Pomerium for detecting pollution levels in the air and the effects of degradation on the constituting materials of Colosseum and Pyramid of Cestius. Editorial activity: since September 2021, she has been a member of the Editorial Board of Axioms (MDPI) and a reviewer of many international journals of applied mathematics: https://publons.com/researcher/1306628/gabriella-bretti/.

Conference organization: She organized Minisymposia for IMACS2018, SIMAI2018, SIAMGS21, ICNAAM2021. She was the chair of Workshop Indam MACH2021 and MCHBS 2021 Virtual Workshop. Research production: She has authored more than 40 papers published in international journals in applied mathematics: https://www.researchgate.net/profile/Gabriella-Bretti. She has been invited to give communications in international conferences. https://orcid.org/0000-0001-5293-2115. Her research interests are focused on numerical methods for nonlinear PDEs, differential models in mathematical biology and flow in heterogeneous media with application to chemical damage of monuments for the conservation of cultural heritage. The highlighted topics imply the application of interdisciplinary methodologies and techniques requiring the integration between mathematics, physics and chemistry.

Editorial

Differential Models, Numerical Simulations and Applications

Gabriella Bretti

Istituto per le Applicazioni del Calcolo "M. Picone", Consiglio Nazionale delle Ricerche, via dei Taurini 19, 00185 Rome, Italy; gab.bretti@gmail.com

Abstract: Differential models, numerical methods and computer simulations play a fundamental role in applied sciences. Since most of the differential models inspired by real world applications have no analytical solutions, the development of numerical methods and efficient simulation algorithms play a key role in the computation of the solutions to many relevant problems. Moreover, since the model parameters in mathematical models have interesting scientific interpretations and their values are often unknown, estimation techniques need to be developed for parameter identification against the measured data of observed phenomena. In this respect, this Special Issue collects some important developments in different areas of application.

Keywords: applied mathematics; numerical methods; computational mathematics; differential and integro-differential models; inverse problems

Citation: Bretti, G. Differential Models, Numerical Simulations and Applications. *Axioms* **2021**, *10*, 260. https://doi.org/10.3390/axioms10040260

Received: 12 October 2021
Accepted: 13 October 2021
Published: 19 October 2021

Publisher's Note: MDPI stays neutral with regard to jurisdictional claims in published maps and institutional affiliations.

Copyright: © 2021 by the author. Licensee MDPI, Basel, Switzerland. This article is an open access article distributed under the terms and conditions of the Creative Commons Attribution (CC BY) license (https://creativecommons.org/licenses/by/4.0/).

1. Special Issue Overview

The Special Issue contains 12 contributions covering a fan of methodology and applications that can be summarized as follows:

1. Numerical methods, simulations and control for particles dynamics [1–4].
2. Modeling and numerical methods for traffic [5,6] and manufacturing problems [7].
3. Inverse problems for biomedical applications [8,9].
4. Theoretical study and numerical solutions for integro-differential equations [10–12].

The main results of the papers are described below.

1.1. Numerical Methods, Simulations and Control for Particles Dynamics

In [1] the authors developed a hybrid PDE–ODE mathematical model mimicking the mechanisms observed in cancer-on-chip experiments, where tumor cells are treated with chemotherapy drugs and secrete chemical signals into the environment, attracting multiple immune cell species. The in silico model proposed here goes towards the construction of a "digital twin" of the experimental immune cells and allows the reconstruction of the chemical gradients in the chip environment in order to better understand the complex mechanisms of immunosurveillance. The development of a trustable simulation algorithm, able to reproduce the dynamics observed in the chip, requires an efficient tool for the calibration of the model parameters. In this respect, the present paper represents a first methodological work to test the feasibility and the soundness of the calibration technique here proposed, based on a multidimensional spline interpolation technique for the time-varying velocity field surfaces obtained from cell trajectories.

The authors in [2] studied a relaxation limit of the so-called aggregation equation with a pointy potential in one-dimensional space. The aggregation equation is today widely used to model the dynamics of a density of individuals attracting each other through a potential. When this potential is pointy, solutions are known to blow up in final time. For this reason, measure-valued solutions have been defined. The convergence of this approximation was studied and a rigorous estimate of the speed of convergence in one dimension with the Newtonian potential was obtained; moreover, the numerical discretization of this relaxation limit by uniformly accurate schemes was investigated.

In [3] a Mean Field Games model where the dynamics of the agents is given by a controlled Langevin equation, and the cost is quadratic, was addressed. An appropriate change of variables transforms the Mean Field Games system into a system of two coupled kinetic Fokker–Planck equations and an existence result for the latter system, obtaining consequently a solution for the Mean Field Games system.

In [4] a tailored version of the Cellular Potts model, a grid-based stochastic approach where cell dynamics are established by a Metropolis algorithm for energy minimization, was developed. The proposed model allowed for quantitatively analyzing selected cell migratory determinants (e.g., the cell and nuclear speed and deformation, and forces acting at the nuclear membrane) in the case of different experimental setups. Most of the numerical results show a remarkable agreement with the corresponding empirical data.

1.2. Modeling and Numerical Methods for Traffic and Manifacturing Problems

In [5], two models describing the dynamics of heavy and light vehicles on a road network were introduced, taking into account the interactions between the two classes. Such models are tailored for two-lane highways where heavy vehicles cannot overtake. The first model couples two first-order macroscopic LWR models, while the second model couples a second-order microscopic follow-the-leader model with a first-order macroscopic LWR model. Numerical results show that both models are able to catch some second-order (inertial) phenomena such as stop and go waves. Models are calibrated by means of real data measured by fixed sensors placed along the A4 Italian highway Trieste–Venice and its branches, provided by Autovie Venete.

The authors of [6] use empirical traffic data collected from three locations in Europe and the US to reveal a three-phase fundamental diagram with two phases located in the uncongested regime. Model-based clustering, hypothesis testing and regression analyses are applied to the speed–flow–occupancy relationship represented in the three-dimensional space to rigorously validate the three phases and identify their gaps. Accordingly, a three-phase macroscopic traffic-flow model and a characterization of solutions to the Riemann problems are proposed. In this work, critical structures in the fundamental diagram that are typically ignored in first- and higher-order models are identified, which could significantly impact travel-time estimation on highways.

In [7], the input-to-state stability (ISS) of an equilibrium for a scalar conservation law with nonlocal velocity and measurement error arising in a highly re-entrant manufacturing system was studied. A numerical discretization of the scalar conservation law with non-local velocity and measurement error was introduced and a suitable discrete Lyapunov function was analyzed to provide ISS of a discrete equilibrium for the proposed numerical approximation.

1.3. Inverse Problems for Biomedical Applications

In [8] a new framework for optimal design was developed, by introducing new protocols for estimating soft tissue parameters in biaxial experiments. This framework is based on the information-theoretic measures of mutual information, and conditional mutual information and their combination is proposed. In particular, the information gain about the parameters from the experiment as the key criterion to be maximized is considered and directly used for optimal design. Information gain is computed through k-nearest neighbor algorithms applied to the joint samples of the parameters and measurements produced by the forward and observation models. For biaxial experiments, the results show that low angles have a relatively low information content compared to high angles. The results also show that a smaller number of angles with suitably chosen combinations can result in higher information gains when compared to a larger number of angles which are poorly combined.

The authors of [9] study the problem of functional connectivity by quantifying the statistical dependencies among time series describing the activity of different neural sources from the magnetic field recorded with magnetoencephalographic (MEG) exam. This problem can be addressed by utilizing connectivity measures whose computation in the frequency domain often relies on the evaluation of the cross-power spectrum of the neural time series, estimated by solving the MEG inverse problem. Recent studies have focused on the optimal determination of the cross-power spectrum in the framework of regularization theory for ill-posed inverse problems, providing indications that, rather surprisingly, the regularization process that leads to the optimal estimate of neural activity does not lead to the optimal estimate of the corresponding functional connectivity. Along these lines, the present paper utilizes synthetic time series, simulating the neural activity recorded by a MEG device to show that the regularization of the cross-power spectrum depends on the spectral complexity of the neural activity.

1.4. Theoretical Study and Numerical Solutions for Integro-Differential Equations

In [10] a prey–predator system with logistic growth of prey and hunting cooperation of predators is studied. The introduction of fractional time derivatives and the related persistent memory strongly characterize the model behavior, as many dynamical systems in the applied sciences are well described by such fractional-order models. Mathematical analysis and numerical simulations are performed to highlight the characteristics of the proposed model. The existence, uniqueness and boundedness of solutions is proved; the stability of the coexistence equilibrium and the occurrence of Hopf bifurcation is investigated. Some numerical approximations of the solution are finally considered; the obtained trajectories confirm the theoretical findings.

The work in [11] is devoted to the study of dynamical models, such as the Rothamsted Carbon (RothC) model, used predict the long-term behavior of soil carbon content for the achievement of land degradation neutrality as measured in terms of the Soil Organic Carbon (SOC), a key indicator of land degradation. Indeed, a reduction in the SOC stock of soil results in degradation and it may also have potential negative effects on soil-derived ecosystem services. In this paper, continuous and discrete versions of the RothC model were compared, especially to achieve long-term solutions. The original discrete formulation of the RothC model was then compared with a novel nonstandard integrator that represents an alternative to the exponential Rosenbrock–Euler approach in the literature.

The authors in [12] studied the asymptotic behavior of the numerical solution to the Volterra integral equations. In particular, a technique based on an appropriate splitting of the kernel is introduced, which allows one to obtain vanishing asymptotic (transient) behavior in the numerical solution, consistent with the properties of the analytical solution, without having to operate restrictions on the integration steplength.

Funding: This research received no external funding.

Conflicts of Interest: The author declares no conflict of interest.

References

1. Bretti, G.; De Ninno, A.; Natalini, R.; Peri, D.; Roselli, N. Estimation Algorithm for a Hybrid PDE–ODE Model Inspired by Immunocompetent Cancer-on-Chip Experiment. *Axioms* **2021**, *10*, 243. [CrossRef]
2. Fabrèges, B.; Lagoutière, F.; Tran Tien, S.; Vauchelet, N. Relaxation Limit of the Aggregation Equation with Pointy Potential. *Axioms* **2021**, *10*, 108. [CrossRef]
3. Camilli, F. A Quadratic Mean Field Games Model for the Langevin Equation. *Axioms* **2021**, *10*, 68. [CrossRef]
4. Scianna, M.; Preziosi, L. A Cellular Potts Model for Analyzing Cell Migration across Constraining Pillar Arrays. *Axioms* **2021**, *10*, 32. [CrossRef]
5. Briani, M.; Cristiani, E.; Ranut, P. Macroscopic and Multi-Scale Models for Multi-Class Vehicular Dynamics with Uneven Space Occupancy: A Case Study. *Axioms* **2021**, *10*, 102. [CrossRef]
6. Delle Monache, M.L.; Chi, K.; Chen, Y.; Goatin, P.; Han, K.; Qiu, J.-M.; Piccoli, B. A Three-Phase Fundamental Diagram from Three-Dimensional Traffic Data. *Axioms* **2021**, *10*, 17. [CrossRef]
7. Göttlich, S.; Herty, M.; Weldegiyorgis, G. Input-to-State Stability of a Scalar Conservation Law with Nonlocal Velocity. *Axioms* **2021**, *10*, 12. [CrossRef]
8. Aggarwal, A.; Lombardi, D.; Pant, S. An Information-Theoretic Framework for Optimal Design: Analysis of Protocols for Estimating Soft Tissue Parameters in Biaxial Experiments. *Axioms* **2021**, *10*, 79. [CrossRef]

9. Vallarino, E.; Sorrentino, A.; Piana, M.; Sommariva, S. The Role of Spectral Complexity in Connectivity Estimation. *Axioms* **2021**, *10*, 35. [CrossRef]
10. Carfora, M.; Torcicollo, I. A Fractional-in-Time Prey–Predator Model with Hunting Cooperation: Qualitative Analysis, Stability and Numerical Approximations. *Axioms* **2021**, *10*, 78. [CrossRef]
11. Diele, F.; Marangi, C.; Martiradonna, A. Non-Standard Discrete RothC Models for Soil Carbon Dynamics. *Axioms* **2021**, *10*, 56. [CrossRef]
12. Messina, E.; Vecchio, A. Analysis of the Transient Behaviour in the Numerical Solution of Volterra Integral Equations. *Axioms* **2021**, *10*, 23. [CrossRef]

Article

Estimation Algorithm for a Hybrid PDE–ODE Model Inspired by Immunocompetent Cancer-on-Chip Experiment

Gabriella Bretti [1,*], Adele De Ninno [2], Roberto Natalini [1], Daniele Peri [1] and Nicole Roselli [3]

1. Istituto per le Applicazioni del Calcolo "M.Picone", 00185 Rome, Italy; roberto.natalini@cnr.it (R.N.); d.peri@iac.cnr.it (D.P.)
2. Istituto di Fotonica e Nanotecnologie, 00156 Rome, Italy; adele.deninno@cnr.it
3. Dipartimento di Scienze di Base e Applicate per l'Ingegneria, Sapienza Università di Roma, 00161 Rome, Italy; nicole.roselli@uniroma1.it
* Correspondence: gabriella.bretti@cnr.it

Citation: Bretti, G.; De Ninno, A.; Natalini, R.; Peri, D.; Roselli, N. Estimation Algorithm for a Hybrid PDE–ODE Model Inspired by Immunocompetent Cancer-on-Chip Experiment. *Axioms* **2021**, *10*, 243. https://doi.org/10.3390/axioms10040243

Academic Editor: Bin Han

Received: 13 June 2021
Accepted: 22 September 2021
Published: 28 September 2021

Publisher's Note: MDPI stays neutral with regard to jurisdictional claims in published maps and institutional affiliations.

Copyright: © 2021 by the authors. Licensee MDPI, Basel, Switzerland. This article is an open access article distributed under the terms and conditions of the Creative Commons Attribution (CC BY) license (https://creativecommons.org/licenses/by/4.0/).

Abstract: The present work is motivated by the development of a mathematical model mimicking the mechanisms observed in lab-on-chip experiments, made to reproduce on microfluidic chips the in vivo reality. Here we consider the Cancer-on-Chip experiment where tumor cells are treated with chemotherapy drug and secrete chemical signals in the environment attracting multiple immune cell species. The in silico model here proposed goes towards the construction of a "digital twin" of the experimental immune cells in the chip environment to better understand the complex mechanisms of immunosurveillance. To this aim, we develop a tumor-immune microfluidic hybrid PDE–ODE model to describe the concentration of chemicals in the Cancer-on-Chip environment and immune cells migration. The development of a trustable simulation algorithm, able to reproduce the immunocompetent dynamics observed in the chip, requires an efficient tool for the calibration of the model parameters. In this respect, the present paper represents a first methodological work to test the feasibility and the soundness of the calibration technique here proposed, based on a multidimensional spline interpolation technique for the time-varying velocity field surfaces obtained from cell trajectories.

Keywords: differential equations; mathematical biology; cell migration; microfluidic chip

MSC: 65M06; 92B05; 92C17; 82C22

1. Introduction

Recruitment of immune cells to a tumor is a key parameter in cancer prognosis and response to therapy and the complex relationship between cellular, noncellular components and secreted chemotactic factors plays an essential role in directing the migration of both activating and suppressive immune cell types. In recent years with the development of the highly multidisciplinary Organ-on-Chip field (OOC) [1,2], microfluidic technologies are employed as valuable in vitro platform tools to build tumor microenvironments, with modular degree of complexity and to visualize and quantify immune infiltration in response to anticancer therapies.

The work that set the stage for Organs-on-Chip (OOC) was published in 2010 [3] and thereafter OOC constituted a dynamic field of research, and substantial effort has been devoted to creating realistic mimics of different organs in recent years [4–7].

The coupling with live-cell imaging may enable extraction of single-cell tracking profiles which can be processed with advanced mathematical tools. In this context, the present study is inspired by the modeling of the complex mechanisms behind the cell dynamics and interactions between immune (ICs) and treated tumor (TCs) cells in microfluidic chips. In this framework, several studies [1,7,8] were conducted on immunocompetent cancer-on-chips to assess the effects of therapeutic drugs on TCs and on the possible reactions of the immune system. One of the first studies addresses the role of formyl peptide receptor 1/annexin a1 axis in anti-tumor response to anthracycline-based chemotherapy [7]. Timelapse recordings were performed in a microfluidic platform designed for oncoimmunology [1] research to assess physical and chemical contacts between malignant and immune populations. Some of the earliest applications of microfluidic cell culture technology focused on modeling specific steps in the cancer cascade, including tumor growth [9] and expansion [10], angiogenesis [11–16], progression from early to late stage lesions involving an epithelial–mesenchymal transition [17,18], tumor cell invasion [19] and metastasis [20]. Although Organs-on-Chip represent an increasing in importance field of research because it allows experimentalists to have major control on the objects of investigation, resulting in more accurate measurements, some aspects still remain difficult to understand. In particular, a quantification of the average chemical gradients present in the chip environment is difficult to be estimated.

Motivated by these laboratory experiments, we present an in silico mathematical model to describe ICs migration in the case of an efficient interaction with TCs treated with chemotherapeutic drug.

The main goal of this work is to build a bridge between experimental data coming from Cancer-on-Chip experiment given as particles and macroscopic mathematical models, such as the one proposed in reference [21], in order to gain further insights on the dynamics and short-range interactions between cells. Indeed, the present works goes towards the direction of the mean-field limit to unveil the statistical properties of the model. Recently, one of the authors of the present paper has derived rigorously in Wasserstein's type topologies the mean-field limit (and propagation of chaos) to the Vlasov-type equation, in the framework of generalizations of the kinetic model given by Cucker–Smale dynamical system, see reference [22]. Here we propose a simulation algorithm based on a hybrid macroscopic-microscopic chemotaxis model able to reproduce the main features of phenomena observed in microfluidic chip, such as migration of ICs and short-range interactions with cancer cells representing the sources of chemical gradients.

Existing methods used in the literature so far were dealing with the description of particle trajectories and are essentially represented by statistics on cell trajectories, as in the work by Agliari et al. [23]. Another possibility, already applied in other papers dealing with particle trajectories, see for instance reference [24], is to compute the distance between immune cells and tumor cells across time. However, we would like to stress that here our main concern is to show the feasibility of the proposed approach and to create a connection with macroscopic modeling of the same problem, in order to see what happens when we go to the limit (corresponding to the situation in which millions of cells move in the environment). Our final aim is indeed to be able in the future to simulate the behavior of immune cells in the tissue of an organ. In this framework, a novel strategy for the estimation of model parameters to perform the validation against real data and the calibration of the mathematical model is introduced. Such strategy involves from one hand the analysis of ICs trajectories coming from experimental data to describe realistically the average speed and the pathways of immune cells and the creation of a synthetic dataset for evaluating the effectiveness of the calibration technique for the estimation of model parameters. Moreover, in order to determine the velocity distribution of ICs, a stochastic component is obtained analyzing real trajectories of cells falling in the area under examination and then it is added to the deterministic velocity field.

The modeling of dynamics and interactions in microfluidic chip environment was already studied in [18,25–27] with particle models and with cellular automata models in [28]. In [21] a fully macroscopic mathematical (PDE) model was considered, both for the chemical gradient which is seen as an average field and for the density of immune cells. With the PDE model we were able to describe long-range interactions in the chip environment and a first estimate of the chemical gradient driving ICs movement was obtained in [29].

Here, instead, we are interested in describing efficient short-range interactions between treated tumor cells and immune cells. As an example of this phenomenon, we considered the study and the experimental setting in [7], where tumor cells treated with chemotherapy drug release a chemical stimulus sensed by healthy immune cells, thus promoting their migration towards the tumor cells. For the construction of the model, we used a *discrete in continuous* approach where we coupled a reaction-diffusion partial differential equation (PDE) describing the evolution of the average substances released by the TCs in the tumor microenvironment with a particle model, where every IC in the system was considered to be a single entity and provided with specific properties, as previously done in [30] in a different experiment. Here the hybrid approach was revealed to be crucial in providing a proper description of the multiscale phenomenon that with a traditional macroscopic model would not be possible to fully decipher. In particular, our model describes the dynamics of each IC by means of ordinary differential equations (ODE) in such a way that every cell can be followed individually. Since the motion of ICs is driven by chemotaxis, the migratory activity is regulated by a chemotactic term which allows them to sense the gradient of the chemicals in the tumor neighborhood and of specific forces such as adhesion–repulsion which establish between cells.

The calibration of the mathematical model is supported by the development of ad-hoc parameter estimation procedure based on an interpolation technique for the approximation of the velocity field of ICs across time, i.e., multidimensional spline method previously introduced in [31] and described in Section 4.1. Such procedure, applied to synthetic cell trajectories dataset, provides accurate estimates for the values of unknown model parameters and demonstrates to be a promising approach for the validation of the proposed model against real data.

1.1. The Geometry of the Microfluidic Chip and the Related Computational Domain

The immune-oncology chip designed for the experiment consists of three main culture chambers for plating adherent TCs and floating ICs connected by a bridge of microcapillaries allowing chemical and physical contacts. In Figure 1 a picture of the two boxes is shown. With regards to the dimensions, the capillaries have, respectively, width and length of 12 µm and 500 µm, while the height is of 10 µm; however, since in the video footage the experiment is recorded at a fixed height, the third spatial dimension in our framework is neglected.

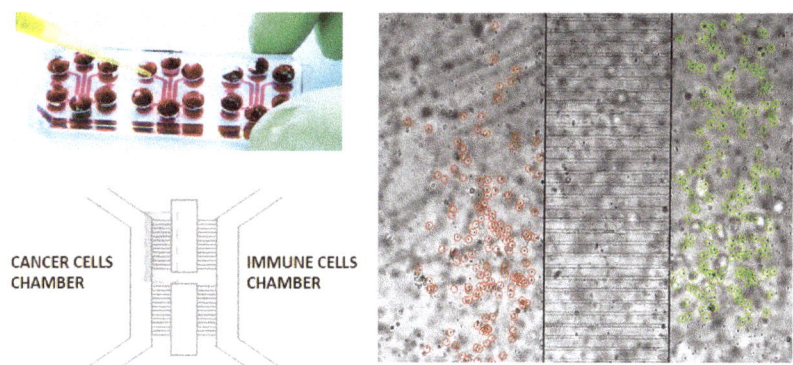

Figure 1. Microfluidic chip environment. On the (**left**): real microphotograph of microfluidic devices filled with food dye and schematics of the 2D layout with three main culture chambers connected by arrays of microchannels. On the (**right**): Timelapse video frame with detected immune cells in the left and intermediate chambers surrounded, respectively, by red and green circles. Credit: Vacchelli et al. (2015) edited by AAAS.

The cross-sectional dimensions of culture chambers are 1 mm (width) × 100 μm (height). Further details about the chip design are illustrated in [7]. We also underline that here we are considering a 2D culture in the liquid with dying cancer cells adherent to the glass slide and mainly static, and immune cells floating.

In the experimental set-up, the timelapse observation area comprises left TCs culture chamber and central one with loaded ICs. Our goal consists of modeling the migration of ICs towards the TCs taking into account the different forces acting on the cells. To achieve such a result, we neglected the migration of ICs in the right chamber and through the microchannels, and we only focused on motility patterns of infiltrated ICs in the TCs left chamber.

Moreover, in order to better analyze the short-range dynamics and interactions between tumor and immune cells, we restricted our study on a subarea of the left chamber of length and height equal to 200 μm. This area corresponds to the subset [400, 600] × [200, 400] μm, and we defined it as $\Omega = [0, L_x] \times [0, L_y]$. In Figure 2 we present the setting of the experiment at time $t = 24$ h, with a focus on the area under analysis. The main reason for considering only a portion of the chip is motivated by:

- the reduction of the computational cost in view of the optimization procedure for the parameter estimation;
- the focus on short-range interactions.

Moreover, Ω contains four treated TCs, and this makes the area a good representative of cell dynamics, since in the laboratory experiment there are about 60 TCs in the entire chamber.

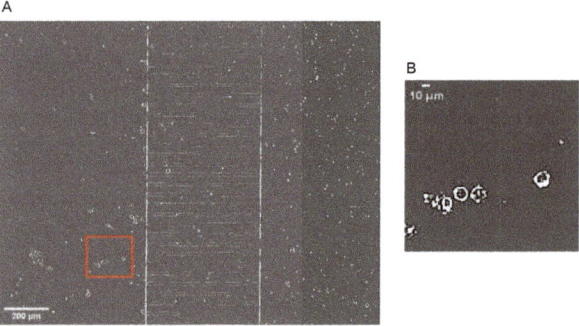

Figure 2. Timelapse pre-processed image of microfluidic chip environment at time t = 24 h. (**A**) Full chip, in red the domain Ω; (**B**) Focus on the area Ω.

1.2. Original Contribution of the Present Paper

To describe such complex mechanisms with a multiscale nature, the model here proposed belongs to the category of hybrid models, involving the coupling between macroscopic and microscopic models.

As mentioned above, we assume the presence of forces between cells, which are established due to the chemoattractant and due to the nature of the cells itself. For this reason, in the current work, we refer to the discrete–continuous model (1)–(2) proposed in [30] to describe the morphogenesis of the posterior lateral line system in zebrafish. Such a model includes a classical reaction-diffusion partial differential equation (PDE) used to describe the evolution of the concentration of chemicals released by TCs in the tumor milieu, coupled with ordinary differential equations (ODEs) for the inter-cellular dynamics of ICs, which establishes both due to the presence of the chemoattractant and to the forces generated between the cells. In particular, the use of a particle model permits the highlighting of the role of single cells in the space, to analyze short-range interactions and to attribute specific characteristics to cells behavior that a macroscopic model would not allow. As mentioned in Section 1.1, for computational reasons and for the datatype at our disposal, cell tracks extracted from video recorded at a fixed level on the z-axis, we neglect the third dimension and consider only the 2D case. However, the third dimension can be easily taken into account by the model, but we do not expect important changes in the overall dynamics.

Here, chemical signals are described as an average field by means of a reaction-diffusion equation

$$\partial_t f = D\Delta f + S(t, \mathbf{Y}, f) \quad (1)$$

with a possible source or degradation term given in S term. Equation (1) was endowed with Robin boundary conditions to state the exchange of chemicals between the internal and the external environment. The gradient field f influences the evolution of cell positions, according to a second order equation of the form:

$$\ddot{\mathbf{X}}_i = F(t, \mathbf{X}, \dot{\mathbf{X}}, \mathbf{Y}, f, \nabla f) - \mu \dot{\mathbf{X}}_i \quad (2)$$

where $\mathbf{X}_i$, $i = 1, \ldots, N_{tot,I}$, is the position vector of the i-th IC, $N_{tot,I}$ is the total number of ICs, $\mathbf{X} = (\mathbf{X}_1, \ldots, \mathbf{X}_{N_{tot,I}})$ contains all the positions, and $\dot{\mathbf{X}} = (\dot{\mathbf{X}}_1, \ldots, \dot{\mathbf{X}}_{N_{tot,I}})$ the velocities. The vector $\mathbf{Y} = (\mathbf{Y}_1, \ldots, \mathbf{Y}_{N_{tot,T}})$, $i = 1, \ldots, N_{tot,T}$, stands for TCs positions, which are taken as constants in the problem, since dying TCs do not migrate as time evolves. The function F includes several effects: from the detection of the chemical signal f (chemotaxis) to mutual interactions between ICs (adhesion and repulsion) and ICs and TCs (adhesion and repulsion). All these effects take into account a non-local sensing radius. The term $\mu \dot{\mathbf{X}}_i$ represents damping due to cell adhesion to the substrate.

We remark that the use of second order equations to describe the cell motion was first used in the seminal work [32] in the framework of Newton's equation of motion for particles. Indeed, even in this case the acceleration is not so high, this formulation describes the effects of the presence of an external force causing velocities and direction changes. Moreover, the friction term for cells immersed in a fluid is also taken into account. In order to have a more complete model able to represent the randomness characterizing cell motility, we also introduced a stochastic component in the velocity field in the spirit of Langevin model, see the review [33] and references therein.

In this context, in order to better reproduce the phenomenon under consideration such as the presence of chemical stimulus of treated cancer cells, we applied the following modification respect to the model in [30]:

- we model the presence of tumor cells, and we also take into account the repulsion forces to avoid overlapping. Eventually, a slight overlay may occur between tumor and immune cell in the case of close interactions. However, it does not seem to occur in the video footage;
- a chemical gradient concentrated around sources represented by cancer cells is considered and diffused in the environment;
- Robin boundary conditions for the inflow of chemoattractant in the area under consideration are applied and adjusted to drive cell migration in a diagonal direction, as observed experimentally;
- the alignment effect between cells is discarded since in this context is not present;
- a chemotactic sensitivity term—i.e., receptor saturation—is added in the drift term of the equation of particles motion, with the effect that chemotaxis of cells is reduced in areas of high chemoattractant concentrations;
- a stochastic component in the particle velocities is added to have more realistic cell trajectories in terms of randomness.

The equations are solved in a square domain Ω, introduced in Section 1.1, subset of the chamber where the experiment was performed and the approximation was carried out using a classical central difference scheme in space and the Crank–Nicolson scheme in time, although in Appendix A we prefer to present the general form of the scheme based on the θ-method.

The main novelty of our approach here is mainly represented by:

- the modeling study on ICs behavior by considering different scenarios, see Section 3;
- the construction of a calibration algorithm to find a trustable estimate of the most significant parameters of the model to assess the effects of chemical gradients on the ICs movement, see Section 4.

In particular, such calibration algorithm—based on the mathematical model—is built to test the possibility of validating the model on real data provided by experimentalists. The main goal of the present work is indeed the introduction of a robust procedure for the estimate of model parameters by means of the paths taken by ICs in the chip subarea under examination. It is worth noting that this is not an easy task, since to succeed in the model calibration we need to extract a common behavior from the time-varying trajectories of immune cells located at different points of the examined area. For this reason, in this first work based on this framework we propose a calibration strategy taking into account this variability and applying it to a synthetic dataset, in order to assess the effectiveness of our methodology.

We underline that the synthetic dataset of ICs pathways has been created reproducing qualitatively the real trajectories extracted from the video footage of the experiment, in order to have a "realistic" dataset. In more detail, a set of trajectories of cells sharing an average behavior has been produced by suitably tuning model parameters to have an upper and lower bound of cell speeds taken from the experimentally observed ones. With these synthetic trajectories, we have computed ICs velocity field at every observation time as a surface produced by the model itself, supposing to have a single population of immune cells thus showing an average behavior. Such velocity field has been then approximated using a multidimensional spline interpolation technique, described in Section 4.1 to have smoothing in time and space on the dataset to be compared.

Finally, this average field has been used as our target solution to be used in the calibration procedure, as described in detail in Section 4.2. The methodology for the error quantification is based on the minimization of a cost functional depending on the difference between the target velocity field of cell in the whole domain and the velocity field obtained for another choice of the model parameters during the optimization procedure performed with a local search method and we compared them at every time step. In addition to this, we also studied the introduction of further terms in functionals with the aim of improving the optimization results and show the soundness of our algorithm representing a starting point for investigations on experimental data in the next future.

1.3. Main Contents and Plan of the Paper

We introduce a hybrid model composed of a reaction diffusion equation, describing the time and space evolution of signaling substances released by treated TCs in a subarea of the main chamber, and of an agent-based model made up of second order differential equations for each IC. The boundary conditions associated with the partial differential equation are of Robin type, so that we can have a control on the amount of chemical exchanged with the surrounding microenvironment. In addition, based on experimental observation, an ad hoc parameter estimation technique was developed. To summarize the main contents of the present work, the mathematical issues faced in this study are identified into:

- the development of a mathematical model describing the behavior of ICs in short-range interactions in the microfluidic chip environment;
- the development of ad-hoc parameter estimation techniques for time-varying velocity fields.

The plan of the paper is as follows. In Section 2 we describe the biological framework that inspired our study, and we introduce the mathematical formulation of biologically inspired models, and we present the adopted model. Section 3 is devoted to the modeling study on the scenarios suggested from the observation of the dynamics in the laboratory experiment performed with the numerical simulation algorithm reported in the Appendix. Section 4 contains the parameter estimation techniques here developed and the results obtained with the calibration algorithm. A sensitivity analysis is also performed and reported in the Appendix. Finally, in Section 5 we discuss the presented results, and the future aims of our work.

2. Materials and Methods

2.1. Biological Framework

The OOC technology allow the design and recreation of more sophisticated in vitro cellular microenvironments under physiological or pathological scenarios; as potential candidates for better prediction of human responses than animal testing cannot. This success is also due thanks to compatibility with a variety of microscopy techniques including live-cell high-content imaging and to the wide variety of cells and tissues that the chips can host, see [1].

One of the most challenging scenarios for the application of these devices is represented by cancer-immune relations due to very complex and not still completely discovered signaling modalities between ICs and TCs. Some attempts have been presented with the aim of modeling the effect of drugs on TCs, see for instance [34,35] where the microfluidic devices helped to control and evaluate the magnitude of cancer responses.

In particular, here we refer to the study in Vacchelli et al. [7], where timelapse imaging of microfluidic co-cultures were performed to investigate the motility patterns and crosstalk between ICs and TCs in the context of chemotherapy-induced anticancer immune responses.

Setting of the Laboratory Experiments

Details on the in vitro microfluidic experiments, type of cells involved and loading procedures can be found in [7,36]. Briefly, human breast cancer cells were previously treated with anthracycline-based chemotherapy and were cultivated in the left chamber. This treatment triggers the process of immunogenic cell death [37], according to which TCs release danger signals. The right chamber is loaded with unlabeled human peripheral blood mononuclear cells (PBMCs) from healthy donors (with normal expression of FPR1). PBMCs start migrating biased by the detection of chemical signal produced by TCs and after crossing microchannels enter in contact with dying cancer cells, engaging in stable interactions leading to TCs killing events.

In the chip, the chosen culture medium is neutral, which means that no exogenous substance is introduced. Timelapse imaging was performed using a microscope placed directly inside the CO_2 incubator for all the duration of the recordings. Images were taken every 2 min over a period of 72 h of migration. Immune cells tracks (<400) were extracted in left chamber (interval time: 24–48 h) using TrackMate plugin(ref) available in the Fiji/ImageJ software (https://imagej.nih.gov/ij/, accessed on 1 August 2020).

Our aim consists of designing a basic mathematical model able to capture the main aspects related to the migratory movement of the ICs with respect to TCs. Therefore, we only focus on the case of efficient interaction between dying cancer cells exposed to ICD inducers and immune cells from healthy donors. Concurrently, we want to develop an algorithmic calibration procedure to derive model parameter values as close to empirical observations as possible.

2.2. Mathematical Framework

In this context, it is important to remark that late years have experienced an increasing interest in the direction of developing techniques to combine experimental data and mathematical models, in order to produce systems, i.e., in silico models, whose solutions could be as close as possible to the experimental outcomes. Indeed, the success of informed models is mainly due to the consistent improvements in computational abilities of the machines and in imaging techniques that allowed a wider access to data.

The involvement of the immune system in all stages of the tumor life cycle, including prevention, maintenance and response to therapy is now recognized as central to understanding cancer development from a systemic point of view. The increasing availability of experimental data and of treatment options, have represented a breeding ground to construct always more precise models. From a mathematical point of view, equations of different nature can be considered, depending on the type of analysis to be carried out. In macroscopic models, consisting of PDEs, any reference to the single constituents, the cells, are neglected, and macroscopic quantities such as the average cell densities are taken into consideration, see for instance the classical models [38,39] and their application to the immunocompetent microfluidic chip experiments in the recent work [21], allowing the simulation of long-range dynamics of immune cells driven by the chemical gradients secreted by cancer cells in the environment. The evolution of chemical stimuli in the tumor microenviroment can be also described by means of ODE models, see for instance the work [25] consisting of three ODEs that model the dynamics of effector and tumor cells and the cytokine IL-2, is one of the first to describe tumor-immune cells interactions. Another model in the same direction is due to Lee et al. [18], where changes in velocity and directionality of ICs are modeled with a system of ODE considering the combined effects of the interstitial flow, the tumor mass, the cytokines IL-8 and CCL-2, and the related receptors. Moreover, ODE models describing cell responses heterogeneity to death ligand (cytotoxic drugs) observed in laboratory experiments with single-cell techniques is proposed in [40]. In [28] the immune system is depicted as a network in the framework of cellular automata, where the nodes are molecules and cells and the arches connecting them are the influences each node has on others. Another kind of modeling consists of combining the *macro* approach expressed by a PDE, with the *micro* approach expressed by an ODE.

Hybrid models, developed including the macroscopic and microscopic scales, have been deeply studied in recent years, see [41–44]. In particular, in [41] a hybrid discrete–continuous model of metastatic cancer cell migration was developed, while in [45] a hybrid discrete-continuum approach was applied to model Turing pattern formation. In [43] a 3D agent-based model is proposed using an off-lattice approach to simulate vasculogenesis. For a comprehensive literature review of hybrid models sees the book chapter [46].

Here, in the framework of hybrid deterministic PDE–ODE models, we propose a *discrete–continuous* approach previously applied in [30], where immune cell motion is governed by ODE and the diffusion of chemoattractant released by cancer cells in the environment is described by PDE. The present work is complementary to the fully macroscopic approach proposed in [21], since it allows a zooming on immune system dynamics. Moreover, we enriched the equations of cell motion with a Brownian component to reproduce zigzag pathways of ICs.

The Model

Here we develop a model that can mimic cell migration under the effects of chemicals released by TCs on immune system dynamics. The considered TCs are experiencing apoptosis during which they liberate damaged-associated molecular patterns together with tumor antigens that elicit an immune response. For the sake of simplicity, the alarm concentrations are not differentiated, and we indicate them by means of the chemoattractant f, whose evolution is described by a PDE. At the cellular level, the model is discrete and includes the equation of the motion of each IC. We also underline that since we refer to experiment of treated cancer cells, according to the experimental setting we do not need to include TCs migration or duplication in our model. Let us summarize the main ingredients which compose the model.

For the IC motion we use a second order dynamic equation, which takes into account the forces acting on the cells that arise from the presence of the chemical signals and from the mechanical interactions between cells. The effect of cancer is described by a chemotactic term produced by the gradient of the f. The cell–cell mechanical interactions due to filopodia, consist of a radial attraction and repulsion depending on the relative positions of the cells, see [47] for experimental results in this direction. To describe this effect we refer to the mechanism introduced in [48]. The action of ICs regarding TCs is described by a radial repulsion term, depending on cells positions. Finally, we introduce a damping term, proportional to the velocities, which models cell adhesion to the substrate [49–51].

About the concentration of the chemoattractant, we associate a diffusion equation including a source term, given by the chemoattractant released by TCs, and a natural degradation term.

In the current context, the species under examination are TCs and ICs, but we underline that the setting can be made more complex with the introduction of a greater number of cell species and with the presence of an exogenous substance in the environment. Furthermore, in the present framework we are not taking into account the killing activity of ICs towards TCs as we already did in [21], since we do not have a quantitative data regarding the killing rate of immune cells in the chip environment. However, in the future also this feature will be included in the model.

Let $\mathbf{X}_i$ be the position of the i-th IC and $f(\mathbf{x}, t)$ the total chemoattractant concentration, the following equations are introduced:

$$\begin{cases} \partial_t f = D \Delta f + \xi F_4(\mathbf{Y}) - \eta f & (3) \\ \ddot{\mathbf{X}}_i = \gamma F_1(\chi(f) \nabla f) + F_2(\mathbf{X}, \mathbf{Y}) + F_3(\mathbf{X}) - \mu \dot{\mathbf{X}}_i & (4) \end{cases}$$

where γ, μ, ξ and η are given constants and $F_n(\cdot)$, $n = 1, 2, 3, 4$ are suitable functions.

Function F_1

The term F_1 is the chemotactic term and therefore it is related to the detection of a chemical signal by i-th cell in its neighborhood and it is taken to be a weighted average over a ball of radius $\bar{R}$ and centered in $\mathbf{X}_i$:

$$F_1(g(\mathbf{x}, t)) = \frac{1}{\mathcal{W}} \int_{\mathbf{B}(\mathbf{X}_i, \bar{R})} g(\mathbf{x}, t) w_i(\mathbf{x}) d\mathbf{x}, \quad (5)$$

where

$$\mathbf{B}(\mathbf{X}_i, \bar{R}) := \{\mathbf{x} : ||\mathbf{x} - \mathbf{X}_i|| \leq \bar{R}\},$$

$||\cdot||$ being the Euclidean norm,

$$w_i(\mathbf{x}) = \begin{cases} 2 \exp\left(-||\mathbf{x} - \mathbf{X}_i||^2 \frac{\log 2}{\bar{R}^2}\right), & \text{if } ||\mathbf{x} - \mathbf{X}_i|| \leq \bar{R}, \\ 0, & \text{otherwise}, \end{cases} \quad (6)$$

is a truncated Gaussian weight function, and

$$\mathcal{W} := \int_{\mathbf{B}(\mathbf{X}_i, \bar{R})} w_i(\mathbf{x}) d\mathbf{x},$$

independently on i. A similar definition holds for the vector quantity F_1.

In addition, the function F_1 contains the gradient of the chemical substance and a chemotaxis function $\chi(f)$, representing the chemotactic sensitivity of ICs. Such chemotactic function, suggested in [52] by Lapidus and Schiller (1976) has the form:

$$\chi(f) = \frac{k_1}{(k_2 + f)^2}, \quad (7)$$

where k_1 represents the cellular drift velocity, while k_2 is the receptor dissociation constant, which indicates how many molecules are necessary to bind the receptors. Such function is known as receptor saturation function and has the effect of reducing chemotaxis of cells in areas of high chemoattractant concentrations. In the integral, $\bar{R}$ will be chosen larger than the IC radius R_I, thus relation (5) describes a chemical signal that is sensed more in the center of the cell and less at the edge of the cell extensions.

Function F_2

Function F_2 includes a repulsion effect among TCs and ICs. In particular, repulsion occurs at a distance between the centers of two cells less than R_1, where R_1 is defined as the sum of the two radii, R_I and R_T, of the immune and tumor cells, respectively. With this formulation, repulsion occurs when an IC and a TC start being effectively overlapped. We assume

$$\mathbf{F}_2(\mathbf{X},\mathbf{Y}) = \sum_{j:\mathbf{Y}_j \in \mathbf{B}(\mathbf{X}_i,R_2)\setminus\{\mathbf{X}_i\}} \mathbf{K}(\mathbf{Y}_j - \mathbf{X}_i), \quad (8)$$

where the function $\mathbf{K}$ depends on the relative positions $\mathbf{Y}_j - \mathbf{X}_i$, namely:

$$\mathbf{K}(\mathbf{Y}_j - \mathbf{X}_i) = -\omega_{rep_{TI}}\left(\frac{1}{||\mathbf{Y}_j - \mathbf{X}_i||} - \frac{1}{R_1}\right)\frac{\mathbf{Y}_j - \mathbf{X}_i}{||\mathbf{Y}_j - \mathbf{X}_i||}, \quad \text{if } ||\mathbf{Y}_j - \mathbf{X}_i|| \leq R_1, \quad (9)$$

where $\omega_{rep_{TI}}$ is a constant. The quantity $R_2 > R_1$ indicates the radius of the ball centered in $\mathbf{X}_i$, in which the centers of the tumor cells can fall. Thus, we are considering all the tumor cells in proximity of the center of an immune cell.

Function F_3

Function $\mathbf{F}_3$ includes adhesion–repulsion effects between ICs. In particular repulsion occurs at a distance between the centers of two ICs less than R_3 and takes into account the effects of a possible cell deformation. Conversely, adhesion occurs at a distance greater than R_3 and less than $R_4 > R_3$, and it is due to a mechanical interaction between cells via filopodia. We assume

$$\mathbf{F}_3(\mathbf{X}) = \sum_{j:\mathbf{X}_j \in \mathbf{B}(\mathbf{X}_i,R_4)\setminus\{\mathbf{X}_i\}} \mathbf{K}(\mathbf{X}_j - \mathbf{X}_i), \quad (10)$$

where the function $\mathbf{K}$ depends on the relative positions $\mathbf{X}_j - \mathbf{X}_i$, namely:

$$\mathbf{K}(\mathbf{X}_j - \mathbf{X}_i) = \begin{cases} -\omega_{rep}\left(\dfrac{1}{||\mathbf{X}_j - \mathbf{X}_i||} - \dfrac{1}{R_3}\right)\dfrac{\mathbf{X}_j - \mathbf{X}_i}{||\mathbf{X}_j - \mathbf{X}_i||}, & \text{if } ||\mathbf{X}_j - \mathbf{X}_i|| \leq R_3, \quad (11)\\[2ex] \omega_{adh}(||\mathbf{X}_j - \mathbf{X}_i|| - R_3)\dfrac{\mathbf{X}_j - \mathbf{X}_i}{||\mathbf{X}_j - \mathbf{X}_i||}, & \text{if } R_3 < ||\mathbf{X}_j - \mathbf{X}_i|| \leq R_4, \quad (12) \end{cases}$$

where ω_{rep}, ω_{adh} are constants. $R_3 = 2R_I$ will be chosen, so that the repulsion occurs when two cells start being effectively overlapped. Please note that function (11) gives a repulsion which goes as $1/r$, r being the distance between the centers of two cells as we can find in [53,54]. The function (12) is the Hooke's law of elasticity. The last term in the first equation is due to the cell adhesion to the substrate (see for example [49–51]).

Friction

The addend $\mu\dot{\mathbf{X}}_i$ in the equation of the motion is due to cell adhesion to the substrate, with damping coefficient equals μ.

Function F_4

In the diffusion equation, only cancer cells are responsible for the production of the chemoattractant, so that

$$\mathbf{F}_4(\mathbf{Y}) = \sum_{j=1}^{N_{tot,c}} \chi_{\mathbf{B}(\mathbf{Y}_j,R_T)}, \quad (13)$$

where $N_{tot,c}$ is the total number of cancer cells, and

$$\chi_{\mathbf{B}(\mathbf{Y}_j,R_T)} = \begin{cases} 1, & \text{if } \mathbf{x} \in \mathbf{B}(\mathbf{Y}_j, R_T), \\ 0, & \text{otherwise.} \end{cases} \quad (14)$$

In the previous formula R_T is the radius of a cancer cell, considering that the source of chemoattractant is defined by the dimension of a single cell.

Initial Conditions

Initial data for Equation (4) are given by the position and velocity of each IC:

$$\mathbf{X}_i(0) = \mathbf{X}_{i,0}, \mathbf{V}_i(0) = \mathbf{V}_{i,0}.$$

Moreover, since TCs do not migrate and maintain their initial position $\mathbf{Y}$ during the whole time, the initial data for Equation (3) is provided by the chemoattractant produced by TCs at time $t = 0$:

$$f(\mathbf{x}, 0) = F_4(\mathbf{Y}).$$

Boundary Conditions

Now, let $\Omega = [0, L_x] \times [0, L_y]$ our domain, for the chemoattractant we require the inhomogeneous Robin boundary condition:

$$D\frac{\partial f}{\partial \mathbf{n}} + af = b, \text{ on } \partial\Omega, \tag{15}$$

where b signals the similarity with the inhomogeneous Neumann boundary condition and regulates the exchange with the external environment.

The system of Equations (3) and (4) can be now rewritten as:

$$\begin{cases} \partial_t f = D\Delta f + \xi \sum_{j=1}^{N_{tot,c}} \chi_{\mathbf{B}(\mathbf{Y}_j, R_T)} - \eta f, \\ \ddot{\mathbf{X}}_i = \frac{\gamma}{\mathcal{W}} \int_{\mathbf{B}(\mathbf{X}_i, \tilde{R})} \chi(f(\mathbf{x}, t)) \nabla f(\mathbf{x}, t) w_i(\mathbf{x}) d\mathbf{x} + \sum_{j: \mathbf{Y}_j \in \mathbf{B}(\mathbf{X}_i, R_2) \setminus \{\mathbf{X}_i\}} \mathbf{K}(\mathbf{Y}_j - \mathbf{X}_i) \\ + \sum_{j: \mathbf{X}_j \in \mathbf{B}(\mathbf{X}_i, R_4) \setminus \{\mathbf{X}_i\}} \mathbf{K}(\mathbf{X}_j - \mathbf{X}_i) - \mu \mathbf{V}_i. \end{cases} \tag{16}$$

2.3. Stochastic Model

The model (16) is composed of deterministic equations, but a stochastic version for Equation (4) can be formulated. In fact, in recent years, several studies have shown that ICs exhibit an intermittent motion composed of a walking phase and of a zigzag phase. The walk is characterized by pause steps between the run steps, while during the zigzag, cells tend to turn away from their last turn directions and prefer to move forward in a zigzag manner [27,55].

This characteristic walk is here described by Brownian motion [56] as a first approach to the problem, revealing to be effective in reproducing the randomness of cell trajectories. The stochastic equation for ICs motion is:

$$\ddot{\mathbf{X}}_i = \gamma F_1(\chi(f)\nabla f) + F_2(\mathbf{X}, \mathbf{Y}) + F_3(\mathbf{X}) - \mu(\dot{\mathbf{X}}_i - \sigma\psi). \tag{17}$$

Equation (17) contains the stochastic contribution, where $\psi(t)$ is a Gaussian white noise, and σ is the standard deviation of ICs trajectories. With this formulation, ICs are not only subjected to mechanical forces such as adhesion or repulsion, but also to random factors that might be related to unknown cell mechanisms. Some estimates of parameter σ based on experimental data are provided in Section 3.1.3.

3. Study on Different Scenarios: Numerical Tests

In this section, we look at some of the different scenarios the model can produce. We remark that the numerical simulations are performed in MATLAB ©. The computational time for a simulation on the total number of frames $T_f = 681 (N_{\Delta_t} = 680)$ takes about 260 s on an Intel(R) Core(TM) i7-9750H CPU @ 2.60 GHz 2.59 GHz.

3.1. Scenarios Representing Relevant Features of ICs Dynamics and Interactions

One of the motivations of the present study is to show the features of the proposed hybrid model in describing Cancer-on-Chip experiment. To this aim, we explore the different dynamics in three significant situations represented by the scenarios here proposed. During apoptosis TCs release alarm substances, which remain localized in proximity of the cells, that become *attractors* for ICs. In absence of an incoming chemical flow through the boundary, if the immune cells fall in the basin of attractions of the tumor cells, would remain trapped in their proximity all the time. From the experiments, however, we observed that ICs migrate also towards the bottom (where a reservoir of TCs was located) and to the left side of Ω, due to the presence of other cells in the surrounding microenvironment, generating a diagonal motion, as illustrated in Figure 2.

To allow ICs migration towards the sides of the domain, we added an incoming chemical flow through the boundaries, which caused changes in ICs orientation.

Moreover, to have control over the changes in orientation and direction of the migratory activity, it is necessary to find proper parameter values to attain a balance between the internal chemical concentration due to TCs and the concentration present at the boundary. For instance, if the chemotactic concentration at the boundary is higher than the internal concentration, ICs will sense the resulting gradient and move towards the boundary. On the opposite, ICs will not sense the gradient at the boundary. Thus, a balance between the internal and the boundary concentrations must be reached to have an IC attracted by the TCs and by the concentration on the sides.

For the numerical simulations we consider the domain Ω where four TCs are present (see Figure 2). The numerical positions of the TCs and the initial concentration of the chemoattractant released by the TCs are shown in Figure 3. We numbered the TCs from 1 to 4 to distinguish them.

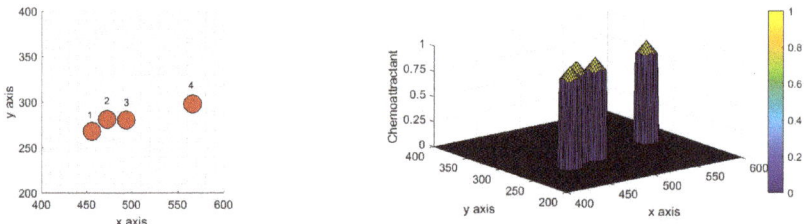

Figure 3. TCs locations (**left**) and TCs chemoattractant (**right**) at time $t = 0$. The grid dimensions are in μm.

In our preliminary study, we also changed the number of TCs and their positions, with the aim of assessing the effect of their presence and their localization on the overall dynamics of ICs.

We assume the coefficients a and b in the Robin boundary condition (15) in such a way to have an incoming flux through the sides $y = 0$, $x = L_x$ and $x = 0$.

The laboratory experiment recording immune-cancer cells interaction has an overall duration of 24 h, but after data analysis we identified the presence of 97 ICs crossing the area in a time interval of 681 frames, which correspond to 1360 min (22 h 40 min) of video recording. To provide an idea of how ICs dynamics changes in time, we divided the time interval in six homogeneous parts, so that we can take a picture of ICs positions at six different instants T_k for $k = 0, \ldots, 5$ ($T_0 = 1$ frame, $T_1 = 136$ frames (4 h 32 min), $T_2 = 272$ frames (9 h 4 min), $T_3 = 408$ frames (13 h 36 min), $T_4 = 544$ frames (18 h 8 min), $T_5 = 681$ frames (22 h 40 min)).

The initial conditions for Equation (4) are assumed as the initial positions and speeds of the experimental cells.

In the following we consider three different scenarios that describe the interactions between TCs and ICs in a subarea Ω of the chip. Specifically, we consider:

1. Deterministic Motion,
2. Deterministic Motion including Cell Death;
3. Stochastic Motion.

The first scenario is assumed to be a prototypal case, where the reciprocal mechanical forces between cells and the chemotactic stimuli are the only guiding forces of ICs migration, the second case takes also into account the possible effects of TCs death over ICs motion, and the third case contemplates the addition of the stochastic component Equation (A11), which modifies the deterministic IC trajectories.

3.1.1. Scenario 1: Deterministic Motion

In Figure 4 we show the final concentration of the chemoattractant, plotted as a 2D surface with a focus on the contour lines (left), and at the boundary of the domain and in correspondence of the centers of the TCs as a function of the longitudinal (center) and transverse distance (right). Cutting the surface this way, the maximum concentrations related to each TC are highlighted.

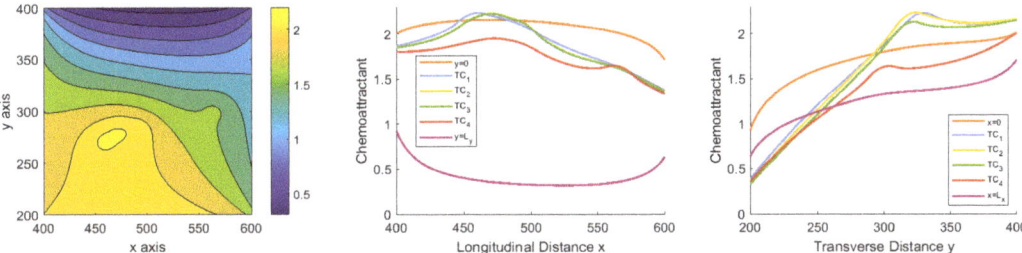

Figure 4. Final Concentration. (**Left**) Concentration plotted in 2D, with contour lines. (**Center**) Concentration profiles as a function of the longitudinal distance. (**Right**) Concentration profiles as a function of the transverse distance. The grid dimensions are in μm.

In Figure 5 the time evolution of the migratory activity of ICs at six different times is depicted. At the initial time, only one IC enters the domain, while at time T_1 the number of ICs increases, and they are directed towards the tumor. As the time grows, most of the ICs approach the TCs and stay nearby, accumulating around them; while the others, guided by the inflow of chemical signal, move towards the left-bottom boundaries of the domain.

3.1.2. Scenario 2: Deterministic Motion including Cell Death

The biological experiment inspiring our work is related to the phase of immunogenic cell death, during which the TCs, previously treated with a drug, release alarm molecules sensed by the immune system. This latter reacts, attacking the tumor.

In the current scenario we focus on the description of the effects of cell death on the ICs dynamics. We remark that in this preliminary study, we did not insert a specific term in the model to describe the death of TCs due to ICs, but we directly *turned off* some TCs. Of course, this represents a simplification of what happens in reality, but is here applied for illustrative purposes. However, in the next future we will include in our modeling the death of cancer cells as a consequence of killing activity of ICs, similarly to the modeling proposed in [21].

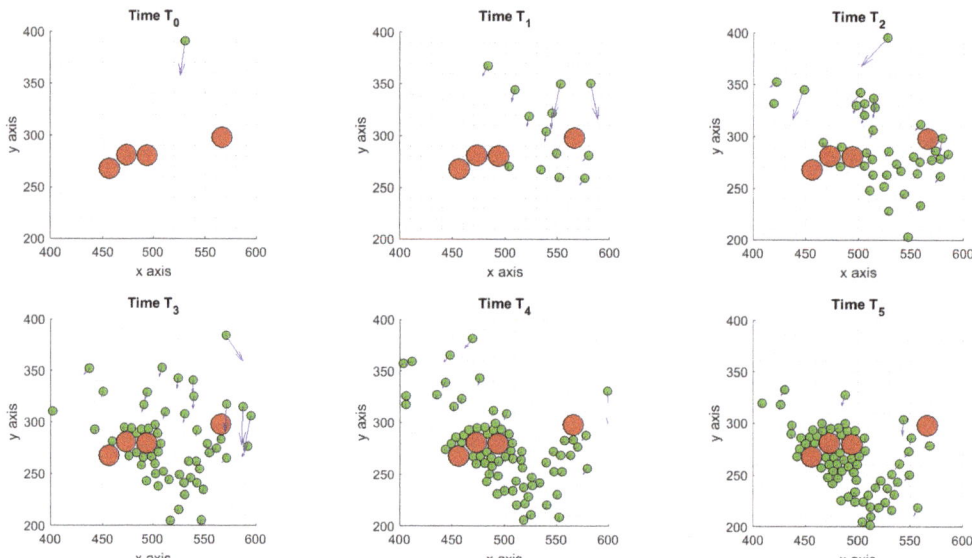

Figure 5. ICs dynamics photographed at six consecutive times. ICs meet all TCs and move towards the left and the right sides of Ω. The grid dimensions are in µm.

Please note that in this simplified setting we let dead TCs physically remain in the domain, thus the effects of repulsion with ICs are still effective, but they do not release alarm molecules anymore. This provokes a decrease of the internal concentration due to the absence of previous sources. In the following, we turn off the third TC during the time interval $[T_2, T_3]$, and then we turn off also the fourth TC in the time interval $[T_3, T_4]$. Figure 4 shows the level of the initial concentration (interval $[T_0, T_2]$), while Figure 6 presents the evolution of the concentration after cell death. Comparing Figure 6 with Figure 4, we can observe how the overall concentration decreases after the two cells are killed.

Figure 7 shows the ICs dynamics. At time T_2 both the third and the fourth cells are approached by ICs, while at time T_3 ICs start moving away from the third cell, which was previously killed and not releasing chemicals anymore. Between time T_3 and T_4 also the fourth cell is turned off and from time T_4 onwards it is not approached anymore. The effect of cell death of some tumor cells (the cells 3 and 4 depicted in Figure 3) on ICs dynamics is depicted in Figure 7. Please note that in the present case a different behavior respect to the one depicted in Figure 5 is observed, since here the accumulation around dead cancer cells does not occur, as expected.

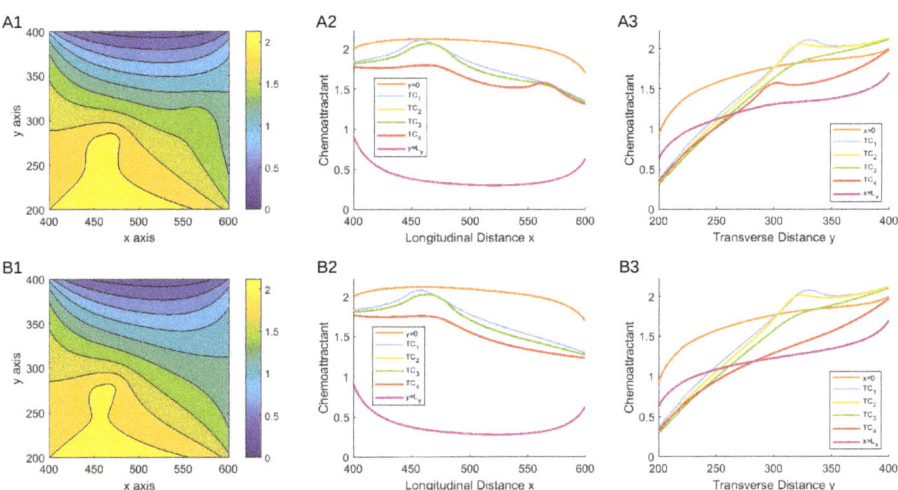

Figure 6. Chemoattractant concentration in the time interval $[T_2, T_3]$ (**Top**) and at the final time (**Bottom**). (**A1,A2**) Concentration plotted in 2D, concentration profiles as a function of the longitudinal distance (**A2,B2**) and transverse (**A3,B3**) distance.

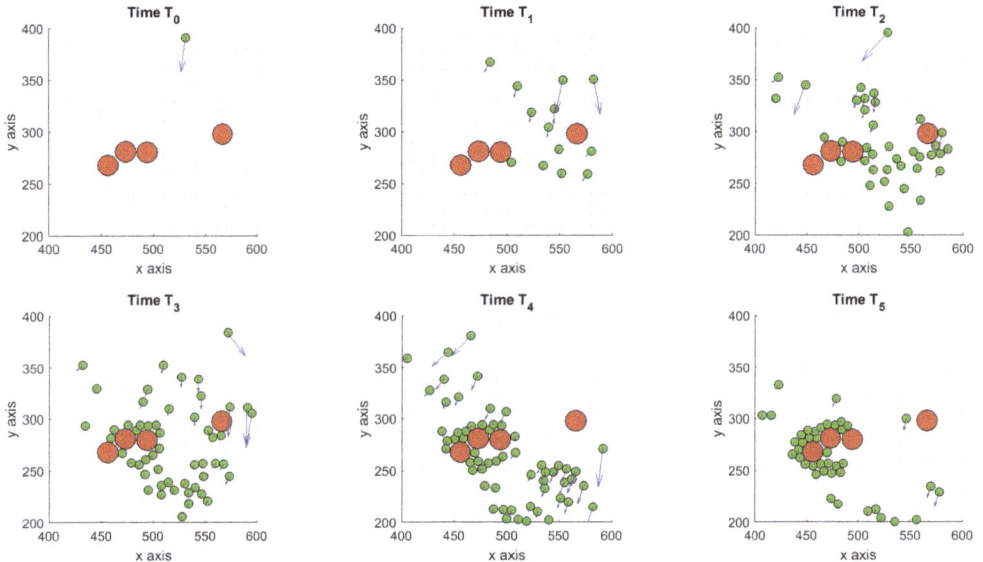

Figure 7. ICs dynamics photographed at six consecutive times. Cell 3 dies during the interval $[T_2, T_3]$, while cell 4 dies during the interval $[T_3, T_4]$. The grid dimensions are in μm.

3.1.3. Scenario 3: Stochastic Motion

Several chemical signals are produced at sites of dying TCs and then diffuse into the surrounding environment. ICs sense these chemoattractants and move in the direction where their concentration is greatest, therefore locating the source of the chemoattractants and their associated targets. If on one hand the deterministic model allows us to control the direction of motion of the ICs acting of the concentration of chemicals, on the other hand the complex migratory activity of ICs is better described by a stochastic model, that takes into account the variability of cells, over the preset deterministic mechanisms. For this reason, the features of this last scenario seem to be more suitable to qualitatively describe the real problem.

Estimates of standard deviations. Here we preliminary determine an appropriate value for the standard deviation σ in Equation (17). To this end, we consider the trajectory P_i of the i-th IC, for $i = 1, \ldots, N_{tot,I}$ and its corresponding smoothing S_i, obtained with a moving average (the Matlab © *smooth* function), and we compute the variance of each IC trajectory:

$$\sigma_i^2 = \frac{1}{T_i} \sum_{k=1}^{T_i} (P_i^k - S_i^k)^2, \tag{18}$$

where T_i is the number of frames the i-th IC spends on the domain Ω and we obtain the sequence $\Sigma = \{\sigma_i^2\}_{i=1,\ldots,N_{tot,I}}$. Successively, we average the values of Σ and divide by 120 s, which corresponds to the time interval between two consecutive frames:

$$\sigma^2 = \left(\frac{1}{N_{tot,I}} \sum_{i=1}^{N_{tot,I}} \sigma_i^2 \right) / 120, \tag{19}$$

obtaining a variance expressed in seconds. As a last step, we apply the square root to relation (19) and then obtain the standard deviation σ of real trajectories. From the analysis conducted on the total number of ICs, we observed cells with variances very far from the average, probably because of the presence of different cell species among the group of ICs. We remark that the immune cell population is heterogeneous, since it is composed by T-lymphocites, monocites, dendritic cells; however, a clear distinction among cell species is not possible currently. For this reason, in order to obtain homogeneous data, we neglect these cell trajectories in the computation of the standard deviation. Indeed, our mathematical model aims at describing an average behavior, and therefore the trajectories whose variance is too far from the average are discarded.

In the presence of more data, we will try to make a classification of cells based on their pathways (length, stops, tortuosity, etc.). This will be the subject of a future study.

The cleaned standard deviations computed from experimental trajectories are reported in Table 1.

ICs dynamics. Figure 8 captures ICs at six different times, while Figure 4 shows the evolution of the chemoattractant. Differences with the deterministic dynamics shown in Figure 5 can be highlighted. In the stochastic case, ICs are more spread.

The computations for this case are performed with a finer time spacing of $\Delta t = 10$ s, with the aim of better describing by the model the complex mechanisms really happening to cells at smaller time scales, and then, the plots are taken every 2 min, which is the timeframe of video recordings.

As an example, in order to evaluate the effects of the stochastic component on the dynamics and establish a qualitative comparison with the deterministic model, we depict the trajectories of six randomly selected cells. In Figure 9 we assume the starting point of a given experimental cell and show the trajectories obtained with the ODE model (blue) and the SDE model (red), while the corresponding trajectory extracted from the video footage is plotted in black. As can be observed, the deterministic trajectories are smooth and tend, in most cases, to stop in correspondence of the TCs, showing that once the immune cells have fallen in the basin of attraction of the tumor, they tend to be trapped there. The stochastic trajectories, instead, have a more similar behavior to the real ones, showing that after some time spent near the tumor, the immune cells move to the boundary.

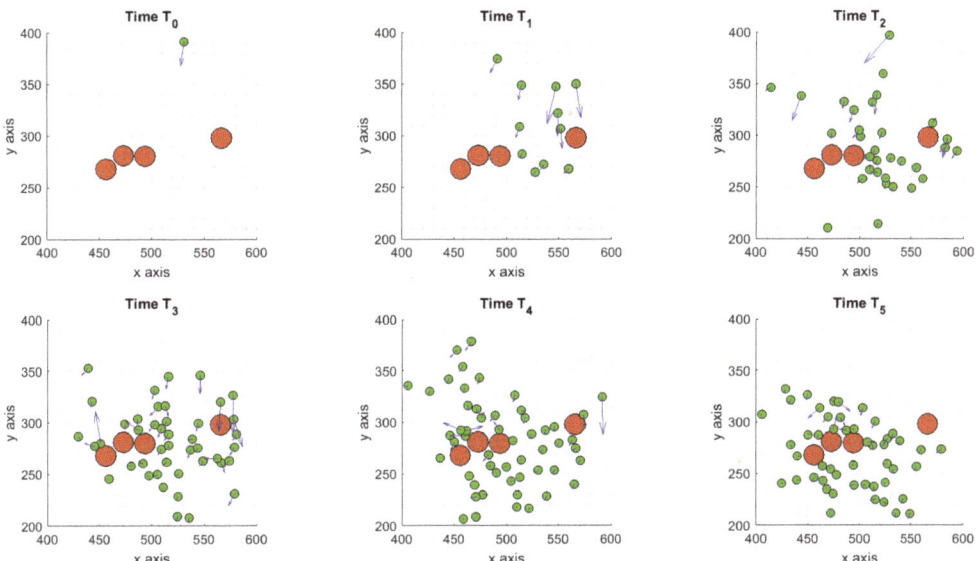

Figure 8. Brownian motion. ICs dynamics photographed at six consecutive times. ICs meet all TCs and move towards the left and the right sides of Ω. The grid dimensions are in µm.

So far the Brownian motion yields satisfactory results as shown in Figure 9, but we do not exclude in the next future to deal with more homogeneous and richer dataset to better identify the probability distribution and then apply a general Levý walk.

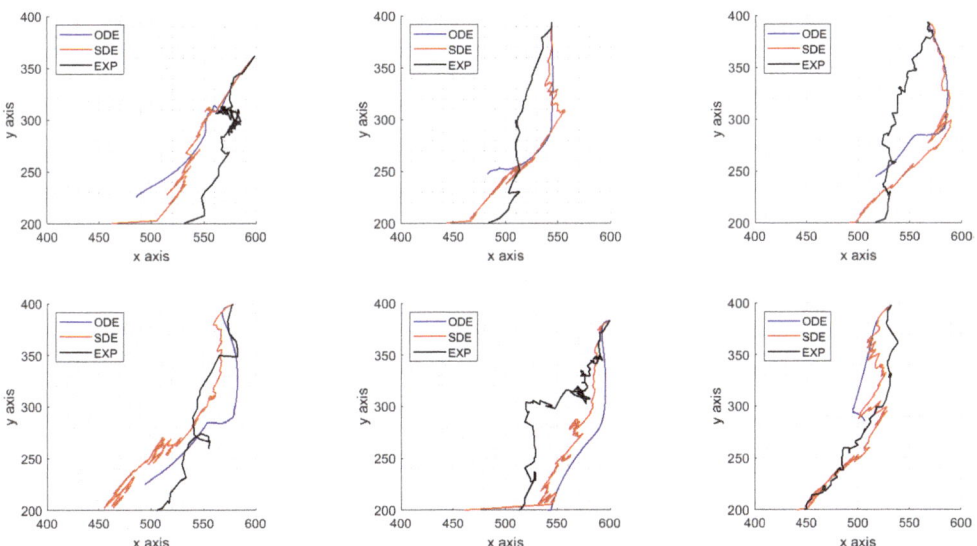

Figure 9. ICs Dynamics. (Blue) Deterministic Motion. (Red) Brownian Motion. (Black) Experimentally observed motion.

Table 1. Estimates of physical parameter values.

Parameter	Description	Units	Value	Ref.
$N_{tot,I}$	Number of ICs in Ω during 24 h		97	Experimental Data
$N_{tot,C}$	Number of TCs in Ω during 24 h		4	Experimental Data
L_x	Horizontal size of Ω	μm	200	Experimental Data
L_y	Vertical size of Ω	μm	200	Experimental Data
R_I	IC radius	μm	4	[57]
R_T	TC radius	μm	10	[58,59]
$\tilde{R}$	Detection radius of chemicals	μm	7	Biological Assumption
R_3	Radius of action of repulsion between ICs	μm	8	Biological Assumption
R_4	Radius of action of adhesion between ICs	μm	10	Biological Assumption
R_1	Radius of action of Repulsion between ICs and TCs	μm	14	Biological Assumption
D	Diffusion Coefficient	$\mu m^2.s^{-1}$	2×10^2	[60]
ξ	Growth Rate of f	s^{-1}/cell	10^{-1}	[61]
η	Consumption Rate of f	s^{-1}	10^{-4}	[61]
ω_{adh}	Coefficient of Adhesion between ICs	s^{-2}	10^{-7}	Biological Assumption
ω_{rep}	Coefficient of Repulsion between ICs	$\mu m^2.s^{-2}$	5×10^{-4}	Biological Assumption
$\omega_{rep_{TI}}$	Coefficient of Repulsion between ICs and TCs	$\mu m^2.s^{-2}$	8.5×10^{-3}	Biological Assumption
μ	Damping Coefficient	s^{-1}	2.1×10^{-3}	Biological Assumption
k_1	Cellular Drift Velocity	$mol.\mu m^2.s^{-1}$	0.39	[60]
k_2	Receptor Dissociation Constant	$mol.\mu m^{-2}$	5×10^{-6}	[60]
a	Rate of exchange of the Chemoattractant with the external environment	$\mu m.s^{-1}$	10	Biological Assumption
γ	Coefficient of Chemotactic Effect. Scenarios 1 and 2	μm^{-1}	5×10^{-3}	Biological Assumption
b_1	Flux condition on $y = 0$. Scenarios 1 and 2	$mol.s^{-1}.\mu m^{-1}$	22	Biological Assumption
b_2	Flux condition on $x = L_x$	$mol.s^{-1}.\mu m^{-1}$	12	Biological Assumption
b_3	Flux condition on $y = L_y$	$mol.s^{-1}.\mu m^{-1}$	0	Biological Assumption
b_4	Flux condition on $y = 0$. Scenarios 1 and 2	$mol.s^{-1}.\mu m^{-1}$	18	Biological Assumption
σ_x	Standard Deviation of x-trajectories	$\mu m.s^{-1/2}$	1.49×10^{-1}	Experimental Data
σ_y	Standard Deviation of y-trajectories	$\mu m.s^{-1/2}$	1.74×10^{-1}	Experimental Data

4. Parameters Estimation on Synthetic Data

In this Section we present the calibration algorithm for the estimation of some model parameters and the results obtained with it. All the model parameters are reported in Table 1. We underline that some of them are given from experimentalists since are taken from the laboratory experiment (size of the cropped area, number of TCs and ICs across time, standard deviation of cell pathways). Other parameters, such as for instance, the diffusion coefficient of chemoattractant, cellular drift velocity, receptor dissociation constant or the radii of cells are taken from the literature. For the model parameters whose values still need to be assigned, we assumed fixed most of them, such as the radii of action between cells or adhesion/repulsion coefficients among cells, setting them in a phenomenological way through the observation of their effects on the overall dynamics.

Then, in order to reduce the computational cost of the optimization procedure, we apply our calibration procedure only to 4 model parameters, i.e., the coefficient of chemotactic effect γ, and the chemical inflow, respectively, at the left, right and bottom boundaries b_1, b_2, b_4, since we observed qualitatively a great effect of such parameters on the overall dynamics and, as a simplification of the model, we assume $b_3 = 0$. In particular, the coefficient γ affects the speed of ICs, while b_1, b_2, b_4 affect the directionality of immune cells.

We underline that the parameter estimation procedure here described is applied to the deterministic version of the model, i.e., (3) and (4). An extension of this procedure to calibrate the stochastic model (3)–(17) will be included in a future work.

In this methodological study we want to show the feasibility and applicability of such estimation technique for some significant model parameters of the hybrid model describing Cancer-on-Chip experiment. Our final goal is indeed not the exact fitting of model parameters, but is to show that we are able to qualitatively reproduce the dynamics observed experimentally, even if the quantification of the chemical gradients in the environment is currently impossible for biologists in this kind of experiments. In the next future, we will complement this methodology with macroscopic models deriving from them to be able to simulate immunocompetent behavior in organs and tissues, where millions of cells are present. To this aim, we applied the proposed procedure to a synthetic dataset produced by the model itself—but strongly inspired by experimental one—to assess the soundness of this strategy. In the future, we aim at having more data in such a way to apply the proposed strategy to a real biological dataset.

Data preparation.

To succeed in the calibration of model parameters, we need to extract a common behavior from the time-varying trajectories of immune cells located at different points of the observed area. Thus, here we propose a calibration algorithm taking into account this variability constructing a "realistic" synthetic dataset of ICs pathways that can reproduce qualitatively the real trajectories extracted from the video footage of the experiment. In particular, a set of trajectories of cells sharing an average behavior has been produced by suitably tuning model parameters to have an upper and lower bound of cell speeds taken from the experimentally observed ones. Then, ICs velocity field given by 2D surface is computed by the model itself at every observation time of 2 min (as the video footage timeframe). It is worth noting that the application of this strategy implies the assumption to have a single population of immune cells showing an average behavior.

The velocity field is then approximated using a multidimensional spline interpolation technique, described in Section 4.1 to have smoothing in time and space on the dataset to be compared.

Main steps of the calibration algorithm.

The calibration algorithm applied to estimate crucial model parameters can be summarized as follows:

- the dataset representing the target solutions of our procedure computed from synthetic ICs trajectories is obtained by the model with fixed parameters;
- a time-space approximation of the velocity fields is carried out with the spline technique presented in Section 4.1 and used as target solution of the calibration algorithm;
- perturbed model parameters are used as initial guess of a global search algorithm minimizing the norm of the difference between the target velocity fields and the estimated ones, as explained in Section 4.2.

The results obtained with the procedure above are reported in Section 4.3.

4.1. Multidimensional Interpolation

The position and speed of the ICs, variable in time, provide punctual information about the velocity field in which the ICs are immersed. As a consequence, they can be considered to be training points for the calibration of an interpolation algorithm, able to provide information about the full space. Here we are considering time as the third dimension, being x and y the first two. As the interpolation scheme we are using a multidimensional spline, described in [31].

Each training point is supposed to influence the value of the interpolating function over a (limited) portion of the space in proximity of it. The degree of influence of the ith training point ($w_i(x,y,t)$) is a function of time and space, driven by a compact-support function radial basis decaying at zero outside the influence area. Then, the interpolated value is obtained as a weighted sum of the values at the training points:

$$\tilde{v}(x,y,t) = \sum_{i=1}^{n} w_i(x,y,t)h(i)c(i), \qquad (20)$$

where $c(i)$ represents the value of the interpolating function at the ith training point and n is the total data-points available, corresponding to $t_1, \ldots, t_n$ discrete times. Specifically, we have two functions $\tilde{v}$, i.e., $\tilde{v}_x(x,y,t)$ and $\tilde{v}_y(x,y,t)$. Moreover, the vector $c(i)$ is different according to the velocities under exam, with training points corresponding alternatively to the x- and y-velocities. Since different influence areas may overlap, the value of $w_i(x,y,t)$ needs to be adjusted to have a correct fit. To this aim, we need to solve a linear system where the n Equations (20) are collocated at the training points, generating an $n \times n$ system. The n coefficients $h(i)$ are computed only once, and then applied in the interpolation procedure. In this specific case, we are using a linear function for the influence coefficients $w_i(x, y, t)$.

4.2. The Calibration Algorithm

With Formula (20) we have generated a velocity field $\tilde{v}$ that from now on we indicate as V^e to point out that the velocity field is produced with the punctual velocities v the immune cells assume at every time step, coming from experimental data (in this specific case from synthetic data). Later we have used this field to define an appropriate functional to be minimized.

To assess the goodness and soundness of our strategy, the analysis which follows is performed on synthetic data, i.e., numerical data produced by the PDE–ODEs system (3) and (4). Future work will be directed towards the application of this methodology to the experimental outcomes.

As a first step we have interpolated immune cell velocities v obtained from synthetic data to the end of generating a velocity field to use as a target, in correspondence of every frame. Thus, we have produced $T_f = 681$ velocity fields V_x^e using the punctual velocities in x and $T_f = 681$ velocity fields V_y^e using the velocities in y. In Figure 10 we present an example of the surfaces resulting from the interpolation at a certain time step. Figure 10A shows the interpolation of the x-velocities, while Figure 10B shows the interpolation of the y-velocities. The green points have coordinates (P_x, P_y, v_x) in (A) and (P_x, P_y, v_y) in (B). For more details about the interpolation technique see Section 4.1.

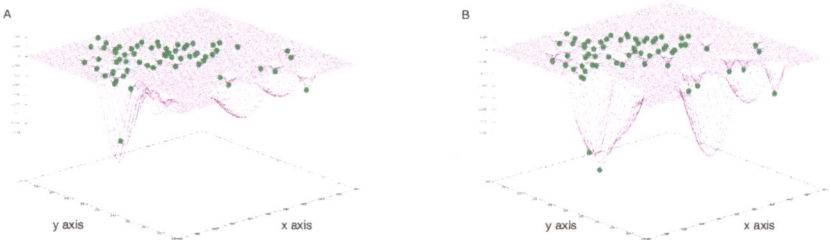

Figure 10. Interpolated surfaces V_x^e (**A**) and V_y^e (**B**) at fixed time. The green points indicate the values of the velocities in the corresponding positions. (**A**) Velocities in the x-direction, (**B**) Velocities in the y-direction.

Successively, we have launched the optimization algorithm and at every run we have computed the numerical solutions of the deterministic model with the approximation scheme described in Appendix A. We have then built a routine in which inserting the numerical positions we are able to find the corresponding velocities on the surface V^e. The resulting interpolated velocities V^n are then used to construct the objective function:

$$J_V(\theta) = \left[\frac{1}{T_f} \sum_{k=1}^{T_f} \left(\frac{1}{N_{tot,I}^k} \sum_{i=1}^{N_{tot,I}^k} \frac{||V_i^{e,k} - V_i^{n,k}(\theta)||}{||V_i^{e,k}||} \right) \right]^2. \qquad (21)$$

In (21) we compare at every time step the punctual interpolated velocities with the punctual target velocities, i.e., the velocities used to generate the field. With θ we indicate the vector whose dimension corresponds to the parameter values we search.

In addition, we have also included a Tikhonov regularization term in the functional, as usually done for the regularization of linear inverse problems. In particular, we consider the following term to be added to the functional to be minimized:

$$P_\lambda(\theta) = \lambda^2 ||\theta - \theta_0||^2, \qquad (22)$$

where θ_0 are the a priori estimate of target parameter values and θ are the parameter values we are optimizing. The constant λ is a regularization parameter that helps the algorithm in the search for optimal values of model parameters by reducing the number of local minima in the functional, see [62]. Tests for different values of λ showed better results of the calibration algorithm for $\lambda^2 = 0.1$. Thus, with relation (22) we want to reduce the error between the target values and the searched ones.

With functionals (21) and (22) we can define the minimization problem:

$$\min_{\theta \in \Theta} J(\theta) = \min_{\theta \in \Theta} (J_V(\theta) + P_\lambda(\theta)), \tag{23}$$

where Θ is the space to explore to search the unknown parameter values. For completeness, we introduce another estimator we have used for our simulations, which is based on the idea of comparing the distances assumed by the tumor and immune cells at every time step. We indicate with C_j the position of the j-th TC (the temporal indicator k is omitted since TC positions do not evolve in time). We call $d_{i,j}^k$ the distance between the i-th IC and the j-th TC at time k:

$$d_{i,j}^k = ||P_i^k - C_j||, \tag{24}$$

and then, fixed the i-th IC, we compare the distances $d_{i,j}^{n,k}$ obtained from the numerical positions with the distances $d_{i,j}^{e,k}$ obtained from the synthetic positions:

$$D_i^k = \frac{1}{N_{tot,c}} \sum_{j=1}^{N_{tot,c}} \frac{||d_{i,j}^{e,k} - d_{i,j}^{n,k}||}{||d_{i,j}^{e,k}||},$$

successively, we sum over all ICs:

$$D^k = \frac{1}{N_{tot,i}^k} \sum_{i=1}^{N_{tot,i}^k} D_i^k,$$

and then we sum over all times. To summarize, we have the functional:

$$J_{D_{TI}} = \left[\frac{1}{T_f} \sum_{k=1}^{T_f} \left(\frac{1}{N_{tot,i}^k} \sum_{j=1}^{N_{tot,i}} \left(\frac{1}{N_{tot,c}} \sum_{j=1}^{N_{tot,c}} \frac{||d_{i,j}^{e,k} - d_{i,j}^{n,k}||}{||d_{i,j}^{e,k}||} \right) \right) \right]^2. \tag{25}$$

In conclusion, to evaluate the error committed by the optimization with respect to synthetic data, we indicate with θ_0 the target value and with θ^* the corresponding optimized parameter and we define:

$$RE = \frac{||\theta^* - \theta_0||}{||\theta^*||} \times 100, \tag{26}$$

the approximation error of each parameter.

The minimization of the functionals above is performed with Particle Swarm Optimization (PSO) [63] as search method using the Matlab © toolbox. For the approximation of target and computed velocity fields resulting in a 2D varying in time surface we use a spline multidimensional interpolation described in Section 4.1.

4.3. Results on Parameters Estimation

Tests on the calibration algorithm are performed to assess the goodness and robustness of the methodology shown in Section 4.2. Please note that the total number of parameters to be assigned is 14, since the other parameters are given from the experimentalists or taken from the literature, as reported in Table 1. Of course, it is possible to make a parameter estimation of all of them. However, in the present work, in order to reduce the computational cost, we apply our calibration procedure only to 4 model parameters that we consider very significant since they strongly affect the dynamics of the ICs, i.e., γ, b_1, b_2 and b_4.

We recall the γ is the coefficient which enhances the effect of the chemotactic function (5), while b_i, for $i = 1, 2, 4$ are the parameters that control the chemical inflow at the boundaries. In order to perform the tests, we produced synthetic solutions choosing specific values for the listed parameters:

$$\gamma = 5 \cdot 10^{-3}, b_1 = 22, b_2 = 12, b_4 = 18, \tag{27}$$

which correspond to those used to create the Scenario 1 (see Section 3.1.1). The synthetic solutions are used to generate the velocity fields V^e to be used as targets.

To assess our strategy, we start testing one parameter and then adding one more at time. The range for the parameter variations is chosen by varying the initial guess by a percentage of ±50%. Specifically, we searched γ in $I_{50\%} = [3.5 \times 10^{-3}, 6.5 \times 10^{-3}]$, b_1 in $I_{50\%} = [11, 33]$, b_2 in $I_{50\%} = [6, 18]$ and b_4 in $I_{50\%} = [9, 27]$. We also choose as initial guess a perturbation of the model parameters by +30%. We tested the algorithm with the following two different functionals:

$$J_1 = J_V + P_\lambda, \tag{28}$$

and

$$J_2 = J_V + J_{D_{TI}} + P_\lambda. \tag{29}$$

Results with the functional J_1 are reported in Table 2, while results with J_2 are in Table 3. In each row of Tables 2 and 3 errors obtained with the parameter estimation algorithm are reported. Please note that errors are reported in increasing order respect to the number of parameters involved in the parameter estimation procedure, from the top to the bottom. We can notice that in both cases the errors in parameters estimation are quite low since their order of magnitude is around 10^{-3} in the worst case, meaning the calibration was successful. We also underline that the norm of the total error of the functional is low, around 10^{-7}, given by the J_V component.

Table 2. J_1. Deterministic Model Calibration. Parameter $\gamma = 5 \times 10^{-3}$, parameter $b_1 = 22$, parameter $b_4 = 18$ and parameter $b_2 = 12$ are varied by +30% and they are optimized in the range ±50%.

Functional	RE % γ	RE % b_1	RE % b_4	RE % b_2	J_V	P_λ	Fval
J_1	4.114×10^{-4}	-	-	-	2.076×10^{-7}	4.231×10^{-17}	2.076×10^{-7}
	2.066×10^{-4}	5.249×10^{-4}	-	-	2.064×10^{-7}	1.334×10^{-9}	2.078×10^{-7}
	7.874×10^{-5}	2.397×10^{-4}	2.283×10^{-4}	-	2.064×10^{-7}	2.633×10^{-10}	2.067×10^{-7}
	1.353×10^{-3}	1.869×10^{-3}	5.18×10^{-4}	7.41×10^{-4}	2.09×10^{-7}	1.857×10^{-8}	2.275×10^{-7}

Table 3. J_2. Deterministic Model Calibration. Parameter $\gamma = 5 \times 10^{-3}$, parameter $b_1 = 22$, parameter $b_4 = 18$ and parameter $b_2 = 12$ are varied by +30% and they are optimized in the range ±50%.

Functional	RE % γ	RE % b_1	RE % b_4	RE % b_2	J_V		P_λ	Fval
J_2	2.915×10^{-6}	-	-	-	2.064×10^{-7}	2.833×10^{-14}	2.134×10^{-21}	2.064×10^{-7}
	2.798×10^{-5}	4.558×10^{-4}	-	-	2.068×10^{-7}	2.037×10^{-8}	1.006×10^{-9}	2.281×10^{-7}
	2.93×10^{-4}	4.335×10^{-4}	7.67×10^{-4}	-	2.066×10^{-7}	6.273×10^{-10}	2.816×10^{-9}	2.1×10^{-7}
	1.102×10^{-4}	5.503×10^{-4}	1.63×10^{-4}	2.65×10^{-4}	2.064×10^{-7}	4.662×10^{-10}	1.653×10^{-9}	2.085×10^{-7}

In conclusion, as a way to quantify the goodness of our reconstruction, we compute the percentage of cells passing through the cropped area of the chip representing the computational domain. In particular, we depict with histograms the statistics obtained from the trajectories reconstructed by the calibrated model and the statistics computed from the trajectories of the synthetic dataset, i.e., the target solutions of the calibration algorithm. The percentage of outgoing cells N_o^k over the number of cells N_p^k present in the domain Ω, is computed by the formula:

$$N^k = \frac{N_o^k}{N_p^k} \times 100, \qquad (30)$$

for each timeframe $k = 1, \ldots, 680$ with a time spacing of 2 min. Then, the optimized parameters (27) obtained with the calibration procedure described in 4.2 are used as the model parameters to be given in input to both the ODE model (3) and (4) and the SDE model (3)–(17) for the computation of the statistics on cell trajectories. More precisely, we perform this computation for the target case, i.e., the synthetic trajectories obtained by the PDE–ODEs system (yellow bars in Figure 11). Then, assuming randomly placed initial positions of ICs, we compute the same statistics for the PDE–ODEs system (red bars in Figure 11) and for the PDE-SDEs system (blue bars in Figure 11). In particular, the initial positions P_i^k, for $i = 1, \ldots, N_{tot,I}$ are randomly perturbed in a range of $(0, 5]$ μm. As can be seen in Figure 11, the statistics of target trajectories (ODE-target) and reconstructed trajectories (ODE-modified and SDE-modified) are quite similar in terms order of magnitude. Please note that the ODE-modified shows the ability to reproduce well also the timing of the exits from the domain, while a slight variability, as expected, can be observed in the SDE-modified case in terms of time occurrence of exits from the domain.

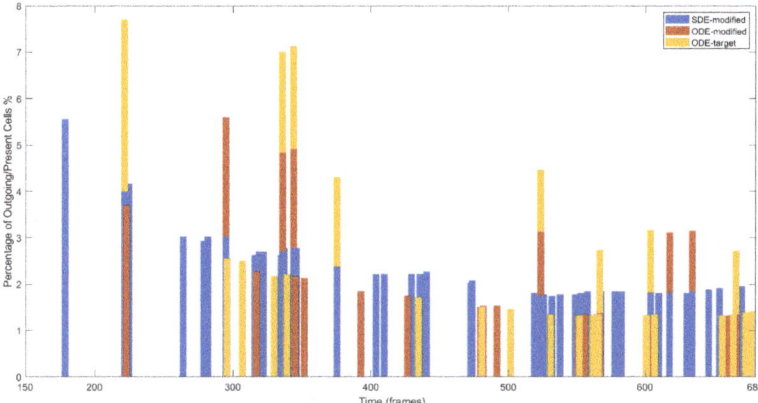

Figure 11. Percentage of outgoing cells over the number of cells present in the domain. The plot shows results starting from frame 150, since no relevant information emerged in the previous frames.

5. Conclusions and Future Work

In the present work we have developed a mathematical model to describe the interactions between ICs and treated TCs in microfluidic chip, together with a strategy to estimate unknown parameter values.

Regarding the modeling part, we have introduced a hybrid model, composed of a reaction-diffusion equation to describe the evolution of chemicals released by TCs, coupled with an equation for the motion of each IC, driven by the chemical gradient. The deterministic system represents a first microscopic model of ICs dynamics in a subarea of the chip.

First, we qualitatively analyzed the overall dynamics of ICs, varying important parameters of the system, such as the chemotactic coefficient γ and the boundary parameters. Indeed, according to their magnitude, the direction of the motion can be controlled and decided *a priori*. Then, in order to reproduce some interesting features characterizing cell behavior, we adjusted suitably the model parameters and we presented three different scenarios that can occur in the chip:

- the deterministic scenario, with some ICs attracted by the tumor and others moving towards to the boundaries of the considered domain;
- the cell death scenario, as a subcase of the previous one, obtained assuming two TCs have died after their interaction with ICs, causing changes in the internal concentration of chemicals thus affecting ICs dynamics;
- the stochastic scenario, obtained adding to the equation of the motion a Brownian walk to mimic the randomness on ICs trajectories.

Regarding the stochastic scenario, it is important to highlight that the Brownian motion is not the most appropriate to describe ICs movement, but in the future we aim at substituting it by the Lévy walk. In recent works [64,65], it has indeed emerged that a heavy-tailed process is more efficient and realistic than Brownian motion. However, in the present work, the shortage of data at our disposal did not provide us information on the nature of the different ICs involved in the experiment. To this end, we are working on a classification strategy in a forthcoming paper to be able to identify the different categories of ICs and then differentiate the model parameters according to the corresponding cell type.

In addition to the pure modeling part, we have developed a model calibration procedure based on the comparison of the velocity fields related to ICs at every time step.

The calibration procedure with synthetic data revealed to be successful, thus representing a first step towards the model calibration on experimental data.

Finally, the main achievements of the present work are represented by:

- the development of a simulation algorithm mimicking short-range dynamics of ICs in the neighborhood of TCs, as observed in Cancer-on-Chip experiment;
- the introduction of an ad-hoc methodology for the calibration of model parameters based on the time-space approximation of synthetic velocity fields computed from immune cell trajectories.

Future perspectives.

Our future aim is to extend this framework to validate our model against the experimental velocity fields computed from the real IC trajectories extracted from the video footage of the experiment.

To face the problem of calibrating the model parameters against real data, we first must produce an interpolation model able to take into account the non-deterministic nature of the cell motions. A possible strategy is to produce a stochastic interpolator, able to determine the expected value and variance of the local speed: the final interpolated value will be then obtained adding a stochastic part, computed using the interpolated variance and the same probability density function of the experimental data, to the interpolated expected value. Using this approach, we have a non-deterministic interpolated value with the same statistical qualities of the interpolating dataset. Implicitly, we are considering as deterministic the local values of the expected value and of the variance of our experimental dataset, and we are also assuming to be able to compute the statistical distribution of the real data (otherwise, we can adopt a prescribed distribution. i.e., Gaussian). It is now evident that the comparison cannot be produced on a single path, but we need to observe the trajectories from a statistical standpoint, i.e., comparing the average path over many simulations, whose output is now non-deterministic. At the same time, in order to derive the statistical properties of the real IC trajectories and speed, a large number of experiments is needed, with a time resolution of the same order of magnitude of the time scale of the physical phenomenon under investigation. This implies a huge experimental and numerical effort, one or two order of magnitude larger than the present work.

Author Contributions: Methodology, G.B. and R.N. (modeling and numerical framework) and D.P. (multidimensional spline interpolation); Software, N.R. (numerical simulation algorithm and calibration procedure) and D.P. (spline based approximation routine); Supervision, G.B. and R.N.; Validation, G.B, N.R. and A.D.N.; Visualization, N.R.; Data curation, A.D.N. and N.R.; Conceptualization, G.B. and R.N.; Investigation, N.R. Writing, G.B., N.R., A.D.N., D.P. All authors have read and agreed to the published version of the manuscript.

Funding: The research of D. Peri was partially funded by Italian Ministry of Education, University and Research (MIUR) to support this research with funds coming from PRIN Project 2017 (No. 2017KKJP4X entitled "Innovative numerical methods for evolutionary partial differential equations and applications".

Institutional Review Board Statement: Not applicable.

Informed Consent Statement: Not applicable.

Data Availability Statement: All data are contained within the article.

Acknowledgments: We are grateful to Francesca Romana Bertani, Luca Businaro, Annamaria Gerardino from IFN-CNR and Antonella Sistigu from Istituto Regina Elena for the interesting and clarifying discussions about the microfluidic chip experiment and to Davide Vergni from IAC-CNR for his support in the comprehension of stochastic behavior of cell trajectories.

Conflicts of Interest: The authors declare no conflict of interest.

Appendix A

In this section we discuss the numerical approximation scheme employed for Equations (3) coupled with (4) or (17) in the numerical simulations. The methods used in the numerical simulations employ a 2D finite difference scheme. We consider the spatial domain $\Omega = [0, L_x] \times [0, L_y]$ and we introduce a discretization on L_x in $N-1$ subintervals of length $\Delta x = \dfrac{L_x}{N-1}$ and a discretization on L_y in $M-1$ subintervals of length $\Delta y = \dfrac{L_y}{M-1}$. Then we introduce a Cartesian grid Ω_Δ consisting of grid points (x_n, y_m), where $x_n = n\Delta x$, for $n = 0, \ldots, N-1$ and $y_m = m\Delta y$, for $m = 0, \ldots, M-1$. The same can be done for the time interval $[0, T]$, in this case if Δt is the time step, t_k will be the k-th temporal step, i.e., $t_k = k\Delta t$, for $k = 0, \ldots, N_{\Delta t}$. Note that the spatial steps were chosen as half radius of an IC: $\Delta x = \Delta y = 2\,\mu\text{m}$, and the temporal step as the 1/6 of the video footage timeframe (of 2 min): $\Delta t = 10\,\text{s}$. With the notation $u_{n,m}^k$ we denote the approximation of a function $u(x, y, t)$ at the grid point (x_n, y_m, t_k).

Moreover, since experimentally it was observed that ICs leave the domain Ω, to manage the entrance and exit of cells and avoid numerical instabilities, we added a ghost grid to Ω_d, where the cells lie after having left the main domain. To construct the grid, we considered the extended x interval $[-L_{x^*}, L_x + L_{x^*}]$, with $L_{x^*} > 0$, with nodes $x_n = n\Delta x$, for $n = -N^*, \ldots, 0, \ldots, N^*$, and the extended y interval $[-L_{y^*}, L_y + L_{y^*}]$, with $L_{y^*} > 0$, with nodes $y_m = m\Delta y$, for $m = -M^*, \ldots, 0, \ldots, M^*$. The equations were solved only on Ω_d.

Appendix A.1. Discretization of the PDE

Regarding the approximation of the parabolic Equation (3). The right hand side is composed of the diffusion term, the source term, and the stiff degradation term $-\eta f$.

To eliminate this last quantity, we perform the classical exponential transformation:

$$f(\mathbf{x}, t) = e^{-\eta t} u(\mathbf{x}, t), \tag{A1}$$

which leads to the diffusion equation with source for $u(\mathbf{x}, t)$:

$$\partial_t u = D\Delta u + e^{\eta t}\xi \sum_{j=1}^{N_{tot,c}} \chi_{\mathbf{B}(\mathbf{Y}_j, R_T)}. \tag{A2}$$

For this equation we apply a central difference scheme in space, i.e., the 5-point stencil for the Laplacian, and the parabolic Crank–Nicolson scheme in time.

The numerical scheme can be written as:

$$\frac{u_{n,m}^{k+1} - u_{n,m}^{k}}{\Delta t} = \frac{D}{2}\left(D_x^2 u^{k+1} + D_y^2 u^{k+1}\right) + \frac{D}{2}\left(D_x^2 u^k + D_y^2 u^k\right) +$$
$$+ \frac{1}{2}e^{\eta(k+1)\Delta t}\zeta \sum_{j=1}^{N_{tot,c}} \chi_{\mathbf{B}(\mathbf{Y}_j^{k+1}, R_T)} + \quad \text{(A3)}$$
$$+ \frac{1}{2}e^{\eta k \Delta t}\zeta \sum_{j=1}^{N_{tot,c}} \chi_{\mathbf{B}(\mathbf{Y}_j^{k}, R_T)},$$

where the second finite difference $D_x^2 u$ is defined as the central difference:

$$D_x^2 u^k = \frac{u_{n-1,m}^k - 2u_{n,m}^k + u_{n+1,m}^k}{\Delta x^2}$$

and $D_y^2 u$ is defined analogously.

Appendix A.2. Boundary Conditions

Using the transformation (A1), the associated boundary conditions are:

$$D\frac{\partial u}{\partial \mathbf{n}} + au = e^{\eta t}b, \quad \text{(A4)}$$

that we rewrite as $\frac{\partial u}{\partial \mathbf{n}} + pu = q(t)$, with $p = \frac{a}{D}$ and $q(t) = e^{\eta t}\frac{b}{D}$. Moreover, to distinguish the values of p and q on the different sides of Ω, we number them as follows: p_1 and q_1 are assumed on $y = 0$, p_2 and q_2 on $x = L_x$, p_3 and q_3 on $y = L_y$ and p_4 and q_4 on $x = 0$.

For the discretization of the boundary conditions, we use a central finite difference scheme. On $y = 0$ and $y = L_y$, we have:

$$\frac{\partial u}{\partial y} + p_s u - q_s = \frac{u_{n,m+1}^k - u_{n,m-1}^k}{2\Delta y} + r_s^k u_{n,m} - h_s^k, \quad \text{(A5)}$$

with $s = 1, 3$, while on $x = 0$ and $x = L_x$, we have:

$$\frac{\partial u}{\partial x} + p_s u - q_s^k = \frac{u_{n+1,m}^k - u_{n-1,m}^k}{2\Delta x} + r_s^k u_{n,m} - h_s^k, \quad \text{(A6)}$$

for $s = 2, 4$ and $k \geq 0$.

The signs of r_s and h_s depend on the incoming/outgoing flow. For instance on $y = 0$, we have $r_1 = -p_1$ and $h_1 = -q_1$, on $y = L_y$, we have $r_3 = p_3$ and $h_3 = q_3$, on $x = L_x$, we have $r_4 = -p_4$ and $h_4 = -q_4$ and on $x = 0$ we have $r_2 = p_2$ and $h_2 = q_2$ in order to have an incoming chemical flux through the boundary. The previous discretizations (A5) and (A6) are used with the well-known ghost nodes technique [66].

Appendix A.3. Discretization of the ODE

The equation of the motion (4) is reduced to the first order system

$$\begin{cases} \dot{\mathbf{V}}_i = \frac{\gamma}{\mathcal{W}} \int_{\mathbf{B}(\mathbf{X}_i, R)} \chi(f(\mathbf{x},t))\nabla f(\mathbf{x},t)w_i(\mathbf{x})d\mathbf{x} + \sum_{j:\mathbf{Y}_j \in \mathbf{B}(\mathbf{X}_i, R_2)\setminus\{\mathbf{X}_i\}} \mathbf{K}(\mathbf{Y}_j - \mathbf{X}_i) \text{ (A7)} \\ + \sum_{j:\mathbf{X}_j \in \mathbf{B}(\mathbf{X}_i, R_4)\setminus\{\mathbf{X}_i\}} \mathbf{K}(\mathbf{X}_j - \mathbf{X}_i) - \mu \mathbf{V}_i, \\ \dot{\mathbf{X}}_i = \mathbf{V}_i, \end{cases} \quad \text{(A8)}$$

for $i = 1, \ldots, N_{tot,I}$. Equation (A7) is discretized with a one-step IMEX method, putting in implicit the term containing $\dot{\mathbf{V}}_i$ and in explicit the other addends, see [67]. Equation (A8) is solved with the forward Euler method. The two-dimensional integral in (A7) can be computed by a 2D quadrature formula, which due to the truncated Gaussian weight function $w_i(\mathbf{x})$ given in (6), is reduced to a sum of the discretized integrand functions on the grid points belonging to the ball $\mathbf{B}(\mathbf{X}_i, \bar{R})$. For an integrand function $g(\mathbf{x}, t)$ holds:

$$\int_{\mathbf{B}(\mathbf{X}_i, \bar{R})} g(\mathbf{x}, t) w_i(\mathbf{x}) d\mathbf{x} \approx \sum_{n,m \text{ s.t. } (x_n, x_m) \in \mathbf{B}(\mathbf{X}_i^n, \bar{R})} g_{n,m}^k (w_i)_{n,m}^{(k)},$$

where $(w_i)^{(k)}$ is the weight function centered in $\mathbf{X}^k$, at time step t_k. The same holds for the integral $\mathcal{W}$ defined in (5), which is approximated by:

$$\widetilde{\mathcal{W}} := \sum_{n,m \text{ s.t. } (x_n, x_m) \in \mathbf{B}(\mathbf{X}_i^k, \bar{R})} (w_i)_{n,m}^{(k)}.$$

The gradients of Equation (A7) are approximated with the first order difference:

$$\nabla g(x_n, y_m, t_k) \approx \left(\frac{g_{n+1,m}^k - g_{n,m}^k}{\Delta x}, \frac{g_{n,m+1}^k - g_{n,m}^k}{\Delta y} \right).$$

Equation (A7) is discretized as follows:

$$\begin{aligned}
\frac{\mathbf{V}_i^{k+1} - \mathbf{V}_i^k}{\Delta t} = & \frac{\gamma}{\widetilde{\mathcal{W}}} \sum_{n,m \text{ s.t. } (x_n, x_m) \in \mathbf{B}(\mathbf{X}_i^k, \bar{R})} \chi(f^k)(\nabla_{n,m} f^k)(w_i)_{n,m}^{(k)} \\
& + \sum_{j : \mathbf{X}_j^k \in \mathbf{B}(\mathbf{X}_i^k, R_2) \setminus \{\mathbf{X}_i^k\}} \mathbf{K}(\mathbf{X}_j^k - \mathbf{X}_i^k) \\
& + \sum_{j : \mathbf{X}_j^k \in \mathbf{B}(\mathbf{Y}_i^k, R_4) \setminus \{\mathbf{Y}_i^k\}} \mathbf{K}(\mathbf{Y}_j^k - \mathbf{X}_i^k) - \mu \mathbf{V}_i^{k+1},
\end{aligned} \quad (A9)$$

and Equation (A8) is discretized as $\dfrac{\mathbf{X}_i^{k+1} - \mathbf{X}_i^k}{\Delta t} = \mathbf{V}_i^{k+1}$, with $\mathbf{V}_i^{k+1}$ computed before Equation (A8) is solved.

Appendix A.4. Discretization of the SDE

The stochastic equation of the motion (17) can be decoupled as follows:

$$\begin{cases}
\dot{\mathbf{V}}_i = \dfrac{\gamma}{\mathcal{W}} \displaystyle\int_{\mathbf{B}(\mathbf{X}_i, \bar{R})} \chi(f(\mathbf{x},t)) \nabla f(\mathbf{x},t) w_i(\mathbf{x}) d\mathbf{x} + \sum_{j : \mathbf{Y}_j \in \mathbf{B}(\mathbf{X}_i, R_2) \setminus \{\mathbf{X}_i\}} \mathbf{K}(\mathbf{Y}_j - \mathbf{X}_i) \text{ (A10)} \\
+ \displaystyle\sum_{j : \mathbf{X}_j \in \mathbf{B}(\mathbf{X}_i, R_4) \setminus \{\mathbf{X}_i\}} \mathbf{K}(\mathbf{X}_j - \mathbf{X}_i) - \mu \mathbf{V}_i, \\
\dot{\mathbf{X}}_i = \mathbf{V}_i + \sigma \psi(t).
\end{cases} \quad (A11)$$

for $i = 1, \ldots, N_{tot,I}$. The discretization of Equation (A10) coincides with the one for Equation (A7), while Equation (A11) requires the application of the Euler-Maruyama method [68].

Equation (A11) can be written in the differential form and use $dW(t) = \psi(t)dt$ where $dW(t)$ denotes the differential form of the Brownian motion:

$$d\mathbf{X}_i(t) = \mathbf{V}_i dt + \sigma dW(t), \quad (A12)$$

where $\mathbf{X}_i(t)$ is a one-dimensional Wiener process with drift $\mathbf{V}_i$ and diffusion σ. This equation is discretized with the Euler-Maruyama scheme, which is the stochastic version of the deterministic Euler scheme. The increments of the Wiener process are defined as:

$$\Delta W = W^{k+1} - W^k,$$

with $0 \leq k \leq N_{\Delta t} - 1$. The increment ΔW is a random variable with zero mean and variance equal to Δt:

$$\Delta W \sim \mathcal{N}(0, \Delta t),$$

and with this increment we can construct approximations by drawing normally distributed numbers from a random generator. We approximate the process (A12) at the discrete time points t_k, $0 \leq k \leq N_{\Delta t} - 1$ by

$$\mathbf{X}_i^{k+1} = \mathbf{X}_i^k + \mathbf{V}_i^{k+1} \Delta t + \sigma \Delta W, \tag{A13}$$

where $\Delta W = \sqrt{\Delta t} Z^k$, with Z^k being standard normal variables with mean 0 and variance 1 for all k.

References

1. Businaro, L.; de Ninno, A.; Schiavoni, G.; Lucarini, V.; Ciasca, G.; Gerardino, A.; Belardelli, F.; Gabriele, L.; Mattei, F. Cross talk between cancer and immune cells: Exploring complex dynamics in a microfluidic environment. *Lab Chip* **2013**, *13*, 229–239. [CrossRef]
2. Gori, M.; Simonelli, M.C.; Giannitelli, S.M.; Businaro, L.; Trombetta, M.; Rainer, A. Investigating Nonalcoholic Fatty Liver Disease in a Liver-on-a-Chip Microfluidic Device. *PLoS ONE* **2016**, *11*, e0159729. [CrossRef] [PubMed]
3. Huh, D.; Matthews, B.D.; Mammoto, A.; Montoya-Zavala, M.; Hsin, H.Y.; Ingber, D.E. Reconstituting Organ-Level Lung Functions on a Chip. *Science* **2010**, *328*, 1662–1668. [CrossRef] [PubMed]
4. Graney, P.L.; Tavakol, D.N.; Chramiec, A.; Ronaldson-Bouchard, K.; Vunjak-Novakovic, G. Engineered models of tumor metastasis with immune cell contributions. *iScience* **2021**, *24*, 102179. [CrossRef] [PubMed]
5. Mattei, F.; Andreone, S.; Mencattini, A.; De Ninno, A.; Businaro, L.; Martinelli, E.; Schiavoni, G. Oncoimmunology Meets Organs-on-Chip. *Front. Mol. Biosci.* **2021**, *8*, 627454. [CrossRef]
6. Maulana, T.I.; Kromidas, E.; Wallstabe, L.; Cipriano, M.; Alb, M.; Zaupa, C.; Hudecek, M.; Fogal, B.; Loskill, P. Immunocompetent cancer-on-chip models to assess immuno-oncology therapy. *Adv. Drug Deliv. Rev.* **2021**, *173*, 281–305. [CrossRef]
7. Vacchelli, E.; Ma, Y.; Baracco, E.E.; Sistigu, A.; Enot, D.P.; Pietrocola, F.; Yang, H.; Adjemian, S.; Chaba, K.; Semeraro, M.; et al. Chemotherapy-induced antitumor immunity requires formyl peptide receptor 1. *Science* **2015**, *350*, 972–978. [CrossRef] [PubMed]
8. Parlato, S.; Grisanti, G.; Sinibaldi, G.; Peruzzi, G.; Casciola, C.M.; Gabriele, L. Tumor-on-a-chip platforms to study cancer–immune system crosstalk in the era of immunotherapy. *Lab Chip* **2020**, *21*, 234–253. [CrossRef]
9. Montanez-Sauri, S.I.; Sung, K.E.; Berthier, E.; Beebe, D.J. Enabling screening in 3D microenvironments: Probing matrix and stromal effects on the morphology and proliferation of T47D breast carcinoma cells. *Integr. Biol.* **2013**, *5*, 631–640. [CrossRef]
10. Hassell, B.; Goyal, G.; Lee, E.; Sontheimer-Phelps, A.; Levy, O.; Chen, C.; Ingber, D.E. Human Organ Chip Models Recapitulate Orthotopic Lung Cancer Growth, Therapeutic Responses, and Tumor Dormancy In Vitro. *Cell Rep.* **2017**, *21*, 508–516. [CrossRef]
11. Baker, B.; Trappmann, B.; Stapleton, S.C.; Toro, E.; Chen, C. Microfluidics embedded within extracellular matrix to define vascular architectures and pattern diffusive gradients. *Lab Chip* **2013**, *13*, 3246–3252. [CrossRef]
12. Bischel, L.L.; Young, E.W.; Mader, B.R.; Beebe, D.J. Tubeless microfluidic angiogenesis assay with three-dimensional endothelial-lined microvessels. *Biomaterials* **2012**, *34*, 1471–1477. [CrossRef]
13. Chen, M.B.; Whisler, J.A.; Jeon, J.; Kamm, R.D. Mechanisms of tumor cell extravasation in an in vitro microvascular network platform. *Integr. Biol.* **2013**, *5*, 1262–1271. [CrossRef]
14. Moya, M.L.; Hsu, Y.-H.; Lee, A.; Hughes, C.C.; George, S.C. In Vitro Perfused Human Capillary Networks. *Tissue Eng. Part C Methods* **2013**, *19*, 730–737. [CrossRef]
15. Nguyen, D.-H.T.; Stapleton, S.C.; Yang, M.T.; Cha, S.S.; Choi, C.K.; Galie, P.; Chen, C.S. Biomimetic model to reconstitute angiogenic sprouting morphogenesis in vitro. *Proc. Natl. Acad. Sci. USA* **2013**, *110*, 6712–6717. [CrossRef]
16. Wang, X.; Phan, D.T.T.; Sobrino, A.; George, S.C.; Hughes, C.C.W.; Lee, A.P. Engineering anastomosis between living capillary networks and endothelial cell-lined microfluidic channels. *Lab Chip* **2015**, *16*, 282–290. [CrossRef] [PubMed]
17. Jeong, S.-Y.; Lee, J.-H.; Shin, Y.; Chung, S.; Kuh, H.-J. Co-Culture of Tumor Spheroids and Fibroblasts in a Collagen Matrix-Incorporated Microfluidic Chip Mimics Reciprocal Activation in Solid Tumor Microenvironment. *PLoS ONE* **2016**, *11*, e0159013. [CrossRef]
18. Lee, J.-H.; Kim, S.-K.; Khawar, I.A.; Jeong, S.-Y.; Chung, S.; Kuh, H.-J. Microfluidic co-culture of pancreatic tumor spheroids with stellate cells as a novel 3D model for investigation of stroma-mediated cell motility and drug resistance. *J. Exp. Clin. Cancer Res.* **2018**, *37*, 1–12. [CrossRef] [PubMed]

19. Ramaswamy, S.; Ross, K.N.; Lander, E.S.; Golub, T.R. A molecular signature of metastasis in primary solid tumors. *Nat. Genet.* **2002**, *33*, 49–54. [CrossRef] [PubMed]
20. Zervantonakis, I.; Hughes, S.; Charest, J.L.; Condeelis, J.S.; Gertler, F.B.; Kamm, R.D. Three-dimensional microfluidic model for tumor cell intravasation and endothelial barrier function. *Proc. Natl. Acad. Sci. USA* **2012**, *109*, 13515–13520. [CrossRef] [PubMed]
21. Braun, E.; Bretti, G.; Natalini, R. Mass-Preserving Approximation of a Chemotaxis Multi-Domain Transmission Model for Microfluidic Chips. *Mathematics* **2021**, *9*, 688. [CrossRef]
22. Natalini, R.; Paul, T. On the Mean Field limit for Cucker-Smale models. *Discret. Contin. Dyn.-Syst. B* **2021**. [CrossRef]
23. Agliari, E.; Biselli, E.; De Ninno, A.; Schiavoni, G.; Gabriele, L.; Gerardino, A.; Mattei, F.; Barra, A.; Businaro, L. Cancer-driven dynamics of immune cells in a microfluidic environment. *Sci. Rep.* **2014**, *4*, 6639. [CrossRef]
24. Zhong, R.X.; Fu, K.Y.; Sumalee, A.; Ngoduy, D.; Lam, W.H. A cross-entropy method and probabilistic sensitivity analysis framework for calibrating microscopic traffic models. *Transp. Res. Part C Emerg. Technol.* **2016**, *63*, 147–169. [CrossRef] [PubMed]
25. Kirschner, D.; Panetta, J.C. Modeling immunotherapy of the tumor-immune interaction. *J. Math. Biol.* **1998**, *37*, 235–252. [CrossRef]
26. Lee, S.W.L.; Seager, R.J.; Litvak, F.; Spill, F.; Sieow, J.L.; Leong, P.H.; Kumar, D.; Tan, A.S.M.; Wong, S.C.; Adriani, G.; et al. Integrated in silico and 3D in vitro model of macrophage migration in response to physical and chemical factors in the tumor microenvironment. *Integr. Biol.* **2020**, *12*, 90–108. [CrossRef] [PubMed]
27. Yang, T.D.; Park, J.-S.; Choi, Y.; Choi, W.; Ko, T.-W.; Lee, K.J. Zigzag Turning Preference of Freely Crawling Cells. *PLoS ONE* **2011**, *6*, e20255. [CrossRef]
28. Checcoli, A.; Pol, J.G.; Naldi, A.; Noel, V.; Barillot, E.; Kroemer, G.; Thieffry, D.; Calzone, L.; Stoll, G. Dynamical Boolean Modeling of Immunogenic Cell Death. *Front. Physiol.* **2020**, *11*, 1320. [CrossRef]
29. Braun, E.C. Organs-On-Chips: Mathematical Modelling and Parameter Estimation. Ph.D. Thesis, Universitá degli Studi di Roma Tre, Roma, Italy, 2021. [CrossRef]
30. Di Costanzo, E.; Natalini, R.; Preziosi, L. A hybrid mathematical model for self-organizing cell migration in the zebrafish lateral line. *J. Math. Biol.* **2014**, *71*, 171–214.
31. Peri, D. Easy-to-implement multidimensional spline interpolation with application to ship design optimisation. *Ship Technol. Res.* **2017**, *65*, 32–46. [CrossRef]
32. Stevens, A.; Othmer, H.G. Aggregation, Blowup, and Collapse: The ABC's of Taxis in Reinforced Random Walks. *SIAM J. Appl. Math.* **1997**, *57*, 1044–1081. [CrossRef]
33. Pomeau, Y.; Piasecki, J. The Langevin equation. *C. R. Phys.* **2017**, *18*, 570–582. [CrossRef]
34. Bai, J.; Tu, T.-Y.; Kim, C.; Thiery, J.P.; Kamm, R.D. Identification of drugs as single agents or in combination to prevent carcinoma dissemination in a microfluidic 3D environment. *Oncotarget* **2015**, *6*, 36603–36614. [CrossRef]
35. Xu, Z.; Gao, Y.; Hao, Y.; Li, E.; Wang, Y.; Zhang, J.; Wang, W.; Gao, Z.; Wang, Q. Application of a microfluidic chip-based 3D co-culture to test drug sensitivity for individualized treatment of lung cancer. *Biomaterials* **2013**, *34*, 4109–4117. [CrossRef] [PubMed]
36. De Ninno, A.; Bertani, F.R.; Gerardino, A.; Schiavoni, G.; Musella, M.; Galassi, C.; Mattei, F.; Sistigu, A.; Businaro, L. Microfluidic Co-Culture Models for Dissecting the Immune Response in in vitro Tumor Microenvironments. *J. Vis. Exp.* **2021**, *170*, e61895. [CrossRef] [PubMed]
37. Kroemer, G.; Galluzzi, L.; Kepp, O.; Zitvogel, L. Immunogenic Cell Death in Cancer Therapy. *Annu. Rev. Immunol.* **2013**, *31*, 51–72. [CrossRef]
38. Greenberg, J.M.; Alt, W. Stability results for a diffusion equation with functional drift approximating a chemotaxis model. *Trans. Am. Math. Soc.* **1987**, *300*, 235. [CrossRef]
39. Keller, E.F.; Segel, L.A. Initiation of slime mold aggregation viewed as an instability. *J. Theor. Biol.* **1970**, *26*, 399–415. [CrossRef]
40. Péré, M.; Chaves, M.; Roux, J. Core Models of Receptor Reactions to Evaluate Basic Pathway Designs Enabling Heterogeneous Commitments to Apoptosis. In Proceedings of the International Conference on Computational Methods in Systems Biology, Konstanz, Germany, 23–25 September 2020; pp. 298–320. [CrossRef]
41. Edalgo, Y.T.N.; Zornes, A.L.; Versypt, A.N.F. A hybrid discrete–continuous model of metastatic cancer cell migration through a remodeling extracellular matrix. *AIChE J.* **2019**, *65*, e16671. [CrossRef]
42. Othmer, H.G.; Kim, Y. Hybrid models of cell and tissue dynamics in tumor growth. *Math. Biosci. Eng.* **2015**, *12*, 1141–1156. [CrossRef]
43. Perfahl, H.; Hughes, B.; Alarcon, T.; Maini, P.K.; Lloyd, M.C.; Reuss, M.; Byrne, H.M. 3D hybrid modelling of vascular network formation. *J. Theor. Biol.* **2017**, *414*, 254–268. [CrossRef]
44. Rousset, M.; Samaey, G. Simulating individual-based models of bacterial chemotaxis with asymptotic variance reduction. *Math. Model. Methods Appl. Sci.* **2013**, *23*, 2155–2191. [CrossRef]
45. Macfarlane, F.R.; Chaplain, M.A.J.; Lorenzi, T. A hybrid discrete-continuum approach to model Turing pattern formation. *Math. Biosci. Eng.* **2017**, *17*, 7442–7479. [CrossRef]
46. Othmer, H.G. Cell-Based, Continuum and Hybrid Models of Tissue Dynamics. In *Mathematical Models and Methods for Living Systems Book Series: Lecture Notes in Mathematics*; Springer International Publishing Switzerland: Cham, Switzerland, 2016; pp. 1–72. [CrossRef]

47. Mertz, A.F.; Che, Y.; Banerjee, S.; Goldstein, J.M.; Rosowski, K.A.; Revilla, S.F.; Niessen, C.M.; Marchetti, M.C.; Dufresne, E.R.; Horsley, V. Cadherin-based intercellular adhesions organize epithelial cell-matrix traction forces. *Proc. Natl. Acad. Sci. USA* **2012**, *110*, 842–847. [CrossRef]
48. D'Orsogna, M.R.; Chuang, Y.L.; Bertozzi, A.L.; Chayes, L.S. Self-Propelled Particles with Soft-Core Interactions: Patterns, Stability, and Collapse. *Phys. Rev. Lett.* **2006**, *96*, 104302. [CrossRef]
49. Bayly, P.V.; Taber, L.A.; Carlsson, A.E. Damped and persistent oscillations in a simple model of cell crawling. *J. R. Soc. Interface* **2011**, *9*, 1241–1253. [CrossRef] [PubMed]
50. Fournier, M.F.; Sauser, R.; Ambrosi, D.; Meister, J.-J.; Verkhovsky, A. Force transmission in migrating cells. *J. Cell Biol.* **2010**, *188*, 287–297. [CrossRef] [PubMed]
51. Rubinstein, B.; Fournier, M.F.; Jacobson, K.; Verkhovsky, A.; Mogilner, A. Actin-Myosin Viscoelastic Flow in the Keratocyte Lamellipod. *Biophys. J.* **2009**, *97*, 1853–1863. [CrossRef] [PubMed]
52. Lapidus, I.; Schiller, R. Model for the chemotactic response of a bacterial population. *Biophys. J.* **1976**, *16*, 779–789. [CrossRef]
53. Cristiani, E.; Piccoli, B.; Tosin, A. Multiscale Modeling of Granular Flows with Application to Crowd Dynamics. *Multiscale Model. Simul.* **2011**, *9*, 155–182. [CrossRef]
54. Scianna, M.; Tosin, A.; Preziosi, L. From discrete to continuous models of cell colonies: A measure-theoretic approach. *arXiv* **2011**, arXiv:1108.1212. [CrossRef]
55. Weninger, W.; Biro, M.; Jain, R. Leukocyte migration in the interstitial space of non-lymphoid organs. *Nat. Rev. Immunol.* **2014**, *14*, 232–246.
56. Miller, M.J.; Wei, S.H.; Parker, I.; Cahalan, M.D. Two-Photon Imaging of Lymphocyte Motility and Antigen Response in Intact Lymph Node. *Science* **2002**, *296*, 1869–1873. [CrossRef] [PubMed]
57. Wiśniewski, J.R.; Hein, M.; Cox, J.; Mann, M. A "Proteomic Ruler" for Protein Copy Number and Concentration Estimation without Spike-in Standards. *Mol. Cell. Proteom.* **2014**, *13*, 3497–3506. [CrossRef] [PubMed]
58. Boulter, E.; Grall, D.; Cagnol, S.; Van Obberghen-Schilling, E. Regulation of cell-matrix adhesion dynamics and Rac-1 by integrin linked kinase. *FASEB J.* **2006**, *20*, 1489–1491. [CrossRef]
59. Puck, T.T.; Marcus, P.I.; Cieciura, S.J. Clonal growth of mammalian cells in vitro. *J. Exp. Med.* **1956**, *103*, 273–284. [CrossRef]
60. Murray, J.D. *Mathematical Biology. II: Spatial Models and Biomedical Applications*, 3rd ed.; Springer: Berlin, Germany, 2002. [CrossRef] [PubMed]
61. Curk, T.; Marenduzzo, D.; Dobnikar, J. Chemotactic Sensing towards Ambient and Secreted Attractant Drives Collective Behaviour of E. coli. *PLoS ONE* **2013**, *8*, e74878.
62. Engl, H.W.; Hanke, M.; Neubauer, A. *Regularization of Inverse Problems*; Springer: Dordrecht, The Netherlands, 2000; Volume 375. [CrossRef] [PubMed]
63. Kennedy, J.; Eberhart, R. Particle swarm optimization. In Proceedings of the ICNN'95–International Conference on Neural Networks, Perth, WA, Australia, 27 November–1 December 1995; Volume 4, pp. 1942–1948.
64. Harris, T.; Banigan, E.; Christian, D.A.; Konradt, C.; Wojno, E.T.; Norose, K.; Wilson, E.H.; John, B.; Weninger, W.; Luster, A.D.; et al. Generalized Lévy walks and the role of chemokines in migration of effector CD8+ T cells. *Nature* **2012**, *486*, 545–548. [CrossRef]
65. Li, H.; Qi, S.; Jin, H.; Qi, Z.; Zhang, Z.; Fu, L.; Luo, Q. Zigzag Generalized Lévy Walk: The In Vivo Search Strategy of Immunocytes. *Theranostics* **2015**, *5*, 1275–1290. [CrossRef] [PubMed]
66. Strikwerda, J.C. *Finite Difference Schemes and Partial Differential Equations*, 2nd ed.; Society for Industrial & Applied Mathematics: Philadelphia, PA, USA, 2004. [CrossRef]
67. Hundsdorfer, W.; Verwer, J.G. Numerical solution of time-dependent advection–diffusion–reaction equations. In *Computational Mathematics*; Springer: Berlin, Germany, 2003. [CrossRef]
68. Higham, D.J. An Algorithmic Introduction to Numerical Simulation of Stochastic Differential Equations. *SIAM Rev.* **2001**, *43*, 525–546.

Article

Relaxation Limit of the Aggregation Equation with Pointy Potential

Benoît Fabrèges [1], Frédéric Lagoutière [1], Sébastien Tran Tien [1] and Nicolas Vauchelet [2,*]

[1] Univ Lyon, Université Claude Bernard Lyon 1, CNRS UMR 5208, Institut Camille Jordan, 43 Blvd. du 11 Novembre 1918, CEDEX, F-69622 Villeurbanne, France; fabreges@math.univ-lyon1.fr (B.F.); lagoutiere@math.univ-lyon1.fr (F.L.); trantien@math.univ-lyon1.fr (S.T.T.)

[2] Laboratoire Analyse, Géométrie et Applications CNRS UMR 7539, Université Sorbonne Paris Nord, 93430 Villetaneuse, France

* Correspondence: vauchelet@math.univ-paris13.fr

Citation: Fabrèges, B.; Lagoutière, F.; Tran Tien, S.; Vauchelet, N Relaxation Limit of the Aggregation Equation with Pointy Potential. *Axioms* **2021**, *10*, 108. https://doi.org/10.3390/axioms10020108

Academic Editor: Giampiero Palatucci

Received: 9 April 2021
Accepted: 26 May 2021
Published: 28 May 2021

Publisher's Note: MDPI stays neutral with regard to jurisdictional claims in published maps and institutional affiliations.

Copyright: © 2021 by the authors. Licensee MDPI, Basel, Switzerland. This article is an open access article distributed under the terms and conditions of the Creative Commons Attribution (CC BY) license (https://creativecommons.org/licenses/by/4.0/).

Abstract: This work was devoted to the study of a relaxation limit of the so-called aggregation equation with a pointy potential in one-dimensional space. The aggregation equation is today widely used to model the dynamics of a density of individuals attracting each other through a potential. When this potential is pointy, solutions are known to blow up in final time. For this reason, measure-valued solutions have been defined. In this paper, we investigated an approximation of such measure-valued solutions thanks to a relaxation limit in the spirit of Jin and Xin. We study the convergence of this approximation and give a rigorous estimate of the speed of convergence in one dimension with the Newtonian potential. We also investigated the numerical discretization of this relaxation limit by uniformly accurate schemes.

Keywords: aggregation equation; relaxation limit; scalar conservation law; finite volume scheme

MSC: 35L65; 65M12; 35D30

1. Introduction

The so-called aggregation equation has been widely used to model the dynamics of a population of individuals in interaction. Let $W : \mathbb{R} \to \mathbb{R}$, sufficiently smooth, be the interaction potential governing the population. Then, in one dimension in space, the dynamics of the density of individuals, denoted by ρ, is governed by the following equation, for $t > 0$ and $x \in \mathbb{R}$:

$$\partial_t \rho + \partial_x (a[\rho]\rho) = 0, \quad \text{with} \quad a[\rho] = -W' * \rho. \tag{1}$$

Such equations appear in many applications in population dynamics: for instance, to describe the collective migration of cells by swarming, the motion of bacteria by chemotaxis, the crowd motion, the flocking of birds, or fishes school, see, e.g., [1–7]. From a mathematical point of view, these equations have been widely studied. When the potential W is not smooth enough, it is known that weak solutions may blow up in finite time [8,9]. Thus, the existence of weak (measure) solutions has been investigated in, e.g., [10,11].

In this paper, we consider a relaxation limit in the spirit of Jin–Xin [12] of the aggregation equation in one space dimension on $\mathbb{R}$. It is now well-established that such modifications allow regularizing the solutions. For a given $c > \|a\|_\infty$, we introduce the system:

$$\partial_t \rho + \partial_x \sigma = 0, \tag{2a}$$

$$\partial_t \sigma + c^2 \partial_x \rho = \frac{1}{\varepsilon}(a[\rho]\rho - \sigma) \tag{2b}$$

$$a[\rho] = -W' * \rho \tag{2c}$$

This system is complemented by initial data ρ_0 and $\sigma_0 := a[\rho_0]\rho_0$. It is clear, at least formally, that when $\varepsilon \to 0$, the solution ρ of system (2) converges to the one of the aggregation equations (1) (and it is actually only true if $c > \|a\|_\infty$). We mention that the aggregation equation may also be derived thanks to a hydrodynamical limit of kinetic equations [6,7,13].

The aim of this work was to study the convergence as $\varepsilon \to 0$ of the relaxation system (2) towards the aggregation equation. More precisely, we establish a precise estimate of the speed of convergence, and we also illustrate with some numerical simulations. These estimates are obtained only in the case of the Newtonian potential in one dimension $W(x) = \frac{1}{2}|x|$. Indeed, in this particular case, we may link the aggregation equation to a scalar conservation law [14,15]. The same link holds for the relaxation system (2)—denoting:

$$u(t,x) = \frac{1}{2} - \int_{-\infty}^{x} \rho(t,dy), \qquad v(t,x) = \frac{1}{2} - \int_{-\infty}^{x} \sigma(t,dy),$$

where the notation $\int \rho(t,dy)$ stands for the integral with respect to the probability measure $\rho(t)$, then we verify easily that:

$$u = -W' * \rho, \qquad \rho = -\partial_x u,$$

so that $a[\rho] = u$. Then, integrating (2), we deduce that (u,v) is a solution to:

$$\partial_t u + \partial_x vs. = 0 \tag{3a}$$

$$\partial_t vs. + c^2 \partial_x u = \frac{1}{\varepsilon}\left(\frac{1}{2}u^2 - v\right), \tag{3b}$$

which is complemented with the initial data $u_0 = \frac{1}{2} - \int_{-\infty}^{x} \rho_0(dy)$, and $v_0 = \frac{1}{2} - \int_{-\infty}^{x} \sigma_0(dy)$. Clearly, as $\varepsilon \to 0$, we expect that the solution of the above system converges to the solution of the following Burgers equation:

$$\partial_t u + \frac{1}{2}\partial_x u^2 = 0.$$

Introducing the quantities $a = v - cu$ and $b = v + cu$, (3) is equivalent to the diagonalized system:

$$\partial_t a - c\partial_x a = \frac{1}{\varepsilon}\left(\frac{1}{2}\left(\frac{b-a}{2c}\right)^2 - \frac{a+b}{2}\right) \tag{4a}$$

$$\partial_t b + c\partial_x b = \frac{1}{\varepsilon}\left(\frac{1}{2}\left(\frac{b-a}{2c}\right)^2 - \frac{a+b}{2}\right). \tag{4b}$$

We will adapt the techniques developed in [16] to obtain convergence estimates for our system.

In order to illustrate this convergence result, numerical discretizations of the relaxation system (2) are investigated. The schemes we propose are such that they are uniform with respect to ε, that is they satisfy the so-called asymptotic preserving (AP) property [17]. Therefore, such schemes in the limit $\varepsilon \to 0$ must be consistent with the aggregation equation. The numerical simulations of solutions of the aggregation equation for pointy potentials have been studied by several authors, see, e.g., [11,13,18–22]. In particular, some authors pay attention to recover the correct behavior of the numerical solutions after the blow-up time. To do so, particular attention must be paid to the definition of the product $a[\rho]\rho$ when ρ is a measure.

In this article, we propose two discretizations of the relaxation system which satisfy the AP property. In a first approach, we propose a simple splitting algorithm where we split the transport part and the right hand side in system (2). It results in a numerical scheme which is very simple to implement and for which we easily verify the AP property. The second approach relies on a well-balanced discretization in the spirit of [20,23]. This scheme is more expensive to implement than the first scheme, but its numerical solution has less diffusion, as it is illustrated by our numerical results.

The outline of the paper is the following. In Section 2, after recalling some useful notations, we prove our main result: an estimation of the speed of convergence in the Wasserstein W_1 distance with respect to ε of the solutions of the relaxation system (2) towards the solution of the aggregation Equation (1) in the case $W(x) = \frac{1}{2}|x|$. The numerical discretization is investigated in Section 3. Two numerical schemes verifying the AP property are proposed. The first scheme is based on a splitting algorithm, whereas the second scheme relies on a well-balanced discretization. Numerical results and comparisons are provided in Section 4.

2. Convergence Result

2.1. Notations

Before stating and proving our main results, we first recall some useful notations and results. Since we are dealing with conservation laws (in which the total mass is conserved), we will work in some space of probability measures, namely the Wasserstein space of order $p \geq 1$, which is the space of probability measures with a finite order p moment:

$$\mathcal{P}_p(\mathbb{R}^N) = \left\{ \mu \text{ nonnegative Borel measure}, \mu(\mathbb{R}^N) = 1, \int |x|^p \mu(dx) < \infty \right\}.$$

This space is endowed with the Wasserstein distance defined by (see, e.g., [24,25])

$$W_p(\mu,\nu) = \inf_{\gamma \in \Gamma(\mu,\nu)} \left\{ \int |y-x|^p \, \gamma(dx,dy) \right\}^{1/p}, \tag{5}$$

where $\Gamma(\mu,\nu)$ is the set of measures on $\mathbb{R}^N \times \mathbb{R}^N$ with marginals μ and ν, meaning that:

$$\Gamma(\mu,\nu) = \left\{ \gamma \in \mathcal{P}_p(\mathbb{R}^N \times \mathbb{R}^N); \forall \xi \in C_0(\mathbb{R}^N), \int_{\mathbb{R}^{2N}} \xi(y_0)\gamma(dy_0,dy_1) = \int_{\mathbb{R}^N} \xi(y_0)\mu(dy_0), \right.$$
$$\left. \int_{\mathbb{R}^{2N}} \xi(y_1)\gamma(dy_0,dy_1) = \int_{\mathbb{R}^N} \xi(y_1)\nu(dy_1) \right\},$$

with $C_0(\mathbb{R}^N)$, the set of continuous functions on $\mathbb{R}^N$ that vanish at infinity. From a simple minimization argument, we know that in the definition of W_p, the infimum is actually a minimum. A map that realizes the minimum in the definition (5) of W_p is called an optimal transport plan, the set of which is denoted by $\Gamma_0(\mu,\nu)$.

In the one-dimensional framework, we may simplify these definitions. Indeed, any probability measure μ on the real line $\mathbb{R}$ can be described in terms of its cumulative distribution function $F_\mu(x) = \mu((-\infty,x))$, which is a right-continuous and non-decreasing function with $F_\mu(-\infty) = 0$ and $F_\mu(+\infty) = 1$. Then, we can define the generalized inverse F_μ^{-1} of F_μ (or monotone rearrangement of μ) by $F_\mu^{-1}(z) := \inf\{x \in \mathbb{R}/F_\mu(x) > z\}$, it is a right-continuous and non-decreasing function as well, defined on $[0,1]$. We have for every non-negative Borel map ξ:

$$\int_\mathbb{R} \xi(x)\mu(dx) = \int_0^1 \xi(F_\mu^{-1}(z))\,dz.$$

In particular, $\mu \in \mathcal{P}_p(\mathbb{R})$ if and only if $F_\mu^{-1} \in L^p(0,1)$. Moreover, in the one-dimensional setting, there exists a unique optimal transport plan realizing the minimum in (5). More precisely, if μ and ν belong to $\mathcal{P}_p(\mathbb{R})$, with monotone rearrangements F_μ^{-1} and F_ν^{-1}, then $\Gamma_0(\mu,\nu) = \{(F_\mu^{-1}, F_\nu^{-1})_\# \mathbb{L}_{(0,1)}\}$ where $\mathbb{L}_{(0,1)}$ is the restriction of the Lebesgue measure on $(0,1)$. Thus, we have the explicit expression of the Wasserstein distance (see [24,26,27]):

$$W_p(\mu,\nu) = \left(\int_0^1 |F_\mu^{-1}(z) - F_\nu^{-1}(z)|^p \, dz\right)^{1/p}, \tag{6}$$

and the map $\mu \mapsto F_\mu^{-1}$ is an isometry between $\mathcal{P}_p(\mathbb{R})$ and the convex subset of (essentially) non-decreasing functions of $L^p(0,1)$.

2.2. Convergence Estimates

Let us first consider the limit $\varepsilon \to 0$ for the system (3). Compactness methods were used in [28] to get L^1_{loc} convergence in space. However, in order to pass to the aggregation equation, one may want global L^1 convergence, which we prove in the following theorem, along the lines of Katsoulakis and Tzavaras [16].

Theorem 1. Let $u_0 \in L^\infty \cap BV(\mathbb{R})$, $c > \|u_0\|_{L^\infty}$ and set $v_0 = \frac{u_0^2}{2}$. There exists a constant $C > 0$ such that, for any $\varepsilon > 0$, denoting by $(u^\varepsilon, v^\varepsilon)$ the solution to (3) with initial data (u_0, v_0), the following estimate holds:

$$\forall T > 0, \qquad \|u(T) - u^\varepsilon(T)\|_{L^1} \leq CTV(u_0)(\sqrt{\varepsilon T} + \varepsilon),$$

where u is the entropy solution to the Burgers equation with initial datum u_0.

Proof. Denote $(a^\varepsilon, b^\varepsilon)$ the solution to (4), and $G(a,b) = \frac{1}{2}\left(\frac{b-a}{2c}\right)^2 - \frac{a+b}{2}$.

So as to obtain entropy inequalities on $(a^\varepsilon, b^\varepsilon)$, we need monotonicity properties on G. One can check that $G(a^\varepsilon, b^\varepsilon)$ is decreasing with respect to a^ε and b^ε if the so-called subcharacteristic condition $|u^\varepsilon| < c$ holds. Up to a slight modification of the nonlinear term $f(u^\varepsilon) = \frac{(u^\varepsilon)^2}{2}$ in (3), which does not affect the value of $(a^\varepsilon, b^\varepsilon)$:

$$f(u) := \begin{cases} -\|u_0\| u - \frac{\|u_0\|^2}{2}, & \text{if } u \leq -\|u_0\|, \\ \frac{u^2}{2}, & \text{if } -\|u_0\| \leq u \leq \|u_0\|, \\ \|u_0\| u - \frac{\|u_0\|^2}{2}, & \text{if } \|u_0\| \leq u, \end{cases}$$

the choice $c > \|u_0\|_{L^\infty}$ ensures that the subcharacteristic condition and the bound $\|u^\varepsilon(t)\|_{L^\infty} \leq \|u_0\|_{L^\infty}$ holds for all time.

Now, obtaining entropy inequalities on $(a^\varepsilon, b^\varepsilon)$ consists of making a comparison with constant state solutions to (4). Namely, letting $m = \|u_0\|_{L^\infty}\left(\frac{\|u_0\|_{L^\infty}}{2} - c\right)$, $M = \|u_0\|_{L^\infty}\left(\frac{\|u_0\|_{L^\infty}}{2} + c\right)$ and $h(a) = a + 2c^2 - 2c\sqrt{c^2 + 2a}$, we have $G(k, h(k)) = 0$ for all $k \in [m, M]$, and therefore $(k, h(k))$ is a solution to (4). Thus, the following system holds:

$$\partial_t(a^\varepsilon - k) - c\partial_x(a^\varepsilon - k) = \frac{1}{\varepsilon}\Big(G(a^\varepsilon, b^\varepsilon) - G(k, h(k))\Big), \tag{7a}$$

$$\partial_t(b^\varepsilon - h(k)) + c\partial_x(b^\varepsilon - h(k)) = \frac{1}{\varepsilon}\Big(G(a^\varepsilon, b^\varepsilon) - G(k, h(k))\Big). \tag{7b}$$

Multiplying (7a) by $\operatorname{sgn}(a^\varepsilon - k)$, (7b) by $\operatorname{sgn}(b^\varepsilon - h(k))$ and summing yields:

$$\partial_t\Big(|a^\varepsilon - k| + |b^\varepsilon - h(k)|\Big) - c\partial_x\Big(|a^\varepsilon - k| - |b^\varepsilon - h(k)|\Big)$$
$$= \frac{1}{\varepsilon}\Big(\operatorname{sgn}(a^\varepsilon - k) + \operatorname{sgn}(b^\varepsilon - h(k))\Big)\Big(G(a^\varepsilon, b^\varepsilon) - G(k, h(k))\Big).$$

Hence, using the monotonicity of G, we obtain the following entropy inequalities on $(a^\varepsilon, b^\varepsilon)$:

$$\partial_t\Big(|a^\varepsilon - k| + |b^\varepsilon - h(k)|\Big) - c\partial_x\Big(|a^\varepsilon - k| - |b^\varepsilon - h(k)|\Big) \leq 0. \tag{8}$$

We now turn to proving the entropy inequalities on u^ε. Straightforward computations yield the existence of a constant $C > 0$ such that, for all $a, b \in [m, M]$, one has $|h(a) - b| \leq C|G(a,b)|$. We therefore work on the variable $w^\varepsilon := \frac{h(a^\varepsilon) - a^\varepsilon}{2c}$ in the first place. Let $\kappa \in \big[-\|u_0\|_{L^\infty}, \|u_0\|_{L^\infty}\big]$, and $k \in [m, M]$ such that $\kappa = \frac{h(k) - k}{2c}$. We have:

$$|w^\varepsilon - \kappa| = \frac{1}{2c}\Big(|h(a^\varepsilon) - h(k)| + |a^\varepsilon - k|\Big) = \frac{1}{2c}\Big(|a^\varepsilon - k| + |b^\varepsilon - h(k)| + r_1^\varepsilon\Big), \tag{9}$$

where $r_1^\varepsilon = |h(a^\varepsilon) - h(k)| - |b^\varepsilon - h(k)|$ verifies $|r_1^\varepsilon| \leq |h(a^\varepsilon) - b^\varepsilon| \leq C|G(a^\varepsilon, b^\varepsilon)|$. Thus, we are left to control $|G(a^\varepsilon, b^\varepsilon)|$. To do so, we formally differentiate this quantity and use (4):

$$\partial_t|G(a^\varepsilon, b^\varepsilon)| = \Big(\partial_t a^\varepsilon \partial_a G(a^\varepsilon, b^\varepsilon) + \partial_t b^\varepsilon \partial_b G(a^\varepsilon, b^\varepsilon)\Big)\operatorname{sgn}(G(a^\varepsilon, b^\varepsilon)),$$
$$= \frac{1}{\varepsilon}\Big(\partial_a G(a^\varepsilon, b^\varepsilon) + \partial_b G(a^\varepsilon, b^\varepsilon)\Big)|G(a^\varepsilon, b^\varepsilon)|$$
$$\quad - c\operatorname{sgn}(G(a^\varepsilon, b^\varepsilon))\Big(\partial_x a^\varepsilon \partial_a G(a^\varepsilon, b^\varepsilon) + \partial_x b^\varepsilon \partial_b G(a^\varepsilon, b^\varepsilon)\Big),$$
$$\leq \frac{1}{\varepsilon}\sup_{[m,M]^2}\Big(\partial_a G + \partial_b G\Big)|G(a^\varepsilon, b^\varepsilon)| + c\sup_{[m,M]^2}\Big(|\partial_a G| + |\partial_b G|\Big)\Big(|\partial_x a^\varepsilon| + |\partial_x b^\varepsilon|\Big).$$

Integrating in space gives:

$$\frac{\mathrm{d}}{\mathrm{d}t}\|G(a^\varepsilon, b^\varepsilon)\|_{L^1} \leq -\frac{A}{\varepsilon}\|G(a^\varepsilon, b^\varepsilon)\|_{L^1} + B\Big(TV(a_0) + TV(b_0)\Big),$$

where $A = -\sup_{[m,M]^2}(\partial_a G + \partial_b G)$ and $B = c\sup_{[m,M]^2}(|\partial_a G| + |\partial_b G|)$ are positive constants which do not depend on ε nor on time. A Gronwall lemma then gives:

$$\|G(a^\varepsilon(t), b^\varepsilon(t))\|_{L^1} \leq C\Big(TV(a_0) + TV(b_0)\Big)\varepsilon, \tag{10}$$

where we still denote $C = B/A$ as a constant independent of time and of ε.

In addition, since $G(a, h(a)) = 0$, one has $\frac{1}{2}\left(\frac{h(a)-a}{2c}\right)^2 = \frac{1}{2}(h(a) + a)$ and therefore:

$$\operatorname{sgn}(w^\varepsilon - \kappa)\left(\frac{(w^\varepsilon)^2}{2} - \frac{\kappa^2}{2}\right) = \frac{1}{2}\operatorname{sgn}\Big(h(a^\varepsilon) - h(k) - (a^\varepsilon - k)\Big)\Big(h(a^\varepsilon) + a^\varepsilon - (h(k) + k)\Big),$$
$$= \frac{1}{2}\Big(|h(a^\varepsilon) - h(k)| - |a^\varepsilon - k|\Big),$$
$$= \frac{1}{2}\Big(|b - h(k)| - |a^\varepsilon - k| + r_2^\varepsilon\Big), \tag{11}$$

with $|r_2^\varepsilon| \leq C|G(a^\varepsilon, b^\varepsilon)|$. Differentiating (9) in time and (11) in space, and using (8) thus yields:

$$\partial_t|w^\varepsilon - \kappa| + \partial_x \operatorname{sgn}(w^\varepsilon - \kappa)\left(\frac{(w^\varepsilon)^2}{2} - \frac{\kappa^2}{2}\right) \leq \frac{1}{2c}\Big(\partial_t r_1^\varepsilon + c\partial_x r_2^\varepsilon\Big). \tag{12}$$

Then, we estimate $\|u(t) - w^\varepsilon(t)\|_{L^1}$ using Kuznetsov's doubling of variables technique (see, e.g., [29] for scalar conservation laws with viscosity and [30] for a more general formalism) in order to combine (12) with Kruzkov inequalities on the entropy solution u, that read:
$$\partial_t |u - \kappa| + \partial_x \operatorname{sgn}(u - \kappa)(f(u) - f(\kappa)) \leq 0. \tag{13}$$

Writing, respectively, (13) at point (s, x) for $\kappa = w^\varepsilon(t, y)$ and (12) at point (t, y) for $\kappa = u(s, x)$, we obtain:

$$\partial_s |u(s,x) - w^\varepsilon(t,y)| + \partial_x \operatorname{sgn}(u(s,x) - w^\varepsilon(t,y)) \left(\frac{u(s,x)^2}{2} - \frac{(w^\varepsilon(t,y))^2}{2} \right) \leq 0, \tag{14a}$$

$$\partial_t |w^\varepsilon(t,y) - u(s,x)| + \partial_y \operatorname{sgn}(w^\varepsilon(t,y) - u(s,x)) \left(\frac{(w^\varepsilon(t,y))^2}{2} - \frac{u(s,x)^2}{2} \right) \tag{14b}$$
$$\leq \frac{1}{2c} \left(\partial_t r_1^\varepsilon(t,y) + c \partial_y r_2^\varepsilon(t,y) \right).$$

Now, let $\omega_\alpha(t) = \frac{1}{\alpha} \omega\left(\frac{t}{\alpha}\right)$ and $\Omega_\beta(x) = \frac{1}{\beta} \Omega\left(\frac{x}{\beta}\right)$ be two mollyfing kernels. Setting $g(s,t,x,y) = \omega_\alpha(s-t)\Omega_\beta(x-y)$ and testing (14a) and (14b) against $g(\cdot, t, \cdot, y)\mathbb{1}_{[0,T]}$ and $g(s, \cdot, x, \cdot)\mathbb{1}_{[0,T]}$, respectively, and integrating over $[0,T] \times \mathbb{R}$, we obtain on the one hand:

$$\iiiint \partial_s g(s,t,x,y) |u(s,x) - w^\varepsilon(t,y)| \, ds \, dx \, dt \, dy$$
$$+ \iiiint \partial_x g(s,t,x,y) \operatorname{sgn}(u(s,x) - w^\varepsilon(t,y)) \left(\frac{u(s,x)^2}{2} - \frac{(w^\varepsilon(t,y))^2}{2} \right) ds \, dx \, dt \, dy \tag{15}$$
$$- \iiint g(T,t,x,y) |u(T,x) - w^\varepsilon(t,y)| \, dx \, dt \, dy$$
$$+ \iiint g(0,t,x,y) |u(0,x) - w^\varepsilon(t,y)| \, dx \, dt \, dy \geq 0,$$

and on the other hand:

$$\iiiint \partial_t g(s,t,x,y) |w^\varepsilon(t,y) - u(s,x)| \, ds \, dx \, dt \, dy$$
$$+ \iiiint \partial_y g(s,t,x,y) \operatorname{sgn}(w^\varepsilon(t,y) - u(s,x)) \left(\frac{(w^\varepsilon(t,y))^2}{2} - \frac{u(s,x)^2}{2} \right) ds \, dx \, dt \, dy$$
$$- \iiint g(s,T,x,y) |w^\varepsilon(T,y) - u(s,x)| \, ds \, dx \, dy + \iiint g(s,0,x,y) |w^\varepsilon(0,y) - u(s,x)| \, ds \, dx \, dy \tag{16}$$
$$\geq \frac{1}{2c} \Bigg(\iiiint \partial_t g(s,t,x,y) r_1^\varepsilon(t,y) \, ds \, dx \, dt \, dy + c \iiiint \partial_y g(s,t,x,y) r_2^\varepsilon(t,y) \, ds \, dx \, dt \, dy$$
$$- \iiint g(s,T,x,y) r_1^\varepsilon(T,y) \, ds \, dx \, dy + \iiint g(s,0,x,y) r_1^\varepsilon(0,y) \, ds \, dx \, dy \Bigg) =: \text{RHS}.$$

Now, since $|\cdot|$ is even, and $\partial_s g = -\partial_t g$ and $\partial_x g = -\partial_y g$, we deduce by adding (15) and (16):

$$- \iiint g(T,t,x,y) |u(T,x) - w^\varepsilon(t,y)| \, dx \, dt \, dy$$
$$+ \iiint g(0,t,x,y) |u(0,x) - w^\varepsilon(t,y)| \, dx \, dt \, dy \tag{17}$$
$$- \iiint g(s,T,x,y) |u(s,x) - w^\varepsilon(T,y)| \, ds \, dx \, dy$$
$$+ \iiint g(s,0,x,y) |u(s,x) - w^\varepsilon(0,y)| \, ds \, dx \, dy \geq \text{RHS}.$$

Then, we write:

$$\|u(T) - w^\varepsilon(T)\|_{L^1} = \iiint \omega_\alpha(T-t)\Omega_\beta(x-y)|u(T,y) - w^\varepsilon(T,y)|\,dx\,dt\,dy$$
$$+ \iiint \omega_\alpha(s-T)\Omega_\beta(x-y)|u(T,y) - w^\varepsilon(T,y)|\,ds\,dx\,dy, \qquad (18)$$
$$=: I_1 + I_2.$$

A triangle inequality gives for I_1:

$$I_1 \leq \iiint \omega_\alpha(T-t)\Omega_\beta(x-y)|u(T,y) - u(T,x)|\,dx\,dt\,dy$$
$$+ \iiint \omega_\alpha(T-t)\Omega_\beta(x-y)|u(T,x) - w^\varepsilon(t,y)|\,dx\,dt\,dy$$
$$+ \iiint \omega_\alpha(T-t)\Omega_\beta(x-y)|w^\varepsilon(t,y) - w^\varepsilon(T,y)|\,dx\,dt\,dy$$
$$=: T_1 + T_2 + T_3.$$

with $T_1 \leq C\beta \cdot TV(u_0)$, the second term T_2 appearing in (17) and for the last one we write:

$$T_3 \leq \int_\mathbb{R} \Omega_\beta(x-y)\int_0^T \omega_\alpha(T-t)\int_\mathbb{R} |w^\varepsilon(t,y) - w^\varepsilon(T,y)|\,dy\,dt\,dx,$$

and then we use the fact that w^ε is uniformly Lipschitz in $L^1(\mathbb{R})$ with respect to ε. Indeed, one has $\partial_t w^\varepsilon = \frac{\partial_t a^\varepsilon(h'(a^\varepsilon) - 1)}{2c}$ with $h'(a^\varepsilon) - 1$ being uniformly bounded with respect to ε as a^ε stays in the compact set $[m, M]$ for all time. In addition, estimating $\|\partial_t a^\varepsilon(t)\|_{L^1}$ can be done reusing (4) and (10):

$$\|\partial_t a^\varepsilon(t)\|_{L^1} \leq c\|\partial_x a^\varepsilon(t)\|_{L^1} + \frac{1}{\varepsilon}\|G(a^\varepsilon(t), b^\varepsilon(t))\|_{L^1} \leq C(TV(a_0) + TV(b_0)).$$

with $C > 0$ still independent of time and of ε. Hence, $\|\partial_t w^\varepsilon(t)\|_{L^1} \leq C(TV(a_0) + TV(b_0))$ and $T_3 \leq \alpha C(TV(a_0) + TV(b_0))$. All in all, we get for I_1:

$$I_1 \leq \iiint \omega_\alpha(T-t)\Omega_\beta(x-y)|u(T,x) - w^\varepsilon(t,y)|\,dx\,dt\,dy + C\beta \cdot TV(u_0)$$
$$+ \alpha C(TV(a_0) + TV(b_0)).$$

Moreover, similarly, for I_2:

$$I_2 \leq \iiint \omega_\alpha(s-T)\Omega_\beta(x-y)|u(s,x) - w^\varepsilon(T,y)|\,ds\,dx\,dy + C(\alpha + \beta)TV(u_0).$$

Returning to (18), we obtain:

$$\|u(T) - w^\varepsilon(T)\|_{L^1} \leq \iiint \omega_\alpha(t)\Omega_\beta(x-y)|u(0,x) - w^\varepsilon(t,y)|\,dx\,dt\,dy \qquad (19)$$
$$+ \iiint \omega_\alpha(s)\Omega_\beta(x-y)|u(s,x) - w^\varepsilon(0,y)|\,ds\,dx\,dy - RHS \qquad (20)$$
$$+ \alpha C(TV(a_0) + TV(b_0)) + C(\alpha + \beta)TV(u_0). \qquad (21)$$

However, using a triangle inequality, one can show that:

$$\iiint \omega_\alpha(t)\Omega_\beta(x-y)|u_0(x) - w^\varepsilon(t,y)|\,dx\,dt\,dy \leq C\beta \cdot TV(u_0) + \alpha C(TV(a_0) + TV(b_0)),$$

and similarly:

$$\iiint \omega_\alpha(s)\Omega_\beta(x-y)|u(s,x) - w^\varepsilon(0,y)|\,ds\,dx\,dy \leq C(\alpha + \beta)TV(u_0).$$

We then bound from above the term RHS using inequality $\|r_i^\varepsilon(t)\|_{L^1} \leq C(TV(a_0) + TV(b_0))\varepsilon$ for $i = 1, 2$:

$$\begin{aligned}|\text{RHS}| &= \frac{1}{2c}\left|\frac{1}{\alpha}\iiiint \omega'\left(\frac{s-t}{\alpha}\right)\Omega_\beta(x-y)r_1^\varepsilon(t,y)\,ds\,dx\,dt\,dy\right.\\ &+ \frac{c}{\beta}\iiiint \omega_\alpha(s-t)\Omega'(x-y)r_2^\varepsilon(t,y)\,ds\,dx\,dt\,dy \\ &- \iiint \omega_\alpha(s-T)\Omega_\beta(x-y)r_1^\varepsilon(T,y)\,ds\,dx\,dy \\ &+ \left.\iiint \omega_\alpha(s)\Omega_\beta(x-y)r_1^\varepsilon(0,y)\,ds\,dx\,dy\right|, \\ &\leq C\left(\frac{T}{\alpha} + \frac{T}{\beta} + 1\right)\cdot(TV(a_0)+TV(b_0))\varepsilon.\end{aligned}$$

Finally, we obtain:

$$\begin{aligned}\|u(T) - w^\varepsilon(T)\|_{L^1} \leq &C\left(\frac{T}{\alpha}+\frac{T}{\beta}+1\right)(TV(a_0)+TV(b_0))\varepsilon \\ &+ C(\alpha+\beta)TV(u_0) + \alpha C(TV(a_0)+TV(b_0)),\end{aligned}$$

which, after optimizing the values of α and β and noticing that $TV(a_0), TV(b_0) \leq C \cdot TV(u_0)$, gives:

$$\|u(T) - w^\varepsilon(T)\|_{L^1} \leq CTV(u_0)(\sqrt{\varepsilon T} + \varepsilon),$$

and this inequality, along with $|h(a) - b| \leq C|G(a, b)|$ and (10) gives in turn the result. □

Denoting $\rho = -\partial_x u$, the convergence of $u^\varepsilon(t)$ towards $u(t)$ in $L^1(\mathbb{R})$ ensures that $\rho(t)$ is a probability measure. Indeed, since for all $\varepsilon > 0$, $\rho^\varepsilon = -\partial_x u^\varepsilon$ is a non-negative distribution, so is ρ. The Riesz–Markov theorem then ensures that ρ can be represented by a non-negative Borel measure. In addition, almost everywhere, for $t \geq 0$, $u^\varepsilon(t)$ is a non-increasing function taking values in $[0, 1]$ and hence converges to a certain limit when x goes to $+\infty$. The same holds true for the limit function $u(t)$. However, since $u^\varepsilon(t) - u(t) \in L^1(\mathbb{R})$, then $u^\varepsilon(t, x) - u(t, x)$ must vanish as x goes to $+\infty$. Therefore, the total mass of $\rho(t)$ is 1.

Then, passing to the relaxation system (2) for the aggregation Equation (1) can be done by using (6) with $p = 1$. As a consequence, Theorem 1 translates as follows for the aggregation.

Theorem 2. *Let $\rho_0 \in \mathcal{P}_2(\mathbb{R})$, $c > 1/2$ and set $\sigma_0 = a[\rho_0]\rho_0$. There exists a constant $C > 0$ such that, for any $\varepsilon > 0$, denoting $(\rho^\varepsilon, \sigma^\varepsilon)$ the solution to (2) with initial data (ρ_0, σ_0), one has:*

$$\forall T > 0, \qquad W_1(\rho(T), \rho^\varepsilon(T)) \leq C(\sqrt{\varepsilon T} + \varepsilon),$$

where $\rho \in C([0, +\infty), \mathcal{P}_2(\mathbb{R}))$ is the unique solution (1) with initial datum ρ_0.

3. Numerical Discretization

Hereafter, we denote Δt the time step and we introduce a Cartesian mesh of size Δx. We denote $t^n = n\Delta t$ for $n \in \mathbb{N}$ and $x_j = j\Delta x$ for $j \in \mathbb{Z}$. In this section, we extend our framework and consider the aggregation Equation (1) with arbitrary pointy potentials W, which satisfy the following conditions:

(i) W is even and $W(0) = 0$;
(ii) $W \in \mathcal{C}^1(\mathbb{R} \setminus \{0\})$;
(iii) W is λ-convex, i.e., there exists $\lambda \in \mathbb{R}$ such that $W(x) - \lambda\frac{|x|^2}{2}$ is convex;
(iv) W is a_∞-lipschitz continuous for some $a_\infty \geq 0$.

In this framework, the convergence of ρ^ε towards ρ for a slightly different problem has also been studied in [7]. Adapting the argument, the convergence still holds provided the sub-characteristic condition $c > a_\infty$ is verified. However, for such general potentials, the authors were not able to obtain the estimates of the speed of convergence as stated in Theorem 2.

In this section, we propose some numerical schemes able to capture the limit $\varepsilon \to 0$, thus satisfying the so-called asymptotic preserving (AP) property. We consider two approaches, the first one based on a splitting algorithm, and the second one based on a well-balanced discretization.

3.1. A Splitting Algorithm

A first simple approach to discretize the system (2) is to use a splitting method. Such a method is known to be convergent and easy to implement but introduces numerical diffusion.

Notice that the system (2) rewrites, with $\mu = \sigma - c\rho$, $\nu = \sigma + c\rho$, as

$$\partial_t \mu - c \partial_x \mu = \frac{1}{\varepsilon} \left(a \left[\frac{\nu - \mu}{2c} \right] \left(\frac{\nu - \mu}{2c} \right) - \frac{\mu + \nu}{2} \right) \tag{22a}$$

$$\partial_t \nu + c \partial_x \nu = \frac{1}{\varepsilon} \left(a \left[\frac{\nu - \mu}{2c} \right] \left(\frac{\nu - \mu}{2c} \right) - \frac{\mu + \nu}{2} \right). \tag{22b}$$

The idea of the method is to solve in a first step on $(t^n, t^n + \Delta t)$ the system:

$$\partial_t \mu = \frac{1}{\varepsilon} \left(a \left[\frac{\nu - \mu}{2c} \right] \left(\frac{\nu - \mu}{2c} \right) - \frac{\mu + \nu}{2} \right)$$

$$\partial_t \nu = \frac{1}{\varepsilon} \left(a \left[\frac{\nu - \mu}{2c} \right] \left(\frac{\nu - \mu}{2c} \right) - \frac{\mu + \nu}{2} \right),$$

with initial data $(\mu(t^n), \nu(t^n)) = (\mu^n, \nu^n)$. We obtain $\mu_j^{n+\frac{1}{2}} = \mu(t^n + \Delta t, x_j)$ and $\nu_j^{n+\frac{1}{2}} = \nu(t^n + \Delta t, x_j)$. Notice that this system may be solved explicitly. Indeed, by adding and subtracting the two equations, we deduce after an integration:

$$\nu_j^{n+\frac{1}{2}} - \mu_j^{n+\frac{1}{2}} = \nu_j^n - \mu_j^n \tag{23a}$$

$$\mu_j^{n+\frac{1}{2}} + \nu_j^{n+\frac{1}{2}} = (\mu_j^n + \nu_j^n) e^{-\Delta t/\varepsilon} + a \left[\frac{\nu^n - \mu^n}{2c} \right] \left(\frac{\nu^n - \mu^n}{2c} \right) (1 - e^{-\Delta t/\varepsilon}). \tag{23b}$$

Then, in a second step, we discretize by a classical finite volume upwind scheme the system:

$$\partial_t \mu - c \partial_x \mu = 0, \qquad \partial_t \nu + c \partial_x \nu = 0.$$

That is:

$$\mu_j^{n+1} = \mu_j^{n+\frac{1}{2}} + c \frac{\Delta t}{\Delta x} (\mu_{j+1}^{n+\frac{1}{2}} - \mu_j^{n+\frac{1}{2}}), \tag{24a}$$

$$\nu_j^{n+1} = \nu_j^{n+\frac{1}{2}} - c \frac{\Delta t}{\Delta x} (\nu_j^{n+\frac{1}{2}} - \nu_{j-1}^{n+\frac{1}{2}}). \tag{24b}$$

Coming back to the variables ρ and σ, we obtain:

$$\nu_j^{n+\frac{1}{2}} = c \rho_j^n + \sigma_j^n e^{-\Delta x/\varepsilon} + a_j^n \rho_j^n (1 - e^{-\Delta t/\varepsilon}),$$

$$\mu_j^{n+\frac{1}{2}} = -c \rho_j^n + \sigma_j^n e^{-\Delta x/\varepsilon} + a_j^n \rho_j^n (1 - e^{-\Delta t/\varepsilon}),$$

with $a_j^n = -\sum_{k \neq j} W'(x_j - x_k)\rho_k^n$. Then, the splitting algorithm reads:

$$\begin{aligned}
\rho_j^{n+1} &= \rho_j^n - \frac{1}{2}\frac{\Delta t}{\Delta x}(\mu_{j+1}^{n+\frac{1}{2}} + v_j^{n+\frac{1}{2}} - \mu_j^{n+\frac{1}{2}} - v_{j-1}^{n+\frac{1}{2}}) \\
&= \rho_j^n - \frac{1}{2}\frac{\Delta t}{\Delta x}\Big((\sigma_{j+1}^n - \sigma_{j-1}^n)e^{-\Delta t/\varepsilon} \\
&\quad + (1 - e^{-\Delta t/\varepsilon})(a_{j+1}^n \rho_{j+1}^n - a_{j-1}^n \rho_{j-1}^n) - c(\rho_{j+1}^n - 2\rho_j^n + \rho_{j-1}^n)\Big),
\end{aligned}$$ (25)

and:

$$\begin{aligned}
\sigma_j^{n+1} &= \sigma_j^{n+\frac{1}{2}} + \frac{c}{2}\frac{\Delta t}{\Delta x}(\sigma_{j+1}^n - 2\sigma_j^n + \sigma_{j-1}^n)e^{-\Delta t/\varepsilon} \\
&\quad + \frac{c}{2}\frac{\Delta t}{\Delta x}\Big((a_{j+1}^n \rho_{j+1}^n - 2a_j^n \rho_j^n + a_{j-1}^n \rho_{j-1}^n)(1 - e^{-\Delta t/\varepsilon}) - c(\rho_{j+1}^n - \rho_{j-1}^n)\Big) \\
&= \sigma_j^n e^{-\Delta t/\varepsilon} + a_j^n \rho_j^n(1 - e^{-\Delta t/\varepsilon}) + \frac{c}{2}\frac{\Delta t}{\Delta x}(\sigma_{j+1}^n - 2\sigma_j^n + \sigma_{j-1}^n)e^{-\Delta t/\varepsilon} \\
&\quad + \frac{c}{2}\frac{\Delta t}{\Delta x}\Big((a_{j+1}^n \rho_{j+1}^n - 2a_j^n \rho_j^n + a_{j-1}^n \rho_{j-1}^n)(1 - e^{-\Delta t/\varepsilon}) - c(\rho_{j+1}^n - \rho_{j-1}^n)\Big).
\end{aligned}$$ (26)

Lemma 1. *For any $\varepsilon > 0$, if both the CFL condition $\frac{c\Delta t}{\Delta x} \leq 1$ and the subcharacteristic condition $c \geq a_\infty$ hold, then the splitting scheme (23) and (24) is L^1-stable:*

$$\forall n \in \mathbb{N}, \quad \sum_{j \in \mathbb{Z}}\Big(|\mu_j^{n+1}| + |v_j^{n+1}|\Big) \leq \sum_{j \in \mathbb{Z}}\Big(|\mu_j^n| + |v_j^n|\Big).$$

Proof. We have:

$$\mu_j^{n+\frac{1}{2}} = \frac{1}{2}\left(e^{-\Delta t/\varepsilon}\left(1 + \frac{a_j^n}{c}\right) + 1 - \frac{a_j^n}{c}\right)\mu_j^n - \frac{1 - e^{-\Delta t/\varepsilon}}{2}\left(1 - \frac{a_j^n}{c}\right)v_j^n,$$

$$v_j^{n+\frac{1}{2}} = -\frac{1 - e^{-\Delta t/\varepsilon}}{2}\left(1 + \frac{a_j^n}{c}\right)\mu_j^n + \frac{1}{2}\left(e^{-\Delta t/\varepsilon}\left(1 - \frac{a_j^n}{c}\right) + 1 + \frac{a_j^n}{c}\right)v_j^n.$$

Under the condition $c \geq a_\infty$, in the expression of $\mu_j^{n+\frac{1}{2}}$, the coefficient in front of μ_j^n is non-negative and the one in front of v_j^n is non-positive. Similarly, in $v_j^{n+\frac{1}{2}}$, the coefficient of μ_j^n is non-positive and the one in front of v_j^n is non-negative. Taking the absolute value and adding up therefore yields:

$$\left|\mu_j^{n+\frac{1}{2}}\right| + \left|v_j^{n+\frac{1}{2}}\right| \leq \left|\mu_j^n\right| + \left|v_j^n\right|.$$

It remains to remark that, provided the CFL condition $\frac{c\Delta t}{\Delta x} \leq 1$ is verified, (24) gives:

$$\begin{aligned}
\sum_{j \in \mathbb{Z}}\Big(|\mu_j^{n+1}| + |v_j^{n+1}|\Big) &\leq \left(1 - \frac{c\Delta t}{\Delta x}\right)\sum_{j \in \mathbb{Z}}\left(\left|\mu_j^{n+\frac{1}{2}}\right| + \left|v_j^{n+\frac{1}{2}}\right|\right) \\
&\quad + \frac{c\Delta t}{\Delta x}\sum_{j \in \mathbb{Z}}\left|\mu_{j+1}^{n+\frac{1}{2}}\right| + \frac{c\Delta t}{\Delta x}\sum_{j \in \mathbb{Z}}\left|v_{j-1}^{n+\frac{1}{2}}\right|, \\
&\leq \left(1 - \frac{c\Delta t}{\Delta x}\right)\sum_{j \in \mathbb{Z}}\Big(|\mu_j^n| + |v_j^n|\Big) + \frac{c\Delta t}{\Delta x}\sum_{j \in \mathbb{Z}}\left|\mu_j^{n+\frac{1}{2}}\right| + \frac{c\Delta t}{\Delta x}\sum_{j \in \mathbb{Z}}\left|v_j^{n+\frac{1}{2}}\right|, \\
&\leq \sum_{j \in \mathbb{Z}}\Big(|\mu_j^n| + |v_j^n|\Big).
\end{aligned}$$

□

Note that similar schemes have also been studied in [31] and proved convergent at a rate of $\sqrt{\Delta x}$.

Let us now verify the AP property. When $\varepsilon \to 0$, we verify that the equation on ρ (25) converges to the following Rusanov discretization of (1) (see [21] for numerical simulations using the Rusanov scheme):

$$\rho_j^{n+1} = \rho_j^n - \frac{1}{2}\frac{\Delta t}{\Delta x}\left(a_{j+1}^n \rho_{j+1}^n - a_{j-1}^n \rho_{j-1}^n\right) + \frac{c\Delta t}{2\Delta x}(\rho_{j+1}^n - 2\rho_j^n + \rho_{j-1}^n), \tag{27a}$$

$$a_j^n = -\sum_{k \neq j} W'(x_j - x_k)\rho_k^n. \tag{27b}$$

This limiting scheme provides a consistent discretization of (1). Indeed, a similar scheme has been extensively studied in [11] using compactness arguments and the following convergence result was proven:

Lemma 2. *Assume $\rho_0 \in \mathcal{P}_2(\mathbb{R})$ and that the stability conditions $c\frac{\Delta t}{\Delta x} \leq 1$ and $c \geq a_\infty$ are satisfied. Let $T > 0$ and suppose we initialize the scheme (27) with $\rho_j^0 = \frac{1}{\Delta x}\rho_0(C_j)$ where $C_j = [x_{j-\frac{1}{2}}, x_{j+\frac{1}{2}}]$. Then, denoting $\rho_{\Delta x}$ the reconstruction given by the scheme (27), that is:*

$$\rho_{\Delta x}(t) = \sum_{n \in \mathbb{N}} \sum_{j \in \mathbb{Z}} \rho_j^n \mathbb{1}_{[t^n, t^{n+1})}(t)\delta_{x_j},$$

then $\rho_{\Delta x}$ converges weakly in the sense of measures on $[0,T] \times \mathbb{R}$ towards the solution ρ of Equation (1), as Δx goes to 0.

It has been also proven in [32] that the scheme (27) converges at a rate of $\sqrt{\Delta x}$.

3.2. Well-Balanced Discretization

Although the splitting method provides a simple way to obtain a discretization which is uniform with respect to the parameter ε, the resulting scheme has strong numerical diffusion and may not have good large time behavior. Then, well-balanced schemes have been introduced. A scheme is said to be well-balanced when it conserves equilibria. The method proposed in this section comes from [20].

Let us assume that, for some $n \in \mathbb{N}$, the approximation $(\mu_j^n, v_j^n)_{j \in \mathbb{Z}}$ of $(\mu(t^n, x_j), v(t^n, x_j))_{j \in \mathbb{Z}}$ solution of (22) is known. We construct an approximation at time t^{n+1} using a finite volume upwind discretization of (22), with the discretization of the source terms $H_{\mu,j}^n, H_{v,j}^n$ to be prescribed right afterwards:

$$\mu_j^{n+1} = \mu_j^n + c\frac{\Delta t}{\Delta x}(\mu_{j+1}^n - \mu_j^n) + \frac{\Delta t}{\varepsilon}H_{\mu,j}^n, \tag{28a}$$

$$v_j^{n+1} = v_j^n - c\frac{\Delta t}{\Delta x}(v_j^n - v_{j-1}^n) + \frac{\Delta t}{\varepsilon}H_{v,j}^n. \tag{28b}$$

In order to preserve equilibria, we set :

$$H_{\mu,j}^n = \frac{1}{\Delta x}\int_{x_{j-1}}^{x_j} H(\overline{\mu}, \overline{v})\,dx, \qquad H(\mu, v) = a\left[\frac{v-\mu}{2c}\right]\left(\frac{v-\mu}{2c}\right) - \frac{\mu+v}{2}, \tag{29}$$

where $(\overline{\mu}, \overline{v})$ solve the stationary system with incoming boundary conditions, on (x_{j-1}, x_j):

$$-c\partial_x \bar{\mu} = \frac{1}{\varepsilon} H(\bar{\mu}, \bar{v}) \tag{30a}$$

$$c\partial_x \bar{v} = \frac{1}{\varepsilon} H(\bar{\mu}, \bar{v}) \tag{30b}$$

$$\bar{\mu}(x_j) = \mu_j^n, \qquad \bar{v}(x_{j-1}) = v_{j-1}^n. \tag{30c}$$

In addition, in the same fashion, $H_{v,j}^n = \frac{1}{\Delta x} \int_{x_j}^{x_{j+1}} H(\tilde{\mu}, \tilde{v}) \, dx$, where $(\tilde{\mu}, \tilde{v})$ is the solution of the stationary system on (x_j, x_{j+1}):

$$-c\partial_x \tilde{\mu} = \frac{1}{\varepsilon} H(\tilde{\mu}, \tilde{v}) \tag{31a}$$

$$c\partial_x \tilde{\mu} = \frac{1}{\varepsilon} H(\tilde{\mu}, \tilde{v}) \tag{31b}$$

$$\tilde{\mu}(x_{j+1}) = \mu_{j+1}^n, \qquad \tilde{v}(x_j) = v_j^n, \tag{31c}$$

Reporting Equations (30b) and (31a) into the discretization of the source term, we obtain $H_{v,j}^n = \frac{c\varepsilon}{\Delta x}(\bar{v}(x_j) - v_{j-1})$ and $H_{\mu,j}^n = -\frac{c\varepsilon}{\Delta x}(\mu_j^n - \tilde{\mu}(x_j))$. Hence, one may rewrite the scheme (28) as

$$\mu_j^{n+1} = \mu_j^n + c\frac{\Delta t}{\Delta x}(\tilde{\mu}(x_j) - \mu_j^n) \tag{32a}$$

$$v_j^{n+1} = v_j^n - c\frac{\Delta t}{\Delta x}(v_j^n - \bar{v}(x_j)). \tag{32b}$$

Remark that the stationary system:

$$-c\partial_x \mu = \frac{1}{\varepsilon} H(\mu, v), \qquad c\partial_x v = \frac{1}{\varepsilon} H(\mu, v), \tag{33}$$

is equivalent to:

$$\partial_x \sigma = 0, \qquad c^2 \partial_x \rho = \frac{1}{\varepsilon}(a[\rho]\rho - \sigma). \tag{34}$$

Therefore, denoting $\sigma_{j+\frac{1}{2}} = \frac{\tilde{\mu} + \tilde{v}}{2}$ and $\sigma_{j-\frac{1}{2}} = \frac{\bar{\mu} + \bar{v}}{2}$, which are constant, respectively, on (x_j, x_{j+1}) and (x_{j-1}, x_j), one has:

$$\tilde{\mu}(x_j) = 2\sigma_{j+\frac{1}{2}} - v_j^n, \qquad \bar{v}(x_j) = 2\sigma_{j-\frac{1}{2}} - \mu_j^n. \tag{35}$$

Thus, it turns out that the scheme can be rewritten only in terms of the discretized unknowns and of $\sigma_{j\pm\frac{1}{2}}$:

$$\mu_j^{n+1} = \mu_j^n - c\frac{\Delta t}{\Delta x}(\mu_j^n + v_j^n) + \frac{2c\Delta t}{\Delta x}\sigma_{j+\frac{1}{2}}, \tag{36a}$$

$$v_j^{n+1} = v_j^n - c\frac{\Delta t}{\Delta x}(\mu_j^n + v_j^n) + \frac{2c\Delta t}{\Delta x}\sigma_{j-\frac{1}{2}}. \tag{36b}$$

Or equivalently:

$$\rho_j^{n+1} = \rho_j^n - \frac{\Delta t}{\Delta x}(\sigma_{j+\frac{1}{2}} - \sigma_{j-\frac{1}{2}}), \tag{37a}$$

$$\sigma_j^{n+1} = \sigma_j^n - c\frac{\Delta t}{\Delta x}(2\sigma_j^n - \sigma_{j+\frac{1}{2}} - \sigma_{j-\frac{1}{2}}). \tag{37b}$$

However, solving the stationary systems (30) and (31) involves the resolution of a nonlinear and nonlocal ODE. Instead, we propose an approximation in the spirit of [20].

We replace the nonlinear term in (30a)–(30b) by $a^n_{j-\frac{1}{2}} \cdot \frac{\bar{v}-\bar{\mu}}{2c}$, where $a^n_{j-\frac{1}{2}}$ stands for a fixed and consistent discretization of $a\left[\frac{\bar{v}-\bar{\mu}}{2c}\right]$ on the interval (x_{j-1}, x_j), to be specified afterwards. Similarly, we will replace the nonlinear term in (31a)–(31b) by $a^n_{j+\frac{1}{2}} \cdot \frac{\bar{v}-\bar{\mu}}{2c}$ with $a^n_{j+\frac{1}{2}}$ defined accordingly. In the following, we detail the construction for the problem (30a)–(30b) on (x_{j-1}, x_j).

Obviously, the definition of $a^n_{j-\frac{1}{2}}$ should be taken with care [11,20]. In [32], the authors showed that, when discretizing the product $a[\rho]\rho$, if $a[\rho]$ and ρ were not evaluated at the same point, then the resulting scheme produces the wrong dynamics. To take this into account, we will split ρ into one contribution coming from the left and one contribution coming from the right, i.e., we set $\bar{\rho} = \rho_L + \rho_R$ and $\bar{\sigma} = \sigma_L + \sigma_R$ where $\rho_L(\Delta x) = 0$ and $\rho_R(0) = 0$. This implies that $\bar{\rho}(\Delta x) = \rho_R(\Delta x)$ and $\bar{\rho}(0) = \rho_L(0)$.

More precisely, we solve the two following boundary value problem, on $(0, \Delta x)$:

$$\varepsilon c^2 \frac{d}{dx}\rho_L = a^n_{j-\frac{1}{2},L}\rho_L - \sigma_L, \qquad \rho_L(\Delta x) = 0, \tag{38a}$$

$$\varepsilon c^2 \frac{d}{dx}\rho_R = a^n_{j-\frac{1}{2},R}\rho_R - \sigma_R, \qquad \rho_R(0) = 0, \tag{38b}$$

We may explicitly solve these linear systems, and since $\rho_L(0) = \bar{\rho}(0)$ and $\rho_R(\Delta x) = \bar{\rho}(\Delta x)$, we obtain the relations:

$$\sigma_L = \bar{\rho}(0)\kappa^n_{j-\frac{1}{2},L}, \qquad \sigma_R = \bar{\rho}(\Delta x)\kappa^n_{j-\frac{1}{2},R}. \tag{39}$$

with:

$$\kappa^n_{j-\frac{1}{2},L} = \frac{a^n_{j-\frac{1}{2},L}}{1 - \exp(-a^n_{j-\frac{1}{2},L}\Delta x/(\varepsilon c^2))}, \qquad \kappa^n_{j-\frac{1}{2},R} = \frac{a^n_{j-\frac{1}{2},R}}{1 - \exp(a^n_{j-\frac{1}{2},R}\Delta x/(\varepsilon c^2))}. \tag{40}$$

Notice that we have:

$$\kappa^n_{j-\frac{1}{2},L} \to (a^n_{j-\frac{1}{2},L})_+, \qquad \kappa^n_{j-\frac{1}{2},R} \to -(a^n_{j-\frac{1}{2},R})_-, \quad \text{when } \varepsilon \to 0, \tag{41}$$

where we denote $a_+ = \max(0, a) \geq 0$ and $a_- = \max(0, -a) \geq 0$—the positive and negative negative part of a. Using the boundary conditions in (30), we have:

$$\bar{\rho}(0) = \frac{v^n_{j-1} - \bar{\mu}(0)}{2c}, \qquad \bar{\rho}(\Delta x) = \frac{\bar{v}(\Delta x) - \mu^n_j}{2c}. \tag{42}$$

with (39) and the fact that $\bar{\sigma} = \sigma_L + \sigma_R$ is constant on $[0, \Delta x]$, we obtain the following 2×2 system on the unknowns $\bar{\mu}(0), \bar{v}(\Delta x)$:

$$\mu^n_j + \bar{v}(\Delta x) = \bar{\mu}(0) + v^n_{j-1}, \tag{43a}$$

$$\mu^n_j + \bar{v}(\Delta x) = \frac{v^n_{j-1} - \bar{\mu}(0)}{2c}\kappa^n_{j-\frac{1}{2},L} + \frac{\bar{v}(\Delta x) - \mu^n_j}{2c}\kappa^n_{j-\frac{1}{2},R} \tag{43b}$$

Solving this system yields:

$$\bar{\mu}(0) = -v^n_{j-1}\frac{c - \kappa^n_{j-\frac{1}{2},R} - \kappa^n_{j-\frac{1}{2},L}}{c - \kappa^n_{j-\frac{1}{2},R} + \kappa^n_{j-\frac{1}{2},L}} - \mu^n_j\frac{\kappa^n_{j-\frac{1}{2},R}}{c - \kappa^n_{j-\frac{1}{2},R} + \kappa^n_{j-\frac{1}{2},L}}, \tag{44a}$$

$$\bar{v}(\Delta x) = v^n_{j-1}\frac{\kappa^n_{j-\frac{1}{2},L}}{c - \kappa^n_{j-\frac{1}{2},R} + \kappa^n_{j-\frac{1}{2},L}} - \mu^n_j\frac{c + \kappa^n_{j-\frac{1}{2},R} + \kappa^n_{j-\frac{1}{2},L}}{c - \kappa^n_{j-\frac{1}{2},R} + \kappa^n_{j-\frac{1}{2},L}}. \tag{44b}$$

From which we deduce with (42):

$$\rho_{j-\frac{1}{2},L}^n := \overline{\rho}(0) = \frac{1}{c}\left(\frac{(c-\kappa_{j-\frac{1}{2},R}^n)v_{j-1}^n + \kappa_{j-\frac{1}{2},R}^n \mu_j^n}{c+\kappa_{j-\frac{1}{2},L}^n - \kappa_{j-\frac{1}{2},R}^n}\right) \tag{45a}$$

$$\rho_{j-\frac{1}{2},R}^n := \overline{\rho}(\Delta x) = \frac{1}{c}\left(\frac{\kappa_{j-\frac{1}{2},L}^n v_{j-1}^n - (c+\kappa_{j-\frac{1}{2},L}^n)\mu_j^n}{c+\kappa_{j-\frac{1}{2},L}^n - \kappa_{j-\frac{1}{2},R}^n}\right) \tag{45b}$$

and with (39):

$$\overline{\sigma}_{j-\frac{1}{2}} := \sigma_L + \sigma_R = \rho_{j-\frac{1}{2},L}^n \kappa_{j-\frac{1}{2},L}^n + \rho_{j-\frac{1}{2},R}^n \kappa_{j-\frac{1}{2},R}^n = \frac{v_{j-1}^n \kappa_{j-\frac{1}{2},L}^n - \mu_j^n \kappa_{j-\frac{1}{2},R}^n}{c-\kappa_{j-\frac{1}{2},R}^n + \kappa_{j-\frac{1}{2},L}^n}, \tag{46}$$

(the above quantities are well-defined since $\kappa_{j-\frac{1}{2},L}^n \geq 0$ and $\kappa_{j-\frac{1}{2},R}^n \leq 0$). Injecting into (37), it gives the following scheme:

$$\mu_j^{n+1} = \left(1 - \frac{c\Delta t}{\Delta x}\right)\mu_j^n - \frac{c\Delta t}{\Delta x}\frac{c-\kappa_{j+\frac{1}{2},R}^n - \kappa_{j+\frac{1}{2},L}^n}{c-\kappa_{j+\frac{1}{2},R}^n + \kappa_{j+\frac{1}{2},L}^n}v_j^n - \frac{2c\Delta t}{\Delta x}\frac{\kappa_{j+\frac{1}{2},R}^n}{c-\kappa_{j+\frac{1}{2},R}^n + \kappa_{j+\frac{1}{2},L}^n}\mu_{j+1}^n, \tag{47a}$$

$$v_j^{n+1} = \left(1 - \frac{c\Delta t}{\Delta x}\right)v_j^n - \frac{c\Delta t}{\Delta x}\frac{c+\kappa_{j-\frac{1}{2},R}^n + \kappa_{j-\frac{1}{2},L}^n}{c-\kappa_{j-\frac{1}{2},R}^n + \kappa_{j-\frac{1}{2},L}^n}\mu_j^n + \frac{2c\Delta t}{\Delta x}\frac{\kappa_{j-\frac{1}{2},L}^n}{c-\kappa_{j-\frac{1}{2},R}^n + \kappa_{j-\frac{1}{2},L}^n}v_{j-1}^n, \tag{47b}$$

where the coefficients $\kappa_{j-\frac{1}{2},L/R}^n$ are defined in (40). Equivalently, for the variable (ρ,σ), the scheme reads:

$$\rho_j^{n+1} = \rho_j^n - \frac{\Delta t}{\Delta x}\left(\frac{v_j^n \kappa_{j+\frac{1}{2},L}^n - \mu_{j+1}^n \kappa_{j+\frac{1}{2},R}^n}{c-\kappa_{j+\frac{1}{2},R}^n + \kappa_{j+\frac{1}{2},L}^n} - \frac{v_{j-1}^n \kappa_{j-\frac{1}{2},L}^n - \mu_j^n \kappa_{j-\frac{1}{2},R}^n}{c-\kappa_{j-\frac{1}{2},R}^n + \kappa_{j-\frac{1}{2},L}^n}\right) \tag{48a}$$

$$\sigma_j^{n+1} = \sigma_j^n - c\frac{\Delta t}{\Delta x}\left(2\sigma_j^n - \frac{v_j^n \kappa_{j+\frac{1}{2},L}^n - \mu_{j+1}^n \kappa_{j+\frac{1}{2},R}^n}{c-\kappa_{j+\frac{1}{2},R}^n + \kappa_{j+\frac{1}{2},L}^n} - \frac{v_{j-1}^n \kappa_{j-\frac{1}{2},L}^n - \mu_j^n \kappa_{j-\frac{1}{2},R}^n}{c-\kappa_{j-\frac{1}{2},R}^n + \kappa_{j-\frac{1}{2},L}^n}\right), \tag{48b}$$

where we recall that $\mu_j^n = \sigma_j^n - c\rho_j^n$ and $v_j^n = \sigma_j^n + c\rho_j^n$.

It remains to define the velocities $a_{j-\frac{1}{2},L/R}^n$ used in (38) and in (40). We take:

$$a_{j-\frac{1}{2},L/R}^n = -\sum_{k \neq j} W'(x_j - x_k)\rho_{k-\frac{1}{2},L/R}^n.$$

However, this discretization implies the resolution of a nonlinear problem, since the quantities $\rho_{k-\frac{1}{2},L/R}^n$ depends nonlinearly on $a_{j-\frac{1}{2},L/R}^n$.

Then, we implement a fixed point method initialized with $a_{j-\frac{1}{2},L}^{n,(0)} := a_{j-1}^n$ and $a_{j-\frac{1}{2},R}^{n,(0)} := a_j^n$. Solving, on each cell (x_{j-1}, x_j), the system of ODEs (38) with these values for the velocities gives two sequences, $(\rho_{j-\frac{1}{2},L}^{(1)})_{j \in \mathbb{Z}}$ and $(\rho_{j-\frac{1}{2},R}^{(1)})_{j \in \mathbb{Z}}$. Then, we assign the next value of the velocity to $a_{j-\frac{1}{2},L/R}^{n,(1)} := -\sum_{k \neq j} W'(x_j - x_k)\rho_{k-\frac{1}{2},L/R}^{(1)}$, which allows us to compute new values for the left and right densities, $(\rho_{j-\frac{1}{2},L}^{(2)})_{j \in \mathbb{Z}}$ and $(\rho_{j-\frac{1}{2},R}^{(2)})_{j \in \mathbb{Z}}$, through (38). We iterate until $W_2(\rho_L^{(i)}, \rho_L^{(i+1)})$ and $W_2(\rho_R^{(i)}, \rho_R^{(i+1)})$ pass below a certain threshold. Notice that the velocities $a_{j-\frac{1}{2},L/R}^{n,(i)}$ always remain bounded by a_∞. In practice, only a few iterations are needed.

The resulting scheme is consistent for any $\varepsilon > 0$ and stable under standard stability conditions, as shown by the following lemmas.

Lemma 3 (L^1 stability). *Under the CFL condition $\frac{c\Delta t}{\Delta x} \leq 1$ and the subcharacteristic condition $c \geq a_\infty$, there holds that the sequence $(\mu_j^n, \nu_j^n)_{j,n}$ defined by the scheme (47), verifies the following L^1 stability property:*

$$\forall n \in \mathbb{N}, \quad \sum_{j \in \mathbb{Z}} \left(|\mu_j^{n+1}| + |\nu_j^{n+1}| \right) \leq \sum_{j \in \mathbb{Z}} \left(|\mu_j^n| + |\nu_j^n| \right).$$

Proof. In each combination of (47), the first coefficient is non-negative under the CFL condition $\frac{c\Delta t}{\Delta x} \leq 1$, and so is the last one since $\kappa_{j\pm\frac{1}{2},L}^n \geq 0$ and $\kappa_{j\pm\frac{1}{2},R}^n \leq 0$. Moreover, under the subcharacteristic condition $c \geq a_\infty$, it holds that $-c \leq \kappa_{j\pm\frac{1}{2},R}^n + \kappa_{j\pm\frac{1}{2},R}^n \leq c$ so the remaining coefficient is non-positive. Thus, applying the triangle inequality and re-indexing the sums appropriately:

$$\sum_{j \in \mathbb{Z}} \left(|\mu_j^{n+1}| + |\nu_j^{n+1}| \right) \leq \sum_{j \in \mathbb{Z}} \left(1 - \frac{c\Delta t}{\Delta x} \right) |\mu_j^n| + \sum_{j \in \mathbb{Z}} \frac{c\Delta t}{\Delta x} \frac{c - \kappa_{j+\frac{1}{2},R}^n - \kappa_{j+\frac{1}{2},L}^n}{c - \kappa_{j+\frac{1}{2},R}^n + \kappa_{j+\frac{1}{2},L}^n} |\nu_j^n|$$

$$- \sum_{j \in \mathbb{Z}} \frac{2c\Delta t}{\Delta x} \frac{\kappa_{j+\frac{1}{2},R}^n}{c - \kappa_{j+\frac{1}{2},R}^n + \kappa_{j+\frac{1}{2},L}^n} |\mu_{j+1}^n| + \sum_{j \in \mathbb{Z}} \left(1 - \frac{c\Delta t}{\Delta x} \right) |\nu_j^n|$$

$$+ \sum_{j \in \mathbb{Z}} \frac{c\Delta t}{\Delta x} \frac{c + \kappa_{j+\frac{1}{2},R}^n + \kappa_{j+\frac{1}{2},L}^n}{c - \kappa_{j+\frac{1}{2},R}^n + \kappa_{j+\frac{1}{2},L}^n} |\mu_{j+1}^n| + \frac{2c\Delta t}{\Delta x} \frac{\kappa_{j+\frac{1}{2},L}^n}{c - \kappa_{j+\frac{1}{2},R}^n + \kappa_{j+\frac{1}{2},L}^n} |\nu_j^n|,$$

$$\leq \left(1 - \frac{c\Delta t}{\Delta x} \right) \sum_{j \in \mathbb{Z}} \left(|\mu_j^n| + |\nu_j^n| \right) + \frac{c\Delta t}{\Delta x} \sum_{j \in \mathbb{Z}} |\mu_{j+1}^n| + \frac{c\Delta t}{\Delta x} \sum_{j \in \mathbb{Z}} |\nu_j^n|,$$

$$\leq \sum_{j \in \mathbb{Z}} \left(|\mu_j^n| + |\nu_j^n| \right).$$

This concludes the proof. □

Lemma 4 (Consistency for smooth solutions). *Assume that, for all $j \in \mathbb{Z}$, we have $a_{j-\frac{1}{2},L/R}^n = -\sum_{k \neq j} W'(x_j - x_k) \rho_{k-\frac{1}{2},L/R}^n$. Then, for any $\varepsilon > 0$, the scheme (37) is consistent with (2) provided that the solutions are smooth enough.*

Proof. For $j \in \mathbb{Z}$, one has, using the Taylor expansions as $\Delta x \to 0$:

$$\frac{\kappa_{j-\frac{1}{2},L}^n}{c - \kappa_{j-\frac{1}{2},R}^n + \kappa_{j-\frac{1}{2},L}^n} = \frac{1}{2} - \frac{1}{4\varepsilon c^2} \left(c - \frac{a_{j-\frac{1}{2},L}^n + a_{j-\frac{1}{2},R}^n}{2} \right) \Delta x + O(\Delta x^2),$$

$$\frac{\kappa_{j-\frac{1}{2},R}^n}{c - \kappa_{j-\frac{1}{2},R}^n + \kappa_{j-\frac{1}{2},L}^n} = -\frac{1}{2} + \frac{1}{4\varepsilon c^2} \left(c + \frac{a_{j-\frac{1}{2},L}^n + a_{j-\frac{1}{2},R}^n}{2} \right) \Delta x + O(\Delta x^2).$$

Thus:

$$\sigma_{j-\frac{1}{2}} = \frac{\sigma_{j-1}^n + \sigma_j^n}{2} + c \frac{\rho_{j-1}^n - \rho_j^n}{2} - \frac{1}{4\varepsilon c^2} \left(\left(c - \frac{a_{j-\frac{1}{2},L}^n + a_{j-\frac{1}{2},R}^n}{2} \right) (\sigma_{j-1}^n + c\rho_{j-1}^n) \right.$$

$$\left. + \left(c + \frac{a_{j-\frac{1}{2},L}^n + a_{j-\frac{1}{2},R}^n}{2} \right) (\sigma_j^n - c\rho_j^n) \right) \Delta x + O(\Delta x^2).$$

In particular, $\sigma_{j-\frac{1}{2}}$ is clearly consistent with $\sigma(t^n, x_{j-\frac{1}{2}})$ as long as the solution (ρ, σ) is smooth enough to perform standard consistency analysis for finite differences. This shows that (37a) is consistent with $\partial_t \rho + \partial_x \sigma = 0$. As for the consistency of (37b) with $\partial_t \sigma + c^2 \partial_x \rho = \frac{1}{\varepsilon}(a[\rho]\rho - \sigma)$, we write:

$$\sigma_{j+\frac{1}{2}} + \sigma_{j-\frac{1}{2}} - 2\sigma_j^n = \frac{\sigma_{j+1}^n - 2\sigma_j^n + \sigma_{j-1}^n}{2} + c\frac{\rho_{j-1}^n - \rho_{j+1}^n}{2} - \frac{\Delta x}{4\varepsilon c^2}\left[c(\sigma_{j-1}^n + 2\sigma_j^n + \sigma_{j+1}^n)\right.$$

$$+ \frac{a_{j-\frac{1}{2},L}^n + a_{j-\frac{1}{2},R}^n}{2}(\sigma_j^n - \sigma_{j-1}^n) + \frac{a_{j+\frac{1}{2},L}^n + a_{j+\frac{1}{2},R}^n}{2}(\sigma_{j+1}^n - \sigma_j^n) + c^2(\rho_{j-1}^n - \rho_{j+1}^n)$$

$$\left. - c\left(\frac{a_{j-\frac{1}{2},L}^n + a_{j-\frac{1}{2},R}^n}{2}\rho_{j-1}^n + \frac{a_{j-\frac{1}{2},L}^n + a_{j-\frac{1}{2},R}^n + a_{j+\frac{1}{2},L}^n + a_{j+\frac{1}{2},R}^n}{2}\rho_j^n + \frac{a_{j+\frac{1}{2},L}^n + a_{j+\frac{1}{2},R}^n}{2}\rho_{j+1}^n\right)\right]$$

$$+ O(\Delta x^2).$$

Using Taylor expansions, we have, for smooth solutions $\sigma(t^n, x_{j+1}) - 2\sigma(t^n, x_j) + \sigma(t^n, x_{j-1}) = O(\Delta x^2)$, $\rho(t^n, x_{j-1}) - \rho(t^n, x_{j+1}) = O(\Delta x)$, $\sigma(t^n, x_j) - \sigma(t^n, x_{j-1}) = O(\Delta x)$ and $\sigma(t^n, x_{j+1}) - \sigma(t^n, x_j) = O(\Delta x)$. Along with the bound $|a_{j\pm\frac{1}{2},L/R}^n| \leq a_\infty$, this implies:

$$\sigma_{j+\frac{1}{2}} + \sigma_{j-\frac{1}{2}} - 2\sigma_j^n = c\frac{\rho_{j-1}^n - \rho_{j+1}^n}{2} - \frac{1}{4\varepsilon c^2}\left[c(\sigma_{j-1}^n + 2\sigma_j^n + \sigma_{j+1}^n)\right.$$

$$- c\left(\frac{a_{j-\frac{1}{2},L}^n + a_{j-\frac{1}{2},R}^n}{2}\rho_{j-1}^n + \frac{a_{j-\frac{1}{2},L}^n + a_{j-\frac{1}{2},R}^n + a_{j+\frac{1}{2},L}^n + a_{j+\frac{1}{2},R}^n}{2}\rho_j^n\right.$$

$$\left.\left. + \frac{a_{j+\frac{1}{2},L}^n + a_{j+\frac{1}{2},R}^n}{2}\rho_{j+1}^n\right)\right]\Delta x + O(\Delta x^2).$$

Clearly, $c\frac{\rho_{j-1}^n - \rho_{j+1}^n}{2}$ and $c(\sigma_{j-1}^n + 2\sigma_j^n + \sigma_{j+1}^n)$ are consistent with an accuracy of $O(\Delta x^2)$ and $O(\Delta x)$, respectively, with $-c\partial_x \rho(t^n, x_j)$ and $4c\sigma(t^n, x_j)$. For the remaining terms, let us recall that, with the notations of (42):

$$\rho_{j-\frac{1}{2},L} = \frac{v_{j-1}^n - \overline{\mu}(0)}{2c} = \frac{v_{j-1}^n - \sigma_{j-\frac{1}{2}}^n}{c}, \quad \rho_{j-\frac{1}{2},R} = \frac{\overline{v}(\Delta x) - \mu_j^n}{2c} = \frac{\sigma_{j-\frac{1}{2}}^n - \mu_j}{c}.$$

Hence, $\rho_{j-\frac{1}{2},L} + \rho_{j-\frac{1}{2},R} = \frac{v_{j-1}^n - \mu_j^n}{c} = \frac{\sigma_{j-1}^n - \sigma_j^n}{c} + \rho_{j-1}^n + \rho_j^n$. Since $\sigma(t^n, x_{j-1}) - \sigma(t^n, x_j) = O(\Delta x)$, and assuming that:

$$a_{j-\frac{1}{2},L/R}^n = -\sum_{k \neq j} W'(x_j - x_k)\rho_{k-\frac{1}{2},L/R}$$

we deduce that $a_{j-\frac{1}{2},L}^n + a_{j-\frac{1}{2},R}^n$ is consistent with $a[\rho(t^n)](x_{j-1}) + a[\rho(t^n)](x_j)$ with accuracy $O(\Delta x)$. It follows that $\sigma_{j+\frac{1}{2}} + \sigma_{j-\frac{1}{2}} - 2\sigma_j^n$ is consistent with $-\partial_x \rho(t^n, x_j) - \frac{1}{\varepsilon}\left(\sigma(t^n, x_j) - a[\rho(t^n)](x_j)\rho(t^n, x_j)\right)$, again with accuracy $O(\Delta x)$, and this concludes the proof. □

The stability conditions in Lemma 3 are independent on ε, we recover in the limit $\varepsilon \to 0$, using (41), the scheme of [20]:

$$\rho_j^{n+1} = \rho_j^n - \frac{\Delta t}{\Delta x}\left(\frac{v_j^n(a_{j+\frac{1}{2},L}^n)_+ + \mu_{j+1}^n(a_{j+\frac{1}{2},R}^n)_-}{c + (a_{j+\frac{1}{2},R}^n)_- + (a_{j+\frac{1}{2},L}^n)_+} - \frac{v_{j-1}^n(a_{j-\frac{1}{2},L}^n)_+ + \mu_j^n(a_{j-\frac{1}{2},R}^n)_-}{c + (a_{j-\frac{1}{2},R}^n)_- + (a_{j-\frac{1}{2},L}^n)_+}\right) \quad (49a)$$

$$\sigma_j^{n+1} = \sigma_j^n - c\frac{\Delta t}{\Delta x}\left(2\sigma_j^n - \frac{v_j^n(a_{j+\frac{1}{2},L}^n)_+ + \mu_{j+1}^n(a_{j+\frac{1}{2},R}^n)_-}{c + (a_{j+\frac{1}{2},R}^n)_- + (a_{j+\frac{1}{2},L}^n)_+}\right.$$

$$\left.- \frac{v_{j-1}^n(a_{j-\frac{1}{2},L}^n)_+ + \mu_j^n(a_{j-\frac{1}{2},R}^n)_-}{c + (a_{j-\frac{1}{2},R}^n)_- + (a_{j-\frac{1}{2},L}^n)_+}\right), \quad (49b)$$

which is stable under the conditions $\frac{c\Delta t}{\Delta x} \leq 1$ and $c \geq a_\infty$. Notice that with the notation in (46), Equation (49a) may be rewritten as

$$\rho_j^{n+1} = \rho_j^n - \frac{\Delta t}{\Delta x}\left(\rho_{j+\frac{1}{2},L}^n(a_{j+\frac{1}{2},L}^n)_+ - \rho_{j+\frac{1}{2},R}^n(a_{j+\frac{1}{2},R}^n)_- \right.$$
$$\left. - \rho_{j-\frac{1}{2},L}^n(a_{j-\frac{1}{2},L}^n)_+ + \rho_{j-\frac{1}{2},R}^n(a_{j-\frac{1}{2},R}^n)_-\right).$$

4. Numerical Experiments

We present some numerical illustrations for the two schemes described in the previous section. In addition to the potential $W(x) = \frac{|x|}{2}$, we also consider the smooth potential $W(x) = \frac{x^2}{2}$.

Numerical tests are conducted on the domain $[-1,1]$ with the inital data $\rho_0 = \frac{1}{2}\delta_{-0.5} + \frac{1}{2}\delta_{0.5}$, $\sigma_0 = a[\rho_0]\rho_0$ and both schemes are initialized with:

$$\rho_j^0 = \frac{1}{\Delta x}\rho_0(C_j), \quad \sigma_j^0 = \frac{1}{\Delta x}\sigma_0(C_j).$$

Figure 1 shows that both schemes recover the correct dynamics in the limit $\varepsilon \to 0$: for the potential $W(x) = \frac{|x|}{2}$, one can compute the exact velocity of both Dirac masses for the aggregation Equation (1) and see that they should be located, respectively, in $x = -0.2$ and $x = 0.2$ in final time $T = 1.2$.

This test is set up with $\varepsilon = 10^{-7}$, on a Cartesian mesh of $[-1,1]$ with 1500 cells, $c = 1$ and the CFL $c\frac{\Delta t}{\Delta x} = 0.9$. Both schemes (27) and (49) display the correct velocity for the Dirac masses, but one can notice that the Rusanov scheme (27) shows more numerical diffusion. Note that both schemes are written in conservation form, they preserve the total mass of ρ, which is also verified numerically.

We then investigated the order of convergence when Δx goes to 0 with ε fixed, in Wasserstein distance W_1 (the numerical results are the same for W_2).

After performing tests for several values of ε, it appears that the convergence rate does not depend on the size of ε. Therefore, as an example, we propose simulations in final time $T = 0.5$, with the same intial data and stability parameters as above, and with $\varepsilon = 2 \times 10^{-6}$ for Figure 2 and with $\varepsilon = 10^{-2}$ for Figure 3:

For a fixed value of ε, both schemes seem to converge with order $1/2$ with respect to Δx for the smooth potential $W(x) = \frac{x^2}{2}$ (see Figure 2) whereas they seem to be of order 1 for the potential $W(x) = \frac{|x|}{2}$ (see Figure 3). This can be explained as both schemes possess some numerical diffusion which is somehow counterbalanced by the aggregation phenomenon in the case of a pointy potential, as already observed in [21]. Due to the link with the Burgers equation, this superconvergence phenomenon is directly linked to the results of Després [33], which should be rigorously extended to our case (the mere extension to the upwind scheme of [11] for the aggregation is not straightforward).

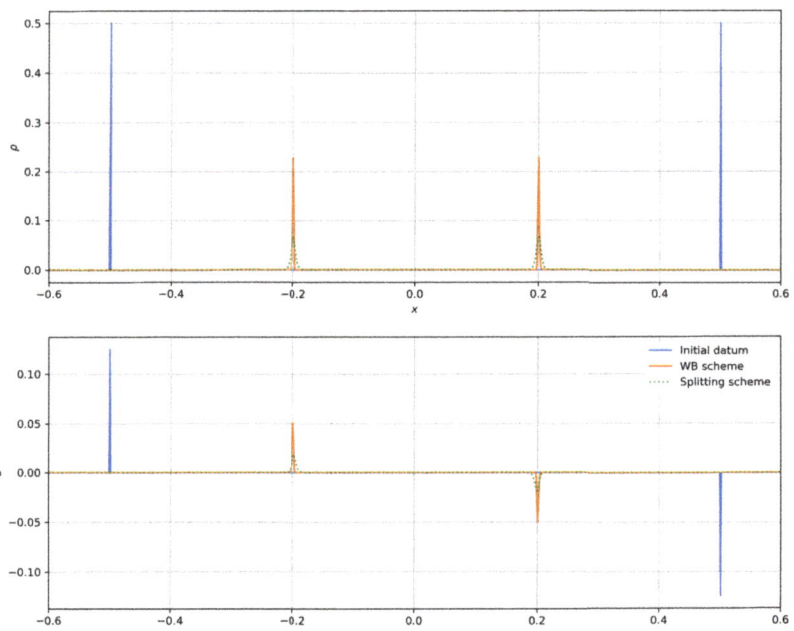

Figure 1. Dynamics of two Dirac masses for the potential $W(x) = \frac{|x|}{2}$ in time $T = 1.2$.

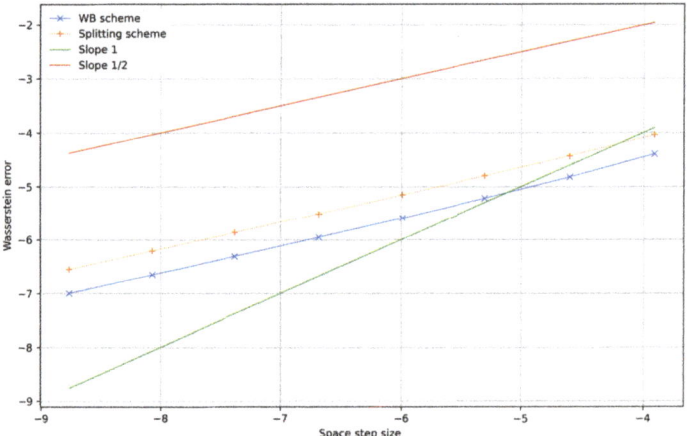

Figure 2. Order of convergence of the splitting scheme and the well-balanced scheme for the smooth potential $W(x) = \frac{x^2}{2}$.

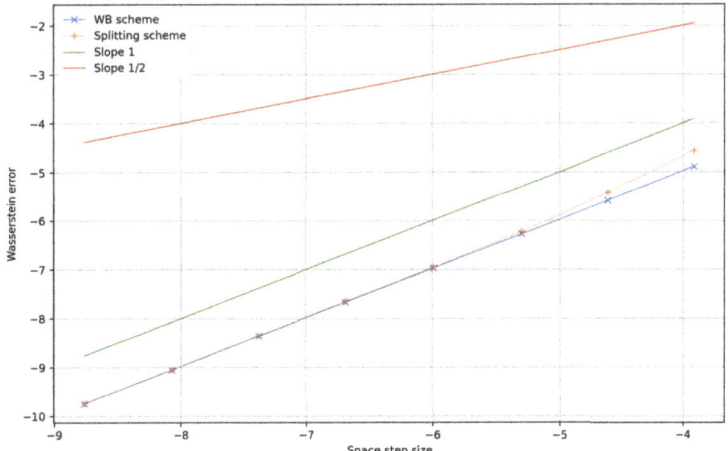

Figure 3. Order of convergence of the splitting scheme and the well-balanced scheme for the pointy potential $W(x) = \frac{|x|}{2}$.

Finally, we also verified the well-balanced property of the scheme (48) by computing the W_1 distance between the approximated solution at time $T = 0.5$ and the stationary solution of (2) given by

$$\rho(t,x) = \rho_0(x) := \frac{1}{8\varepsilon c^2}\left(1 - \tanh^2\left(\frac{x}{4\varepsilon c^2}\right)\right).$$

The test is conducted with $\varepsilon = 2 \times 10^{-4}$, with the exact boundary conditions given by the above formula, and for several values of Δx. As we show in Figure 4, the scheme (48) preserves well the above equilibrium for any Δx (although we replaced the resolution of the systems (30) and (31) with linear systems, see (38)), while for the splitting scheme, we recover the linear convergence towards ρ_0 which is, in this case, the exact solution.

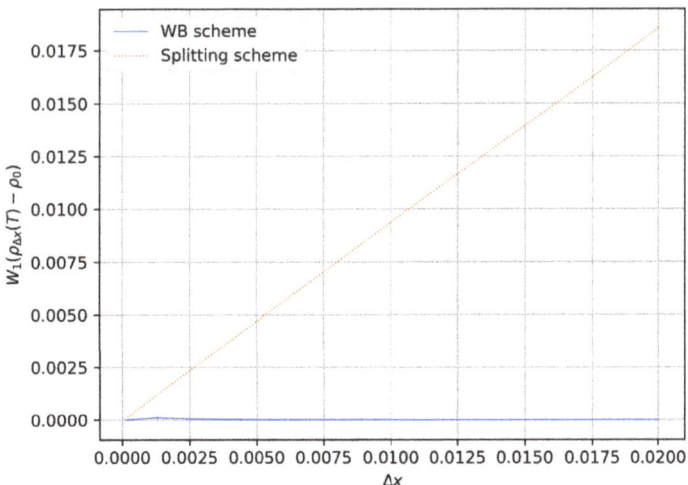

Figure 4. Distance to the equilibrium for the splitting scheme and the well-balanced scheme and for the pointy potential $W(x) = \frac{|x|}{2}$.

Author Contributions: B.F., F.L., S.T.T. and N.V. contributed equally in writing this article. All authors have read and agreed to the published version of the manuscript.

Funding: This research received no external funding.

Conflicts of Interest: The authors declare no conflict of interest.

Abbreviations

The following abbreviations are used in this manuscript:

MDPI Multidisciplinary Digital Publishing Institute
DOAJ Directory of Open Access Journals
AS   asymptotic preserving

References

1. Morale, D.; Capasso, V.; Oelschläger, K. An interacting particle system modelling aggregation behavior: From individuals to populations. *J. Math. Biol.* **2005**, *50*, 49–66. [CrossRef]
2. Burger, M.; Di Francesco, M. Large time behavior of nonlocal aggregation models with nonlinear diffusion. *Netw. Heterog. Media* **2008**, *3*, 749–785. [CrossRef]
3. Burger, M.; Capasso, V.; Morale, D. On an aggregation model with long and short range interactions. *Nonlinear Anal. Real World Appl.* **2007**, *8*, 939–958. [CrossRef]
4. Topaz, C.M.; Bertozzi, A.L.; Lewis, M.A. A nonlocal continuum model for biological aggregation. *Bull. Math. Biol.* **2006**, *68*, 1601–1623. [CrossRef]
5. Topaz, C.M.; Bertozzi, A.L. Swarming patterns in a two-dimensional kinematic model for biological groups. *SIAM J. Appl. Math.* **2004**, *65*, 152–174. [CrossRef]
6. Dolak, Y.; Schmeiser, C. Kinetic models for chemotaxis: Hydrodynamic limits and spatio-temporal mechanisms. *J. Math. Biol.* **2005**, *51*, 595–615. [CrossRef] [PubMed]
7. James, F.; Vauchelet, N. Chemotaxis: from kinetic equations to aggregate dynamics. *NoDEA Nonlinear Differ. Equ. Appl.* **2013**, *20*, 101–127. [CrossRef]
8. Bertozzi, A.L.; Brandman, J. Finite-time blow-up of L^∞-weak solutions of an aggregation equation. *Commun. Math. Sci.* **2010**, *8*, 45–65. [CrossRef]
9. Bertozzi, A.L.; Carrillo, J.A.; Laurent, T. Blow-up in multidimensional aggregation equations with mildly singular interaction kernels. *Nonlinearity* **2009**, *22*, 683–710. [CrossRef]
10. Carrillo, J.A.; Difrancesco, M.; Figalli, A.; Laurent, T.; Slepčev, D. Global-in-time weak measure solutions and finite-time aggregation for nonlocal interaction equations. *Duke Math. J.* **2011**, *156*, 229–271. [CrossRef]
11. Carrillo, J.A.; James, F.; Lagoutière, F.; Vauchelet, N. The Filippov characteristic flow for the aggregation equation with mildly singular potentials. *J. Differ. Equ.* **2016**, *260*, 304–338. [CrossRef]
12. Jin, S.; Xin, Z. The relaxation schemes for systems of conservation laws in arbitrary space dimensions. *Commun. Pure Appl. Math.* **1995**, *48*, 235–276. [CrossRef]
13. James, F.; Vauchelet, N. Numerical methods for one-dimensional aggregation equations. *SIAM J. Numer. Anal.* **2015**, *53*, 895–916. [CrossRef]
14. Bonaschi, G.A.; Carrillo, J.A.; Di Francesco, M.; Peletier, M.A. Equivalence of gradient flows and entropy solutions for singular nonlocal interaction equations in 1D. *ESAIM Control Optim. Calc. Var.* **2015**, *21*, 414–441. [CrossRef]
15. James, F.; Vauchelet, N. Equivalence between duality and gradient flow solutions for one-dimensional aggregation equations. *Discret. Contin. Dyn. Syst.* **2016**, *36*, 1355–1382.
16. Katsoulakis, M.A.; Tzavaras, A.E. Contractive relaxation systems and the scalar multidimensional conservation law. *Commun. Partial. Differ. Equ.* **1997**, *22*, 225–267. [CrossRef]
17. Jin, S. Efficient asymptotic-preserving (AP) schemes for some multiscale kinetic equations. *SIAM J. Sci. Comput.* **1999**, *21*, 441–454. [CrossRef]
18. Carrillo, J.A.; Chertock, A.; Huang, Y. A finite-volume method for nonlinear nonlocal equations with a gradient flow structure. *Commun. Comput. Phys.* **2015**, *17*, 233–258. [CrossRef]
19. Craig, K.; Bertozzi, A.L. A blob method for the aggregation equation. *Math. Comput.* **2016**, *85*, 1681–1717. [CrossRef]
20. Gosse, L.; Vauchelet, N. Numerical High-Field Limits in Two-Stream Kinetic Models and 1D Aggregation Equations. *SIAM J. Sci. Comput.* **2016**, *38*, A412–A434. [CrossRef]
21. Fabrèges, B.; Hivert, H.; Le Balc'h, K.; Martel, S.; Delarue, F.; Lagoutière, F.; Vauchelet, N. Numerical schemes for the aggregation equation with pointy potentials. *ESAIM Proc. Surv.* **2019**. [CrossRef]
22. Carrillo, J.A.; Fjordholm, U.S.; Solem, S. A second-order numerical method for the aggregation equations. *Math. Comput.* **2021**, *90*, 103–139. [CrossRef]
23. Gosse, L. *Computing Qualitatively Correct Approximations of Balance Laws. Exponential-Fit, Well-Balanced and Asymptotic-Preserving*; Springer: Milano, Italy, 2013; Volume 2, p. 340.

24. Villani, C. *Topics in Optimal Transportation*; American Mathematical Society (AMS): Providence, RI, USA, 2003; Volume 58, p. 370.
25. Santambrogio, F. *Optimal Transport for Applied Mathematicians. Calculus of Variations, PDEs, and Modeling*; Birkhäuser/Springer: Cham, Switzerland, 2015; Volume 87, p. 353.
26. Vallender, S.S. Calculation of the Wasserstein Distance Between Probability Distributions on the Line. *Theory Probab. Appl.* **1974**, *18*, 784–786. [CrossRef]
27. Rachev, S.T.; Rüschendorf, L. *Mass Transportation Problems. Vol. 1: Theory. Vol. 2: Applications*; Springer: New York, NY, USA, 1998; p. 430.
28. Natalini, R. Convergence to equilibrium for the relaxation approximations of conservation laws. *Commun. Pure Appl. Math.* **1996**, *49*, 795–823. [CrossRef]
29. Serre, D. *Systems of Conservation Laws 1: Hyperbolicity, Entropies, Shock Waves*; Cambridge University Press: New York, NY, USA, 1999.
30. Bouchut, F.; Perthame, B. Kružkov's Estimates for Scalar Conservation Laws Revisited. *Trans. Am. Math. Soc.* **1998**, *350*, 2847–2870. [CrossRef]
31. Liu, H.; Warnecke, G. Convergence Rates for Relaxation Schemes Approximating Conservation Laws. *SIAM J. Numer. Anal.* **2000**, *37*, 1316–1337. [CrossRef]
32. Delarue, F.; Lagoutière, F.; Vauchelet, N. Convergence analysis of upwind type schemes for the aggregation equation with pointy potential. *Ann. Henri Lebesgue* **2020**, *3*, 217–260. [CrossRef]
33. Després, B. Discrete Compressive Solutions of Scalar Conservation Laws. *J. Hyper. Differ. Equ.* **2004**, *01*, 493–520. [CrossRef]

Article

Macroscopic and Multi-Scale Models for Multi-Class Vehicular Dynamics with Uneven Space Occupancy: A Case Study

Maya Briani [1,*], Emiliano Cristiani [1,*] and Paolo Ranut [2]

[1] Istituto per le Applicazioni del Calcolo, Consiglio Nazionale delle Ricerche, 00185 Rome, Italy
[2] Autovie Venete S.p.A., 34143 Trieste, Italy; Paolo.Ranut@autovie.it
* Correspondence: m.briani@iac.cnr.it (M.B.); e.cristiani@iac.cnr.it (E.C.)

Abstract: In this paper, we propose two models describing the dynamics of heavy and light vehicles on a road network, taking into account the interactions between the two classes. The models are tailored for two-lane highways where heavy vehicles cannot overtake. This means that heavy vehicles cannot saturate the whole road space, while light vehicles can. In these conditions, the creeping phenomenon can appear, i.e., one class of vehicles can proceed even if the other class has reached the maximal density. The first model we propose couples two first-order macroscopic LWR models, while the second model couples a second-order microscopic follow-the-leader model with a first-order macroscopic LWR model. Numerical results show that both models are able to catch some second-order (inertial) phenomena such as stop and go waves. Models are calibrated by means of real data measured by fixed sensors placed along the A4 Italian highway Trieste–Venice and its branches, provided by Autovie Venete S.p.A.

Keywords: LWR model; follow-the-leader model; phase transition; creeping; seepage; fundamental diagram; lane discipline; networks

MSC: 35L65; 35F25; 90B20; 76T99

Citation: Briani, M.; Cristiani, E.; Ranut, P. Macroscopic and Multi-Scale Models for Multi-Class Vehicular Dynamics with Uneven Space Occupancy: A Case Study. *Axioms* **2021**, *10*, 102. https://doi.org/10.3390/axioms10020102

Academic Editor: Angel R. Plastino

Received: 31 March 2021
Accepted: 15 May 2021
Published: 24 May 2021

Publisher's Note: MDPI stays neutral with regard to jurisdictional claims in published maps and institutional affiliations.

Copyright: © 2021 by the authors. Licensee MDPI, Basel, Switzerland. This article is an open access article distributed under the terms and conditions of the Creative Commons Attribution (CC BY) license (https://creativecommons.org/licenses/by/4.0/).

1. Introduction

In this paper, we deal with macroscopic and multi-scale modeling of traffic flow on a road network, focusing on multi-class dynamics which couple light and heavy vehicles (in the following, cars and trucks). The proposed models are characterized by the fact that cars and trucks interact with each other and that trucks are confined to a part of the road space (slow lane) and cannot overtake. As a consequence, when trucks saturate the space and form a queue, cars can still move, although at reduced speed.

1.1. State of the Art

The literature about traffic flow is very large and many different aspects of traffic dynamics were described through mathematical models. Let us start from classic approaches: in a single-lane *microscopic* (agent-based) framework with N vehicles and no overtaking, each vehicle $k \in \{1, \ldots, N\}$ is singularly identified by its position $X_k(t)$ and its velocity $V_k(t)$. By assumption, the $(k+1)$-th vehicle is always in front of the k-th one. Further, each vehicle is assumed to adjust its acceleration based on the difference in positions and velocities between the vehicle itself and the vehicle in front of it. This approach leads to the following system of ordinary differential equations:

$$\begin{cases} \dot{X}_k = V_k \\ \dot{V}_k = A(X_k, X_{k+1}, V_k, V_{k+1}) \end{cases}, \quad k = 1, \ldots, N-1 \qquad (1)$$

where A is a given acceleration function. The first vehicle in the row ($k = N$), called *leader*, has an independent dynamics. Since the whole dynamics is determined by the leader's one through a domino effect, these kinds of models are known as follow-the-leader.

Adopting, instead, a *macroscopic* (fluid dynamics) point of view, we describe the mass of vehicles by means of their density $\rho(x,t)$ only. The celebrated LWR model [1,2] is based on the observation that the density ρ evolves in time, ruled by the following conservation law:

$$\partial_t \rho + \partial_x f(\rho) = 0, \qquad x \in \mathbb{R}, \quad t > 0 \tag{2}$$

where the function $f(\rho)$, called *fundamental diagram*, is given and represents the flux of vehicles as a function of the density itself. The velocity of the vehicles can be recovered from ρ thanks to the relation

$$v(\rho) = \frac{f(\rho)}{\rho}, \qquad (\rho \neq 0). \tag{3}$$

It is important to note that the microscopic model (1) is second-order, i.e., acceleration-based, while the macroscopic model (2) is first-order, i.e., velocity-based. The difference is important because velocity-based models, allowing nonphysical instantaneous accelerations, are not able to catch effects caused by inertia, such as stop and go waves. Due to this difference, it is plain that the model (2) is not the many-particle limit of the model (1). We refer the interested reader to [3] for a review of various many-particle limits (i.e., micro-to-macro correspondences) and the existing multi-scale models.

The first generalization of the models (1) and (2) of our interest is that of *road networks*. While managing junctions in microscopic models is relatively easy, doing the same in a macroscopic setting is more challenging. The reason is that, in general, the conservation of the mass alone is not sufficient to characterize a unique solution at junctions. We refer the reader to the book by Garavello and Piccoli [4] for more details about the ill-posedness of the problem at junctions. Multiple workarounds for such ill-posedness have been suggested in the literature: (i) maximization of the flux across junctions and introduction of priorities among the incoming roads [4–6]; (ii) introduction of a *buffer* to model the junctions by means of an additional ordinary differential equation coupled with (2) [7–9]; (iii) reformulation of the problem on all possible paths on the network rather than on roads and junctions. The last approach has both a global formulation [10–12] and a more manageable local formulation, described in [13], which is the one we will adopt in this paper. All these approaches allow determining a unique solution for the traffic evolution on the network, but the solutions might be different.

The second generalization of our interest is that of *multi-class* dynamics. "Multi-class" is a very generic term used in the literature to refer to the case in which the road is populated by different groups of vehicles/drivers, and tracking each group separately is desired. Again, doing this in a microscopic framework is easy since it is sufficient to label each vehicle on the basis of the class it belongs to. In the macroscopic setting, instead, we need to introduce as many density functions as there are classes and then establish the interactions between classes. This leads to a system of conservation laws of the form

$$\partial_t \rho_c + \partial_x f_c(\rho_1, \ldots, \rho_C) = 0, \qquad c = 1, \ldots, C, \quad x \in \mathbb{R}, \quad t > 0 \tag{4}$$

where C is the number of classes, $c \in \{1, \ldots, C\}$, ρ_c is the density of class c and f_c is the flux of class c which depends on all densities (usually the dependence is on the sum of all densities $\sum_c \rho_c$). Multi-class models are used to describe very different situations, such as the co-presence of vehicles with

- Different driving modes (e.g., autonomous vs. classic);
- Different origins and destinations;
- Different lengths (i.e., space occupied);
- Different velocities/flux functions;

- Reserved roads or reserved entry/exit lanes.

A complete review of multi-class models is out of the scope of this paper. We refer to [14–16] and to the recent books [17,18] for an overview of the most used multi-class models.

Before introducing our contributions, let us introduce the *creeping or seepage effect* [14,19] which will be useful to describe the features of the proposed models. This term denotes the situations where the road space is shared by small and large vehicles, and small vehicles are able to move (at reduced velocity) even if large vehicles have reached the maximal density. This is in contrast to classical models such as the one proposed by Benzoni-Gavage and Colombo [20], in which the saturation of a class of vehicles immediately stops all the other classes. It is useful to note that the creeping phenomenon is typically considered in a context of *disordered traffic*, i.e., traffic with no lane discipline: smaller vehicles (e.g., two wheels) slip into the empty spaces left by large vehicles, similar to motion through porous media. This is not the case considered here, since we assume a strict lane discipline.

Finally, let us recall some important contributions about the fundamental diagram and its properties. It is well known that a single function $f = f(\rho)$ is not able, alone, to describe real data correctly. Indeed, by (3), we deduce that for any given density value ρ, only one velocity $v(\rho)$ is possible. This is not what happens in reality, where a scattered fundamental diagram is observed instead, due to the fact that different drivers respond in a different way to the same traffic conditions. Many papers investigated this phenomenon from different points of view, trying to explain its features, including instabilities; see, e.g., [21–31].

1.2. Case Study

In this paper, we consider the Italian motorway A4 Trieste–Venice and its branches to/from Udine, Pordenone and Gorizia, managed by Autovie Venete S.p.A., see Figure 1.

Figure 1. The Italian motorway A4 Trieste–Venice and its branches to/from Udine, Pordenone and Gorizia, managed by Autovie Venete S.p.A.

At the time of the present study (2019), the motorway had two lanes per direction, except for the leftmost segment near Venice (Venice–San Donà). To avoid heterogeneous conditions, we dropped the three-lane segment of the road to focus exclusively on the parts with *two lanes per direction*. In those segments, cars can use both lanes at any time, while trucks can use only the slow lane and cannot overtake. Due to the large flow of heavy vehicles, it happens sometimes that a queue of trucks is formed. In this case, cars move into the fast lane and keep going, although at moderate speed. When traffic conditions are sustained, the two classes of vehicles interact with each other: on the one hand, trucks act as moving bottlenecks for cars (cf. [32]), which are forced to slow down due to the restricted space; on the other hand, trucks must slow down when cars find it convenient to occupy part of the slow lane.

1.3. Our Contribution

In this paper, we propose two models for describing multi-class traffic flow on networks in which vehicles belonging to different classes share the road space only partially. More precisely, light vehicles can occupy the whole road, while heavy vehicles can occupy only a part of it. To align with the case study, we will assume that the road has two lanes in total and trucks can occupy only the slow one, without overtaking.

1. The first model is purely macroscopic. Both cars and truck are described by two coupled first-order LWR-based models. Fundamental diagrams are shaped in order to allow cars to move even in the presence of fully congested trucks. Considering that the fundamental diagram of each class is influenced by the presence of the other class, in the case of unstable (rapidly varying) traffic conditions of one class, we observe a scattered behavior in the fundamental diagram of the other class. Numerical results will show that this feature allows the model to catch, at least in part, some second-order (inertial) phenomena in traffic behavior, such as stop and go waves.

2. The second model is multi-scale. Cars are described by a first-order LWR-based model, while trucks are described by a second-order microscopic follow-the-leader model. For trucks, we consider the microscopic model used in [3], inspired, in turn, by a model originally proposed in [33] and specifically designed to reproduce stop and go waves. The choice of second-order model for trucks is crucial, since inertia effects are not at all negligible for those vehicles, while they are less important in car dynamics. Finally, note that, since trucks are confined to only one lane and cannot overtake, their dynamics perfectly matches the constituting assumptions of the follow-the-leader model.

Let us finally mention that the idea of coupling first- and second-order models was already exploited in [3] in a single-class scenario.

Remark 1. *Both models distinguish classes, but not lanes. The fact that trucks cannot use the fast lane while cars can occupy both slow and fast lanes is encapsulated in the choice of the fundamental diagrams.*

2. Dataset

Autovie Venete constantly monitors traffic conditions by means of video cameras, mobile sensors and fixed sensors. In this paper, we focus on the latest kind of data. Fixed sensors are located along the motorway, on each lane, and measure the flux and velocity of all vehicles passing in front of them, also distinguishing the class of vehicles. Data are aggregated per minute and are stored in a database for later analysis. For light vehicles, we further aggregated data coming from slow and fast lanes. For heavy vehicles, instead, we considered the slow lane only. In Figures 2–4, we show some flux and velocity data coming from some fixed sensors, used to conceive and calibrate the models presented in this paper. For better readability, flux data are plotted as both raw (as is) and smoothed by a Gaussian filter. Note that the flux data are always a multiple of 60 since they are evaluated every minute, but they are expressed in terms of vehicles per hour.

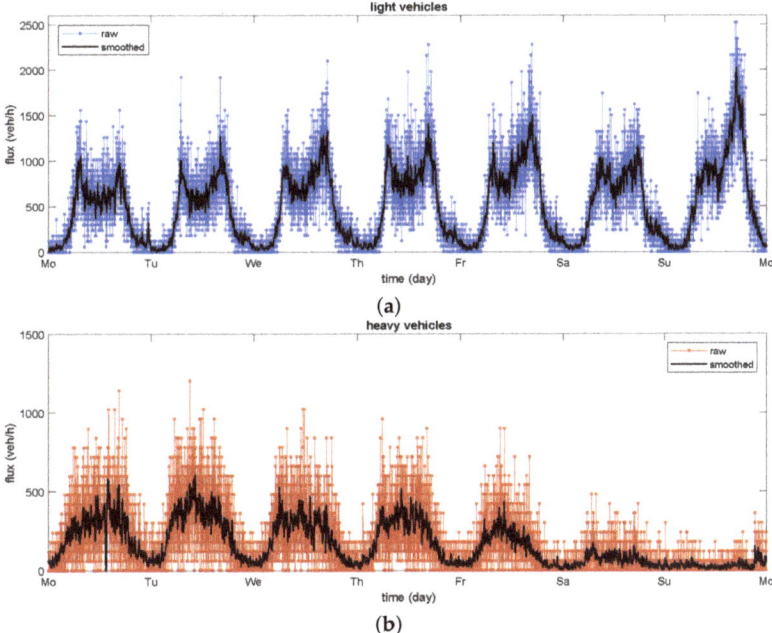

Figure 2. Typical weekly (from Monday to Sunday) flux data on the A4 motorway of (**a**) light and (**b**) heavy vehicles collected on March 2019 near Redipuglia. Smoothed data are plotted in black. Note the flux drop of cars in the middle of the day and of trucks on the weekend.

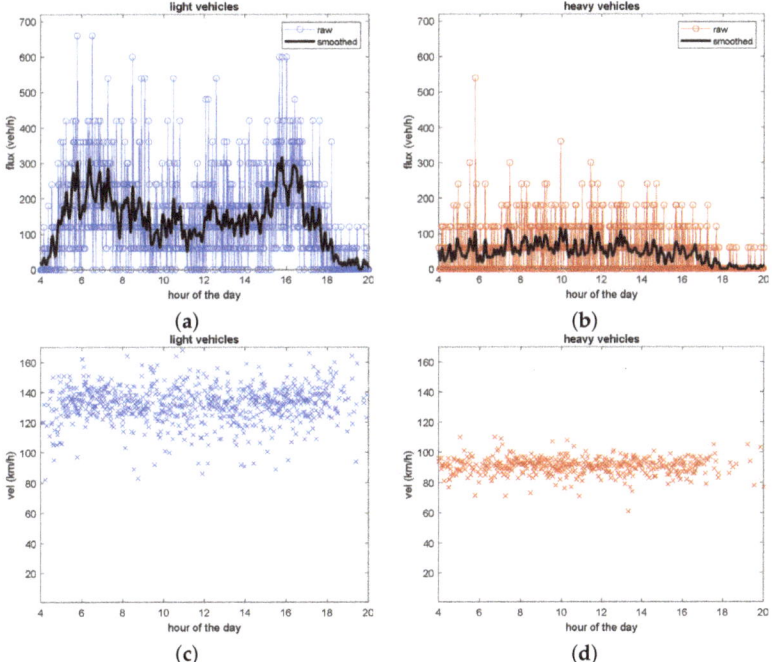

Figure 3. Typical daily (Thursday) flux and velocity data on the A28 motorway of (**a**–**c**) light and (**b**–**d**) heavy vehicles collected in May 2019 near Sesto al Reghena.

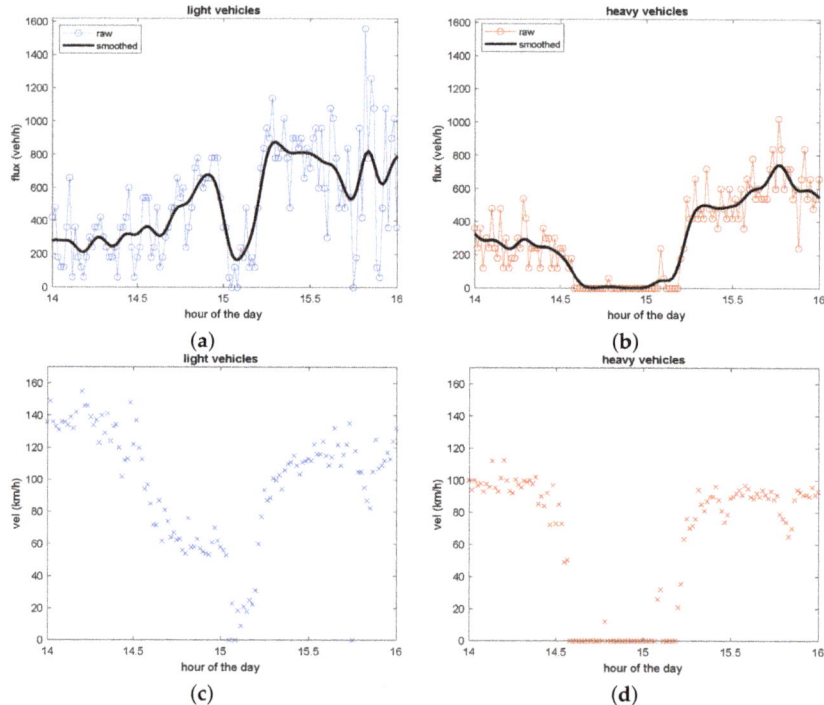

Figure 4. Creeping phenomenon registered in May 2019 near Portogruaro: (**a**) Light vehicles move in the fast lane even if (**b**) heavy vehicles queue in the slow lane. (**c**) Light vehicles' velocity drops from ∼140 to ∼60 km/h and then to ∼20 km/h while (**d**) heavy vehicles are completely stopped.

3. Models

In this section, we present the two models. As already stated in the Introduction, the models are not meant to provide the same results or to be the many-particle limit of the other. Nevertheless, they share the most important constitutive assumptions, and for this reason, they are expected to provide the same qualitative results. The most important common modeling assumption is that the car dynamics is influenced, at any time, by the presence of trucks, while the truck dynamics is affected by cars only if the density of cars exceeds a certain threshold, which corresponds to the fact that cars cannot be confined to the fast lane any longer and must invade the slow lane where trucks live. This assumption comes from an important piece of evidence: cars tend to avoid being trapped between two trucks in the slow lane and prefer moving to the fast lane. Doing this, cars move to the side of trucks (overtaking them if possible) and do not affect their dynamics, unless the density of cars is so high that they must necessarily occupy the slow lane too.

3.1. Macroscopic Model

We denote by ℓ_L and by ℓ_H the average length of light vehicles (cars) and heavy vehicles (trucks), respectively, and we define

$$\beta := \frac{\ell_L}{\ell_H} < 1. \tag{5}$$

We also denote by ρ_L the density of cars and by ρ_H the density of trucks. Similarly, we denote by $\rho_L^{\max}$ and $\rho_H^{\max}$ the maximal densities for cars and trucks, respectively. They are defined as

$$\rho_L^{\max} = \frac{2}{\ell_L} \quad \text{and} \quad \rho_H^{\max} = \frac{1}{\ell_H}, \tag{6}$$

having assumed that there are two available lanes for cars and only one for trucks. Note that density values are expressed in terms of number of vehicles per unit of space. Considering that trucks occupy more space than cars, a direct comparison of the two densities is not meaningful. For this reason, the two classes are typically compared in terms of occupied space.

The two-class dynamics is physically admissible if the two densities fall in the set

$$\mathcal{D} := \left\{ (\rho_L, \rho_H) : 0 \le \rho_L \le \rho_L^{\max}, \ 0 \le \rho_H \le \rho_H^{\max}, \ 0 \le \rho_L + \frac{\rho_H}{\beta} \le \rho_L^{\max} \right\}, \tag{7}$$

which is well defined if $\rho_L^{\max} - \frac{\rho_H^{\max}}{\beta} \ge 0$. In the following, in order to cope with the uneven space occupancy, we assume that the last condition is verified with the strict inequality

$$\rho_L^{\max} - \frac{\rho_H^{\max}}{\beta} > 0. \tag{8}$$

We consider the following two-class model for $(\rho_L, \rho_H) \in \mathcal{D}$:

$$\begin{cases} \partial_t \rho_L + \partial_x f_L(\rho_L, \rho_H) = 0 \\ \partial_t \rho_H + \partial_x f_H(\rho_L, \rho_H) = 0 \end{cases} \quad x \in \mathbb{R}, \ t > 0, \tag{9}$$

where

$$f_L(\rho_L, \rho_H) := \rho_L v_L(\rho_L, \rho_H), \qquad f_H(\rho_L, \rho_H) := \rho_H v_H(\rho_L, \rho_H)$$

define the two fundamental diagrams and v_L, v_H are the speed functions for light and heavy vehicles, respectively. We then have a family of flow–density curves $\rho_L \mapsto f_L(\rho_L, \rho_H)$ for cars, parameterized by the truck density ρ_H, and, analogously, a family of flow–density curves $\rho_H \mapsto f_H(\rho_L, \rho_H)$ for trucks, parameterized by ρ_L.

We assume that the flux and speed functions satisfy the following properties:

(L1) $v_L(\rho_L, \rho_H) \ge 0$ for all $(\rho_L, \rho_H) \in \mathcal{D}$ and $v_L(\rho_L, \rho_H) = 0$ iff $\rho_L = \rho_L^*(\rho_H)$, where

$$\rho_L^*(\rho_H) := \rho_L^{\max} - \rho_H/\beta \tag{10}$$

is the maximum admissible car density given the truck density ρ_H;

(L2) $v_L(\rho_L, \rho_H)$ is a decreasing function with respect to ρ_L and ρ_H;
(L3) $f_L(0, \rho_H) = 0$ and $f_L(\rho_L^*(\rho_H), \rho_H) = 0$ for all $\rho_H \in [0, \rho_H^{\max}]$;
(L4) $f_L(\rho_L, \rho_H)$ is concave with respect to ρ_L for any ρ_H. We define

$$\sigma_L(\rho_H) := \arg\max_{\rho_L} f_L(\rho_L, \rho_H) \tag{11}$$

which represents, as usual, the interface between freeflow and congested regimes;
(L5) $f_L(\rho_L, \rho_H)$ is a decreasing function with respect to ρ_H for any ρ_L.

Similarly,

(H1) $v_H(\rho_L, \rho_H) \ge 0$ for all $(\rho_L, \rho_H) \in \mathcal{D}$ and $v_H(\rho_L, \rho_H) = 0$ iff $\rho_H = \rho_H^*(\rho_L)$, where

$$\rho_H^*(\rho_L) := \min\{\rho_H^{\max}, \beta(\rho_L^{\max} - \rho_L)\} \tag{12}$$

is the maximum admissible truck density given the car density ρ_L;

(H2) $v_H(\rho_L, \rho_H)$ is a decreasing function with respect to ρ_L and ρ_H;
(H3) $f_H(\rho_L, 0) = 0$ and $f_H(\rho_L, \rho_H^*(\rho_L)) = 0$ for all $\rho_L \in [0, \rho_L^{\max}]$;

(H4) $f_H(\rho_L, \rho_H)$ is concave with respect to ρ_H for any ρ_L. We define

$$\sigma_H(\rho_L) := \arg\max_{\rho_H} f_H(\rho_L, \rho_H) \tag{13}$$

which represents, as usual, the interface between freeflow and congested regimes;

(H5) $f_H(\rho_L, \rho_H)$ is a decreasing function with respect to ρ_L for any ρ_H.

To cope with the peculiarities of the dynamics, we consider a *phase transition* (cf. [34–36]) caused by the presence of two states of the system:

- The *partial coupling phase* is in place when

$$(\rho_L, \rho_H) \in \mathcal{D}_1 := \left\{ 0 \le \rho_H \le \rho_H^{\max},\ 0 \le \rho_L \le \rho_L^{\max} - \rho_H^{\max}/\beta \right\}, \tag{14}$$

see Figure 5.

In this phase, we assume that cars are mainly in the fast lane and do not affect the truck dynamics. Trucks are then independent from cars.

For trucks, we choose a triangular fundamental diagram with

$$v_H(\rho_H) = V_H^{\max} \quad \text{for all} \quad \rho_H \le \sigma_H, \tag{15}$$

where $V_H^{\max}$ is the maximum speed of trucks, see Figure 6b.

Cars do not interfere with trucks but adapt their dynamics to the presence of them. Moreover, for cars, we choose (a family of) triangular fundamental diagrams, see Figure 6a. Specifically, we set

$$v_L(\rho_L, \rho_H) = \begin{cases} V_L^*(\rho_H) & \text{if } \rho_L \le \sigma_L(\rho_H), \\ \dfrac{V_L^*(\rho_H)\,\sigma_L(\rho_H)}{\rho_L^*(\rho_H) - \sigma_L(\rho_H)} \left(\dfrac{\rho_L^*(\rho_H)}{\rho_L} - 1 \right) & \text{if } \sigma_L(\rho_H) < \rho_L \le \rho_L^{\max} - \rho_H^{\max}/\beta, \end{cases} \tag{16}$$

where $V_L^*(\rho_H)$ is the maximum speed of cars given the truck density. We also define $V_L^*(0) = V_L^{\max}$ as the maximum speed of cars in the absence of trucks. Then, $V_L^*(\rho_H) \ge 0$ and $\sigma_L(\rho_H) \ge 0$ are continuous linear decreasing functions of ρ_H.

For $(\rho_L, \rho_H) \in \mathcal{D}_1$, the model (9) then becomes

$$\begin{cases} \partial_t \rho_L + \partial_x f_L(\rho_L, \rho_H) = 0 \\ \partial_t \rho_H + \partial_x f_H(\rho_H) = 0 \end{cases} \tag{17}$$

where $f_L(\rho_L, \rho_H) = \rho_L v_L(\rho_L, \rho_H)$ and $f_H(\rho_H) = \rho_H v_H(\rho_H)$, as described in Figure 6.

- The *full coupling phase* is in place when $(\rho_L, \rho_H) \in \mathcal{D}_2 := \mathcal{D} \setminus \mathcal{D}_1$, see Figure 5. In this case, we assume that there are too many cars to find it convenient to be confined to the fast lane. For this reason, they invade the slow lane, thus influencing the dynamics of trucks. The two equations in system (9) are then fully coupled.

As before, we choose for both classes a family of triangular fundamental diagrams which extend, by continuity, those defined in $\mathcal{D}_1$, as shown in Figure 7.

We define the *transition level* as the threshold density of light vehicles which acts as an interface between the two phases, see Figure 5. In our setting, trucks are confined to one of the two available lanes, and then the transition level is equal to $\rho_L^{\max} - \rho_H^{\max}/\beta = \rho_L^{\max}/2$.

Note also that the fundamental diagrams we use in this work verify all the properties (L1)–(L5) and (H1)–(H5).

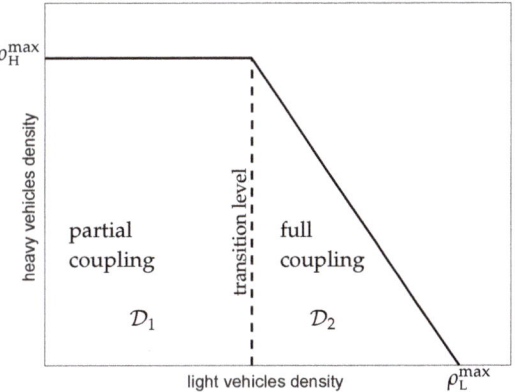

Figure 5. Domains $\mathcal{D}_1$ and $\mathcal{D}_2$ of the macroscopic model (9).

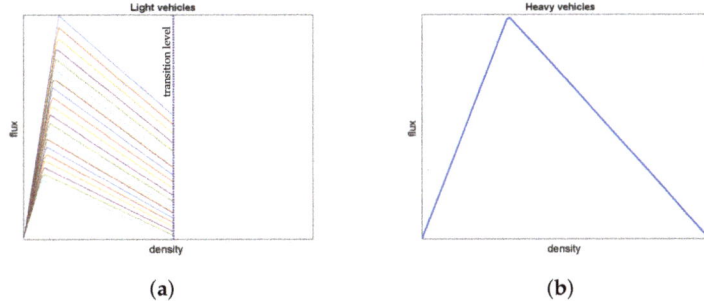

Figure 6. Fundamental diagrams of the macroscopic model in the partial coupling phase, i.e., $(\rho_L, \rho_H) \in \mathcal{D}_1$. (**a**) Light vehicles, (**b**) heavy vehicles.

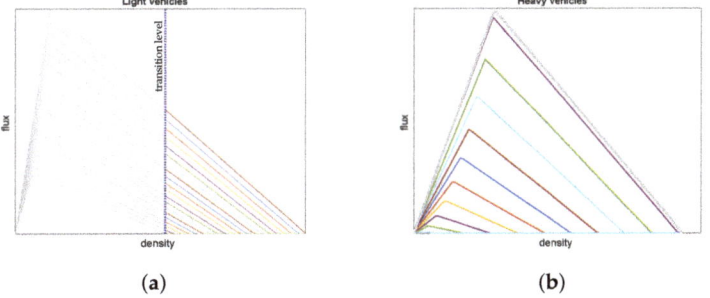

Figure 7. Fundamental diagrams of the macroscopic model in the full coupling phase, i.e., $(\rho_L, \rho_H) \in \mathcal{D}_2$. (**a**) Light vehicles, (**b**) heavy vehicles.

3.2. Multi-Scale Model

In this section, we describe the multi-scale model. Here, cars are described by a first-order LWR model of type (2), and trucks are described by a second-order microscopic follow-the-leader model of type (1). Let us describe the microscopic model first, dropping, for the moment, the coupling with light vehicles.

3.2.1. Microscopic Model for Heavy Vehicles

The microscopic model is the one presented in [3], which is, in turn, inspired by the model originally proposed by Zhao and Zhang in [33].

In the following, we denote by Δ_k the gap between truck k and truck $k+1$ at any time t:
$$\Delta_k(t) := X_{k+1}(t) - X_k(t).$$

It is plain that this gap is inversely proportional to the density of heavy vehicles. We define in (1)

$$A(X_k, X_{k+1}, V_k, V_{k+1}) = \begin{cases} \frac{1}{\tau_{acc}}\left(v^{zz}(\Delta_k) - V_k\right), & \text{if } v^{zz}(\Delta_k) \geq V_k \\ \frac{1}{\tau_{dec}}\left(v^{zz}(\Delta_k) - V_k\right), & \text{if } v^{zz}(\Delta_k) < V_k \end{cases} \quad (18)$$

where the function v^{zz} represents the equilibrium velocity all drivers tend to and depends on the gap Δ_k. Parameters $\tau_{acc}, \tau_{dec} > 0$ are the relaxation times as usual, differentiated for the acceleration and the deceleration phase. Diversifying the relaxation times appeared to be crucial to fit real data.

The velocity function v^{zz} is defined by

$$v^{zz}(\Delta) := \begin{cases} 0, & \text{if } \Delta \leq \Delta_{close} \\ \frac{V_H^{max}}{\Delta_{far} - \Delta_{close}}(\Delta - \Delta_{close}), & \text{if } \Delta_{close} < \Delta < \Delta_{far} \\ V_H^{max}, & \text{if } \Delta \geq \Delta_{far} \end{cases} \quad (19)$$

where $\Delta_{close}, \Delta_{far}, V_H^{max}$ are positive parameters, see Figure 8.

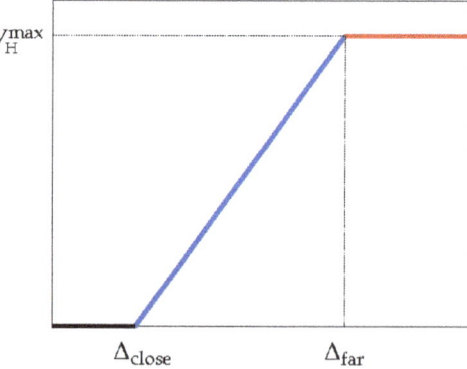

Figure 8. The shape of the velocity function $v^{zz}(\Delta)$ defined in (19).

The plateau in $\Delta \in [0, \Delta_{close}]$ is crucial for correctly reproducing stop and go waves. Indeed, once the relaxation times τ_{acc}, τ_{dec} are fixed, the capability of the model to trigger stop and go waves is ruled precisely by Δ_{close}.

3.2.2. Full Model

First of all, given the parameter $\delta > 0$, we assume that cars located at x are influenced by a truck iff the distance between the truck and x is less than δ. We denote the number of trucks falling in the road interval $[x - \delta, x + \delta]$ at any time t by

$$N_H^\delta(x, t) := \#\{k : X_k(t) \in [x - \delta, x + \delta]\}. \quad (20)$$

Second, we denote by ρ_L the density of light vehicles and by v_L their velocity. To couple the dynamics of the two classes, we assume that v_L depends on both ρ_L (as in the classical LWR model) and N_H^δ. Following standard assumptions, we assume that v_L is decreasing with respect to both arguments.

Finally, we couple the dynamics of heavy vehicles with those of light vehicles. The interaction is obtained by introducing the dependence on ρ_L in the parameters Δ_{close} and Δ_{far}. More precisely, we introduce the increasing functions $\Delta_{\text{close}} = \Delta_{\text{close}}(\rho_L)$ and $\Delta_{\text{far}} = \Delta_{\text{far}}(\rho_L)$, and we denote by $A^C = A^C(X_k, X_{k+1}, V_k, V_{k+1}, \rho_L)$ the coupled acceleration defined as A in (18) and (19), with the new dependence on ρ_L.

We are now ready to present the fully coupled multi-scale model which reads as

$$\begin{cases} \begin{cases} \dot{X}_k = V_k \\ \dot{V}_k = A^C(X_k, X_{k+1}, V_k, V_{k+1}, \rho_L) \end{cases}, \quad k = 1, \ldots, N-1 \\ \partial_t \rho_L + \partial_x \left(\rho_L v_L(\rho_L, N_H^\delta) \right) = 0, \quad x \in \mathbb{R}, \quad t > 0. \end{cases} \quad (21)$$

To be coherent with our modeling assumptions, the functions Δ_{close} and Δ_{far} are constant for car densities below the transition level, i.e., $\rho_L \le \rho_L^{\max}/2$. In this case, the dynamics of trucks is independent from those of cars. Conversely, for $\rho_L > \rho_L^{\max}/2$, we assume that the distances Δ_{close} and Δ_{far} increase linearly with respect to the average number of cars which are positioned between two trucks. This number can be easily computed considering the average number of cars in a road segment of length ℓ (equal to $\rho_L \ell$) and the number of trucks in the same road segment (assuming that all vehicles are uniformly distributed). We were unable to precisely calibrate the shape of the functions Δ_{close} and Δ_{far} from real data because it happens rarely that many cars are found between trucks: indeed, trucks tend to "push" cars into the fast lane rather than reacting to their presence.

3.3. Extension of the Models to General Road Networks

In order to perform a complete simulation on a generic network of highways, some important generalizations are needed.

3.3.1. Any Number of Lanes

Highways often have more than two lanes. Consider a road with n lanes of which n_H can be occupied by trucks. To allow the creeping phenomenon, we assume that $n_H < n$, which corresponds to $\rho_L^{\max} \ell_L - \rho_H^{\max} \ell_H > 0$ in terms of space occupied, cf. (8).

In the macroscopic approach, the model is easy generalized. Fundamental diagrams are modified in such a way that trucks start interacting with cars when the density of cars becomes greater than $\frac{n_H}{n}\rho_L^{\max}$.

In the microscopic model, instead, an important modification is needed if $n_H > 1$. Indeed, in this case, trucks can overtake, and the microscopic model must be able to handle this. Typically, some new parameters are introduced in order to establish when a truck decides to overtake and if the truck can actually overtake, considering suitable safety constraints. From the computational point of view, additional difficulty arises when one has to find the truck in front of any other truck, since the ordering is lost whenever a truck overtakes. To make the search for the preceding vehicle computationally feasible, one can keep track, in a specific list, of all trucks located in each numerical cell and then update the list whenever a truck leaves or enters the cell.

3.3.2. Junctions

In order to perform a full simulation on a network of highways, both theoretical and numerical treatments of junctions are needed. Typically, highways do not have roundabouts, traffic lights or complex junctions; therefore, we can limit ourselves to handle simple merging (2 incoming roads and 1 outgoing road) and diverging (1 incoming road and 2 outgoing roads). We adopted the approach detailed in [13], in which the dynamics reformulated along paths and junctions "disappears". The price to pay is that the number of equations is multiplied by the number of possible paths the drivers can follow at junctions.

In both merging and diverging, we have only two possible paths: for example, in the case of diverging, one can choose among the first and second outgoing roads, while in merging, one can decide to come from the first or the second incoming road.

Following this approach in the macroscopic model, the densities of each class of vehicles are split around every junction, ending up with a system of four conservation laws (two paths for each of the two classes of vehicles) with a discontinuous flux. After the junctions, densities are gathered together again, and the two-equation system (9) is restored.

In the multi-scale model, instead, the path-based approach is applied only for car dynamics since managing trucks is much simpler. Indeed, in the microscopic model, one can just move vehicles from one road to another on the basis of their destination, see [37]. Unfortunately, the ordering of trucks is lost every time a change of road takes place. In order to reduce the computational effort needed for the computation of the preceding truck of every truck, the same solution proposed in Section 3.3.1 can be applied.

4. Numerical Approximation and Calibration

In this section, we describe how the models introduced above can actually be implemented. First, we briefly recall the numerical methods we have adopted, and then we describe how we used real data to set the models' parameters.

4.1. Macroscopic Model

For the numerical approximation of the macroscopic model (9), we employ the extension of the *cell transmission model* (CTM) to the heterogeneous multi-class model proposed in [14]. Let Δx and Δt be the space and time steps, respectively, and let $(\rho_L^{n,i}, \rho_H^{n,i})$ be the traffic densities in the ith cell at the nth time step. The finite volume numerical scheme reads

$$\begin{cases} \rho_L^{n+1,i} = \rho_L^{n,i} + \dfrac{\Delta t}{\Delta x}\left(\mathcal{F}_L^{n,i-1/2} - \mathcal{F}_L^{n,i+1/2}\right) & \text{(22a)} \\ \rho_H^{n+1,i} = \rho_H^{n,i} + \dfrac{\Delta t}{\Delta x}\left(\mathcal{F}_H^{n,i-1/2} - \mathcal{F}_H^{n,i+1/2}\right) & \text{(22b)} \end{cases}$$

where

$$\mathcal{F}_L^{n,i+1/2} := \min\left\{S_L(\rho_L^{n,i}, \rho_H^{n,i}), R_L(\rho_L^{n,i+1}, \rho_H^{n,i+1})\right\}, \tag{23}$$

$$\mathcal{F}_H^{n,i+1/2} := \min\left\{S_H(\rho_L^{n,i}, \rho_H^{n,i}), R_H(\rho_L^{n,i+1}, \rho_H^{n,i+1})\right\}, \tag{24}$$

and (S_L, R_L), (S_H, R_H) represent the sending and receiving functions of the two vehicle classes, respectively, defined by

$$\begin{aligned} S_L(\rho_L, \rho_H) &:= \begin{cases} f_L(\rho_L, \rho_H), & \text{if } \rho_L \leq \sigma_L(\rho_H), \\ f_L(\sigma_L(\rho_H), \rho_H), & \text{if } \rho_L > \sigma_L(\rho_H), \end{cases} \\ R_L(\rho_L, \rho_H) &:= \begin{cases} f_L(\sigma_L(\rho_H), \rho_H), & \text{if } \rho_L \leq \sigma_L(\rho_H), \\ f_L(\rho_L, \rho_H), & \text{if } \rho_L > \sigma_L(\rho_H), \end{cases} \end{aligned} \tag{25}$$

and similarly for (S_H, R_H).

The numerical grid is chosen as $\Delta x = 100$ m and $\Delta t = 2.6$ s. The choice of the space step comes from the fact that the company Autovie Venete finds such granularity convenient for sharing traffic information to drivers, while the time step is dictated by the CFL condition.

Calibration of the fundamental diagrams was performed by fitting real data. We used all data measured in 2019 by one fixed sensor located near Cessalto, see Figures 9 and 10. Note that for high densities, the velocities drop rapidly to zero. Since we have no data for completely stationary vehicles under the sensor, we are not able to reconstruct data on high traffic density. For this reason, the maximal densities $\rho_L^{\max}$ and $\rho_H^{\max}$ are estimated by simply computing the ratio between the number of available lanes for the class and the average length of vehicles of that class, see Equation (6).

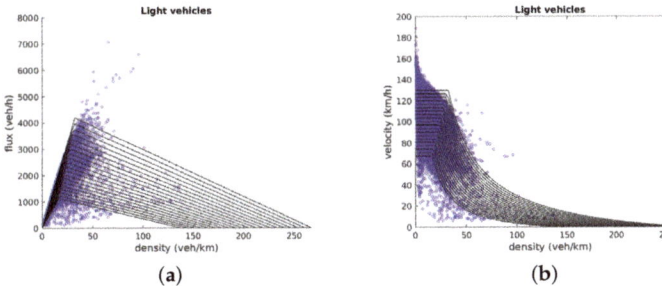

Figure 9. (a) Flux–density and (b) velocity–density relationships for cars with real data superimposed.

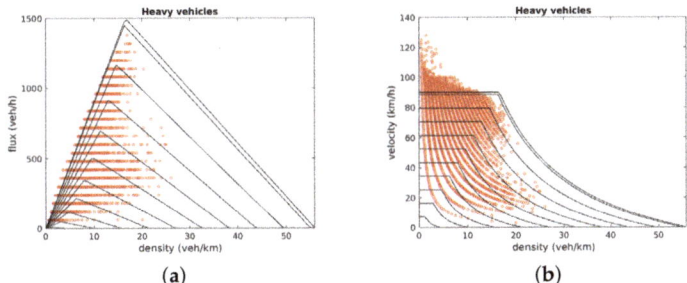

Figure 10. (a) Flux–density and (b) velocity–density relationships for trucks with real data superimposed.

Model parameters are summarized in Table 1. All functions which rule the dependence of ρ, v, f on the density of the other class are linear.

Table 1. Parameters for the macroscopic model.

	Light Vehicles	Heavy Vehicles
veh. length + safety dist. (km)	7.5×10^{-3}	18×10^{-3}
max max density (veh/km)	$\rho_L^*(0) = 267$	$\rho_H^*(0) = 56$
min max density (veh/km)	$\rho_L^*(\rho_H^{max}) = 133$	$\rho_H^*(\rho_L^{max}) = 0$
max max speed (km/h)	$v_L(0,0) = 130$	$v_H(0,0) = 90$
min max speed (km/h)	$v_L(0, \rho_H^{max}) = 65$	$v_H(\rho_L^{max}, 0) = 0$
max max flux (veh/h)	$f_L(\sigma_L(0), 0) = 4200$	$f_H(0, \sigma_H(0)) = 1500$
min max flux (veh/h)	$f_L(\sigma_L(\rho_H^{max}), \rho_H^{max}) = 1200$	$f_H(\rho_L^{max}, \sigma_H(\rho_L^{max})) = 0$

4.2. Multi-Scale Model

For the numerical approximation of the macroscopic part of the multi-scale model (21), we again employ scheme (22a), where N_H^δ plays the role of ρ_H in the obvious manner. The numerical grid is chosen as $\Delta x = 100$ m and $\Delta t = 2$ s.

The dependence of the flux on N_H^δ can generate some issues. For example, consider the case of no trucks and a car density $\hat{\rho}_L$ close to ρ_L^{max}. When a truck enters the road, the maximal density allowed in the cell occupied by the truck drops to $\rho_L^*(N_H^\delta)$ according to (10). Now, if $\rho_L^*(N_H^\delta) < \hat{\rho}_L$, the current density $\hat{\rho}_L$ is found not to be compatible with the new maximal density. Although the entering truck perceives the cars, it is not guaranteed that the compatibility with the maximal density is respected at any time. To avoid this problem, trucks must be prevented from entering cells if the new maximal density caused by the presence of the truck itself is not compatible with current traffic conditions.

For the numerical approximation of the microscopic part, we used a standard Euler scheme with a time step of $\delta t = 0.1$ s. Note that this time step is much smaller than the time step Δt used for the Godunov scheme, meaning that the updates of the trucks and cars are asynchronous.

Regarding the parameters, the macroscopic part of the model is treated as in Section 4.1 (Table 1). For the microscopic model, some parameters are easily calibrated by using real data and considering physical constraints. For example, V_H^{max} was defined as in the macroscopic model. Δ_{far} was set in order to guarantee that trucks do not collide even in the event that a truck suddenly brakes with full power until it stops (note that our model allows, in principle, collisions since deceleration is bounded). Δ_{close}, instead, was set to a distance which guarantees catching the maximal observed density of trucks. In other words, when a queue of trucks is formed, the model predicts the correct maximal density.

Parameters τ_{acc} and τ_{dec} are, instead, more difficult to calibrate since they are not easily measurable. For those values, we considered a real stop and go wave observed by the company staff on 12 June 2017, generated by the slowdown of a truck near a bottleneck. The initial perturbation (slowdown) was amplified and, in a short time, generated a queue which propagated backwards. We run the microscopic model using real inflow data as left boundary conditions, and then we fitted the parameters in order to catch the real queue as measured in the field, see Figure 11.

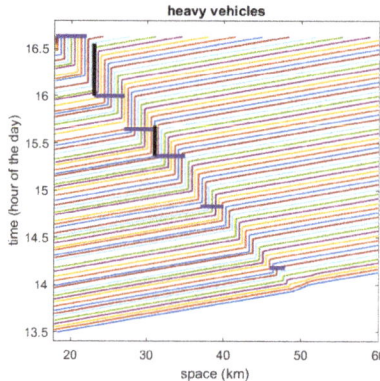

Figure 11. Simulated trajectories obtained with real inflow data as left boundary conditions (not all vehicles are plotted for visualization purposes). Horizontal blue lines and vertical black lines indicate, respectively, the position and the duration of the real queue as measured in the field.

The role of the parameters τ_{acc} and τ_{dec} is to adjust the points/times of the start and the end of the queue. We noted a strong sensitivity of the model to those parameters. As a consequence, it is quite difficult to catch the correct speed of the backward propagation of a queue when inertia comes into play. We summarize the values of the parameters in Table 2.

Table 2. Parameters for the microscopic model.

δ	50×10^{-3}	km
Δ_{close}	25×10^{-3}	km
Δ_{far}	50×10^{-3}	km
V_H^{max}	90	km/h
τ_{dec}	2×10^{-4}	h
τ_{acc}	1.4×10^{-2}	h

5. Numerical Results

In this section, we present the numerical results obtained with models (9) and (21).

5.1. Macroscopic Model

Here, we present three tests which highlight how the macroscopic model reproduces some interesting phenomena arising from the coupled dynamics of cars and trucks. In particular, we focus on the creeping phenomenon, the shared occupancy and the stop and go waves.

5.1.1. Test 1A: Creeping

In this simple test, we observe the creeping phenomenon, see Figure 12. The simulation starts with a constant density $(\rho_L, \rho_H) = (10, 13)$ veh/km all along the road. At the end of the road (right boundary), trucks are stopped by fixing their density at its maximum value $\rho_H^{max} = 56$: a queue of trucks propagates backward from the end of the road, while a constant flux of cars approaches the beginning of the queue. Once cars reach the trucks' queue, they have to slow down but do not stop completely. More precisely, the cars' velocity drops to 65 km/h. Note that the car density remains under the transition level, and then the dynamics is in the partial coupling phase all the time. Moreover, cars are always in the freeflow regime and then move at maximal speed, but the maximal speed changes as a function of the truck density.

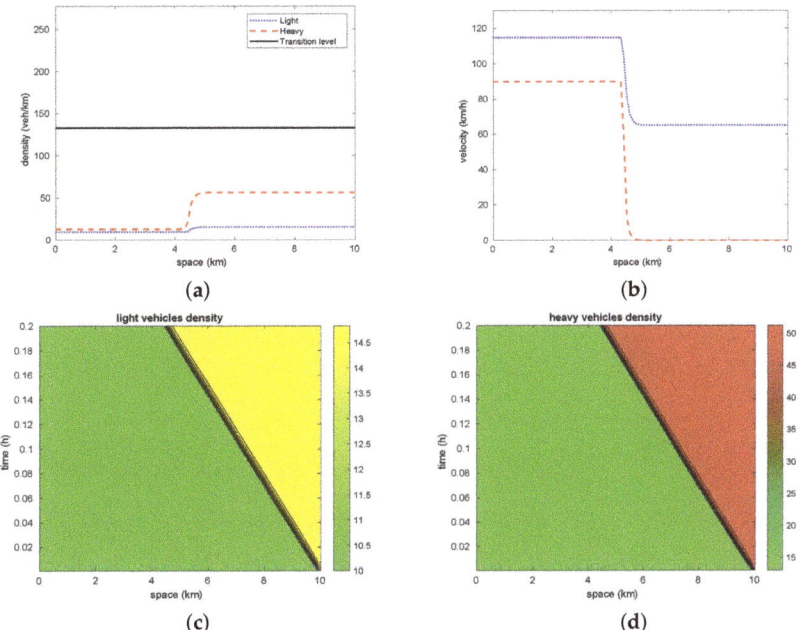

Figure 12. Test 1A: (**a**) Density and (**b**) velocity of light and heavy vehicles as a function of space at final time. (**c**) Density of light and (**d**) heavy vehicles in space–time.

5.1.2. Test 2A: Cars' Congestion Affects Truck Dynamics

In this test, we observe the effect of congestion of cars, see Figure 13. The simulation starts with a constant density $(\rho_L, \rho_H) = (10, 8)$ veh/km all along the road. At the end of the road (right boundary), the density of cars is fixed to 186 veh/km to create the slowdown. The car density is larger than the transition level, so cars have to invade the slow lane. Trucks facing the car congestion slow down but do not just occupy the space left to them by cars; rather, they conquer some extra space, thus decreasing the car density. As a result, both cars and trucks proceed slowly without stopping, and the initial car congestion propagates backward with a density lower than the transition level.

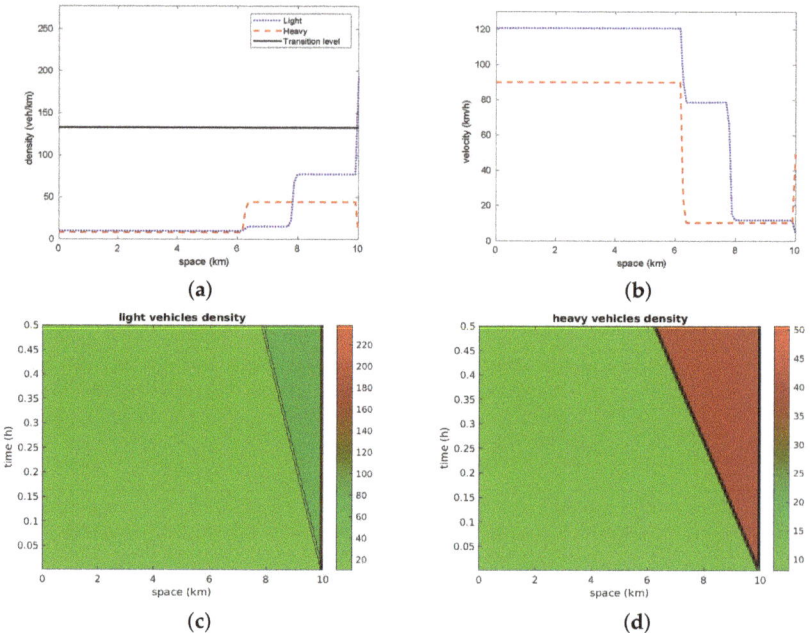

Figure 13. Test 2A: (**a**) Density and (**b**) velocity of light and heavy vehicles as a function of space at final time. (**c**) Density of light and (**d**) heavy vehicles in space–time.

5.1.3. Test 3A: Stop and Go Wave

In this test, we study the evolution of a small perturbation in the truck density, see Figure 14. At the initial time, the truck density is constant and equal to 12 veh/km, except for a small perturbation at the end of the road where the density is equal to 30 veh/km. The car density instead oscillates just above the transition level. It is plain that a single-class LWR model for trucks only would flatten the perturbation in a short time. Conversely, in this case, the coupling with car dynamics causes the perturbation to propagate backward without vanishing. This second-order-type effect is obtained thanks to the fact that the fundamental diagram of trucks is continuously modified by the oscillating car density.

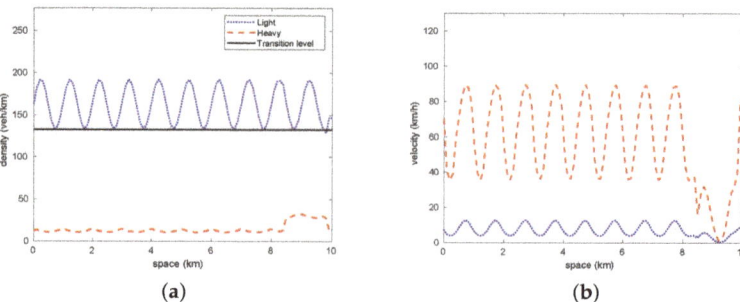

Figure 14. *Cont.*

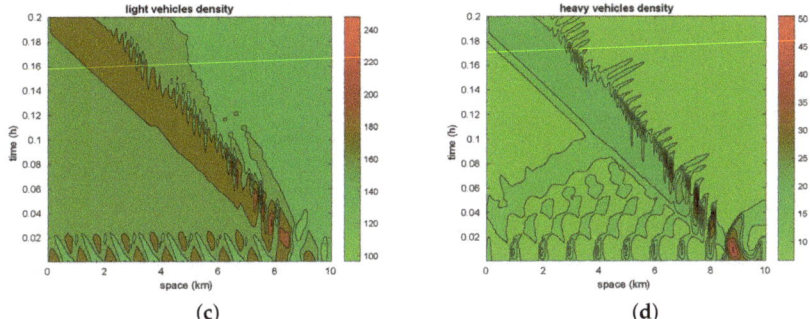

(c) (d)

Figure 14. Test 3A: (**a**) Density and (**b**) velocity of light and heavy vehicles as a function of space at $t = \Delta t$ (i.e., just after the initial time). (**c**) Density of light and (**d**) heavy vehicles in space–time. The evolution of the initial perturbation in the truck density starting at 9 km is perfectly visible, which creates, in turn, a perturbation in the car density.

5.2. Multi-Scale Model

Here, we replicate, with the multi-scale model, the first two scenarios already investigated in Section 5.1. The third scenario was already considered in Figure 11, where the second-order microscopic model is able to reproduce stop and go waves alone, without the need to couple car dynamics. Finally, we consider the case of a merge.

5.2.1. Test 1B: Creeping Effect

Similar to Test 1A in Section 5.1.1, here, one truck stops completely and creates a long queue of trucks behind, which saturates the slow lane. When cars reach the truck queue, they all move to the fast lane staying at the (new, reduced) maximal velocity of 65 km/h, see Figure 15.

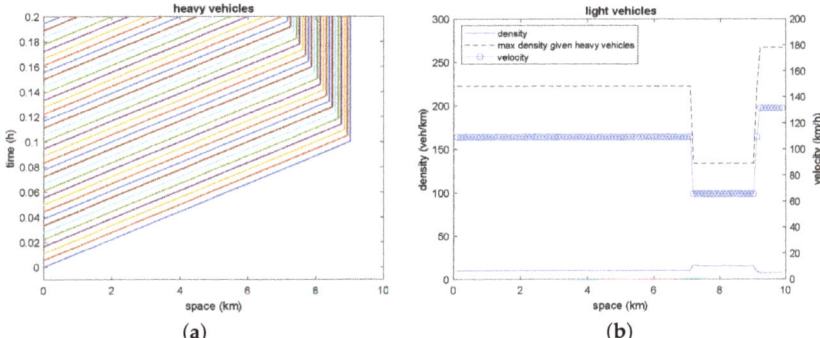

(a) (b)

Figure 15. Test 1B: (**a**) Trajectories of trucks in space–time (for visualization purposes, not all trucks are actually plotted). When the first truck stops, a queue is formed behind. (**b**) Car density, car velocity and car maximal density given the number of trucks at final time. Creeping is visible between 7 and 9 km.

5.2.2. Test 2B: Cars' Congestion Affects Truck Dynamics

Similar to Test 2A in Section 5.1.2, congestion of cars at the end of the road slows down trucks, see Figure 16. The results are similar to those obtained by the macroscopic model, but here, trucks stop completely, forming a queue.

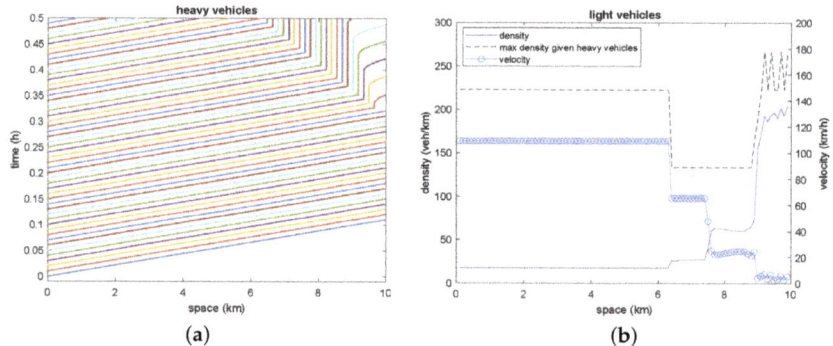

Figure 16. Test 2B: (**a**) Trajectories of trucks in space–time (for visualization purposes, not all trucks are actually plotted). They stop for a while and then accelerate. (**b**) Car density, car velocity and car maximal density given the number of trucks at final time.

5.2.3. Test 3B: Merge

In this test, we consider a merge (two incoming roads and one outgoing road). At time $t = 0$, the three roads are empty. A constant inflow of trucks (one every 4 s) comes from the left boundary of both incoming roads, while a constant density of cars ($\rho_L = 32$) is imposed as a Dirichlet left boundary condition on the second incoming road only. The first incoming road has no cars. When trucks reach the junction and merge, they suddenly break and rapidly form a queue which propagates backward along both incoming roads, see Figure 17a,b.

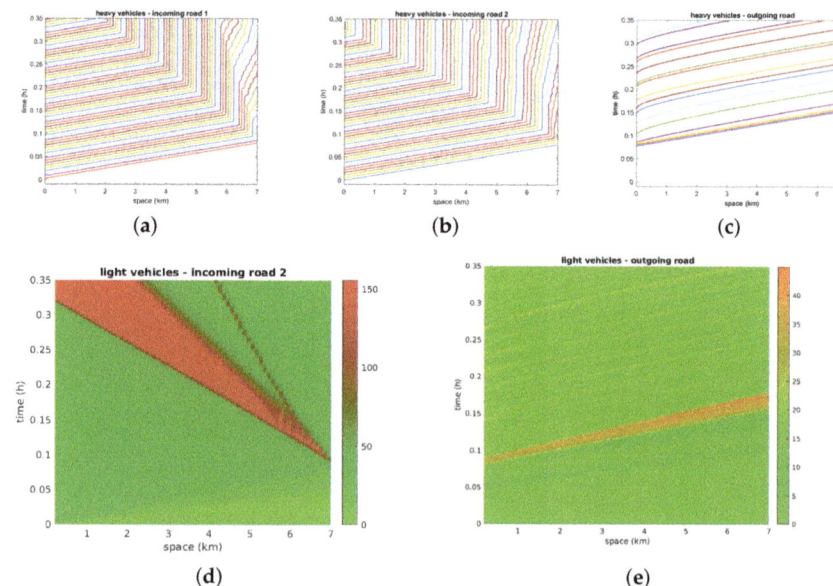

Figure 17. Test 3B: (**a**–**c**) Trajectories of trucks in space–time on the first incoming road, second incoming road and outgoing road, respectively (for visualization purposes, not all trucks are actually plotted). (**d**,**e**) Car density on second incoming road and outgoing road, respectively.

Queues are not identical due to the presence of cars along the second incoming road. One can note that when the trucks downstream of the queue start moving again, their flux is not maximal: indeed, if the flow were maximum, a queue at the junction would immediately reform as it happened in the first place. This is the well-known *capacity drop* phenomenon, ruled by τ_{acc}, cf. [38]. As a consequence, trucks are able to cross the junction without spillback. Cars, instead, move at the maximal flux until they encounter the truck queue. The queue acts as a moving bottleneck and drops the road capacity; therefore, the car traffic immediately enters the congested state, and the density increases. Downstream, the density remains in the freeflow state, and cars cross the junction without spillback, see Figure 17d,e.

6. Conclusions and Future Work

In this paper, we presented two models for two-class traffic flow. Although the models are tailored for a specific case study, they are sufficiently general to be useful in other motorways. Moreover, both models can be easily generalized to more than two classes of vehicles and a different ratio between the number of lanes used by trucks and the number of lanes used by cars.

We have shown that the models are able to reproduce, both qualitatively and quantitatively, some notable traffic phenomena arising from the interactions of the two classes. Interestingly, the macroscopic model, although purely first-order, is able to reproduce stop and go waves thanks to the coupling of the two classes.

After this preliminary analysis, it is possible to sketch some conclusions about the advantages and drawbacks of the two models: The multi-scale model has a greater potential since the second-order microscopic part makes it more realistic and then suitable for quantitative predictions. Nevertheless, the macroscopic model appears to be simpler and more manageable, thus representing a valid alternative if one wants to avoid tracking all single vehicles, especially for saving computational time.

In conclusion, we believe that both the proposed models represent the best compromise between accuracy and implementability. In fact, decoupling the dynamics of different classes excessively simplifies the problem description and does not allow obtaining an accurate forecast; conversely, moving to second-order macroscopic models or including multi-lane features in the models notably increases the complexity of the code as well as the number of parameters to be tuned. These generalizations would allow, in principle, easily catching inertia-based phenomena in all classes of vehicles and tracking the density of *each class* of vehicle in *each lane*, but, in our opinion, they make that model unfeasible for practical applications.

In the future, we plan to improve the models including the possibility that they are fed by both Lagrangian (GPS-like) and Eulerian data coming from mobile and fixed sensors, respectively, cf. [39]. Moreover, we plan to estimate, in real time, the difference between predicted and measured densities using the machinery developed in [13], hopefully creating an algorithm for the auto-calibration of the models in real time.

Author Contributions: Conceptualization, M.B. and E.C.; data curation, P.R.; funding acquisition, M.B., E.C. and P.R.; investigation, M.B., E.C. and P.R.; methodology, M.B. and E.C.; visualization, M.B. and E.C.; writing—original draft, M.B. and E.C.; writing—review and editing, M.B., E.C. and P.R. All authors have read and agreed to the published version of the manuscript.

Funding: This work was partially funded by the company Autovie Venete S.p.A. This work was also carried out within the research project "SMARTOUR: Intelligent Platform for Tourism" (No. SCN_00166) funded by the Ministry of University and Research with the Regional Development Fund of European Union (PON Research and Competitiveness 2007–2013). The authors also acknowledge the Italian Minister of Instruction, University and Research for supporting this research with funds coming from the project entitled *Innovative numerical methods for evolutionary partial differential equations and applications* (PRIN Project 2017, No. 2017KKJP4X). M.B. and E.C. are members of the INdAM Research group GNCS.

Data Availability Statement: Data are not publicly available.

Acknowledgments: The authors want to thank all the Autovie Venete staff as well as Gabriella Bretti, Matteo Piu, Elisa Iacomini, Caterina Balzotti and Elia Onofri for valuable help.

Conflicts of Interest: The authors declare no conflict of interest.

References

1. Lighthill, M.J.; Whitham, G.B. On kinematic waves II. A theory of traffic flow on long crowded roads. *Proc. R. Soc. Lond. Ser. A* **1955**, *229*, 317–345. [CrossRef]
2. Richards, P.I. Shock waves on the highway. *Oper. Res.* **1956**, *4*, 42–51. [CrossRef]
3. Cristiani, E.; Iacomini, E. An interface-free multi-scale multi-order model for traffic flow. *Discret. Contin. Dyn. Syst. Ser. B* **2019**, *24*, 6189–6207. [CrossRef]
4. Garavello, M.; Piccoli, B. *Traffic Flow on Networks*; American Institute of Mathematical Sciences: Springfield, MO, USA, 2006.
5. Coclite, G.M.; Garavello, M.; Piccoli, B. Traffic flow on a road network. *SIAM J. Math. Anal.* **2005**, *36*, 1862–1886. [CrossRef]
6. Holden, H.; Risebro, H. A mathematical model of traffic flow on a network of unidirectional roads. *SIAM J. Math. Anal.* **1995**, *26*, 999–1017. [CrossRef]
7. Bressan, A.; Nguyen, K.T. Conservation law models for traffic flow on a network of roads. *Netw. Heterog. Media* **2015**, *10*, 255–293. [CrossRef]
8. Garavello, M.; Goatin, P. The Cauchy problem at a node with buffer. *Discret. Contin. Dyn. Syst. Ser. A* **2012**, *32*, 1915–1938. [CrossRef]
9. Herty, M.; Lebacque, J.P.; Moutari, S. A novel model for intersections of vehicular traffic flow. *Netw. Heterog. Media* **2009**, *4*, 813–826. [CrossRef]
10. Bretti, G.; Briani, M.; Cristiani, E. An easy-to-use algorithm for simulating traffic flow on networks: Numerical experiments. *Discret. Contin. Dyn. Syst. Ser. S* **2014**, *7*, 379–394. [CrossRef]
11. Briani, M.; Cristiani, E. An easy-to-use algorithm for simulating traffic flow on networks: Theoretical study. *Netw. Heterog. Media* **2014**, *9*, 519–552. [CrossRef]
12. Hilliges, M.; Weidlich, W. A phenomenological model for dynamic traffic flow in networks. *Transp. Res. Part B* **1995**, *29*, 407–431. [CrossRef]
13. Briani, M.; Cristiani, E.; Iacomini, E. Sensitivity analysis of the LWR model for traffic forecast on large networks using Wasserstein distance. *Commun. Math. Sci.* **2018**, *16*, 123–144. [CrossRef]
14. Fan, S.; Work, D.B. A heterogeneous multiclass traffic flow model with creeping. *SIAM J. Appl. Math.* **2015**, *75*, 813–835. [CrossRef]
15. van Wageningen-Kessels, F. Framework to assess multiclass continuum traffic flow models. *Transp. Res. Rec.* **2016**, *2553*, 150–160. [CrossRef]
16. (Sean) Qian, Z.; Li, J.; Li, X.; Zhang, M.; Wang, H. Modeling heterogeneous traffic flow: A pragmatic approach. *Transp. Res. Part B* **2017**, *99*, 183–204. [CrossRef]
17. Ferrara, A.; Sacone, S.; Siri, S. *Freeway Traffic Modelling and Control*; Springer: Berlin/Heidelberg, Germany, 2018.
18. Kessels, F. *Traffic Flow Modelling*; Springer: Berlin/Heidelberg, Germany, 2019.
19. Agarwal, A.; Lämmel, G. Modeling *seepage* behavior of smaller vehicles in mixed traffic conditions using an agent based simulation. *Transp. Dev. Econ.* **2016**, *2*, 8. [CrossRef]
20. Benzoni-Gavage, S.; Colombo, R.M. An *n*-populations model for traffic flow. *Eur. Jnl Appl. Math.* **2003**, *14*, 587–612. [CrossRef]
21. Balzotti, C.; Göttlich, S. A two-dimensional multi-class traffic flow model. *Netw. Heterog. Media* **2021**, *16*, 69–90. [CrossRef]
22. Fan, S.; Seibold, B. Data-fitted first-order traffic models and their second-order generalizations. Comparison by trajectory and sensor data. *Transp. Res. Rec.* **2013**, *2391*, 32–43. [CrossRef]
23. Fan, S.; Herty, M.; Seibold, B. Comparative model accuracy of a data-fitted generalized Aw-Rascle-Zhang model. *Netw. Heterog. Media* **2014**, *9*, 239–268. [CrossRef]
24. Klar, A.; Günther, M.; Wegener, R.; Materne, T. Multivalued fundamental diagrams and stop and go waves for continuum traffic flow equations. *SIAM J. Appl. Math.* **2004**, *64*, 468–483. [CrossRef]
25. Herty, M.; Illner, R. Coupling of non-local driving behaviour with fundamental diagrams. *Kinet. Relat. Models* **2012**, *5*, 843–855. [CrossRef]
26. Ni, D.; Hsieh, H.K.; Jiang, T. Modeling phase diagrams as stochastic processes with application in vehicular traffic flow. *Appl. Math. Model.* **2018**, *53*, 106–117. [CrossRef]
27. Paipuri, M.; Leclercq, L. Bi-modal macroscopic traffic dynamics in a single region. *Transp. Res. Part B* **2020**, *133*, 257–290. [CrossRef]
28. Puppo, G.; Semplice, M.; Tosin, A.; Visconti, G. Fundamental diagrams in traffic flow: The case of heterogeneous kinetic models. *Commun. Math. Sci.* **2016**, *14*, 643–669. [CrossRef]
29. Visconti, G.; Herty, M.; Puppo, G.; Tosin, A. Multivalued fundamental diagrams of traffic flow in the kinetic Fokker–Planck limit. *Multiscale Model. Simul.* **2017**, *15*, 1267–1293. [CrossRef]
30. Wang, H.; Ni, D.; Chen, Q.Y.; Li, J. Stochastic modeling of the equilibrium speed-density relationship. *J. Adv. Transp.* **2013**, *47*, 126–150. [CrossRef]
31. Fan, S.; Sun, Y.; Piccoli, B.; Seibold, B.; Work, D.B. A collapsed generalized Aw-Rascle-Zhang model and its model accuracy. *arXiv* **2017**, arXiv:1702.03624.
32. Bretti, G.; Cristiani, E.; Lattanzio, C.; Maurizi, A.; Piccoli, B. Two algorithms for a fully coupled and consistently macroscopic PDE-ODE system modeling a moving bottleneck on a road. *Math. Eng.* **2018**, *1*, 55–83. [CrossRef]

33. Zhao, Y.; Zhang, H.M. A unified follow-the-leader model for vehicle, bicycle and pedestrian traffic. *Transp. Res. Part B* **2017**, *105*, 315–327. [CrossRef]
34. Colombo, R.M. Hyperbolic phase transitions in traffic flow. *SIAM J. Appl. Math.* **2002**, *63*, 708–721. [CrossRef]
35. Colombo, R.M.; Goatin, P.; Piccoli, B. Road networks with phase transitions. *J. Hyperbolic Differ. Eq.* **2010**, *7*, 85–106. [CrossRef]
36. Delle Monache, M.L.; Chi, K.; Chen, Y.; Goatin, P.; Han, K.; Qiu, J.; Piccoli, B. Three-phase fundamental diagram from three-dimensional traffic data. *Axioms* **2021**, *10*, 17. [CrossRef]
37. Cristiani, E.; Sahu, S. On the micro-to-macro limit for first-order traffic flow models on networks. *Netw. Heterog. Media* **2016**, *11*, 395–413. [CrossRef]
38. Calvert, S.C.; van Wageningen-Kessels, F.L.M.; Hoogendoorn, S.P. Capacity drop through reaction times in heterogeneous traffic. *J. Traffic Transp. Eng.* **2018**, *5*, 96–104. [CrossRef]
39. Colombo, R.M.; Marcellini, F. A traffic model aware of real time data. *Math. Models Methods Appl. Sci.* **2016**, *26*, 445–467. [CrossRef]

Article

An Information-Theoretic Framework for Optimal Design: Analysis of Protocols for Estimating Soft Tissue Parameters in Biaxial Experiments

Ankush Aggarwal [1,†], Damiano Lombardi [2,3,†] and Sanjay Pant [4,*,†]

1. Glasgow Computational Engineering Centre, James Watt School of Engineering, University of Glasgow, Glasgow G12 8LT, UK; Ankush.Aggarwal@glasgow.ac.uk
2. COMMEDIA, Inria Paris, 2 rue Simone Iff, 75012 Paris, France; Damiano.Lombardi@inria.fr
3. Laboratoire Jacques-Louis Lions, Sorbonne Université, 75006 Paris, France
4. Zienkiewicz Centre for Computational Engineering, College of Engineering, Swansea University, Swansea SA1 8EN, UK
* Correspondence: Sanjay.Pant@swansea.ac.uk
† These authors contributed equally to this work.

Abstract: A new framework for optimal design based on the information-theoretic measures of mutual information, conditional mutual information and their combination is proposed. The framework is tested on the analysis of protocols—a combination of angles along which strain measurements can be acquired—in a biaxial experiment of soft tissues for the estimation of hyperelastic constitutive model parameters. The proposed framework considers the information gain about the parameters from the experiment as the key criterion to be maximised, which can be directly used for optimal design. Information gain is computed through k-nearest neighbour algorithms applied to the joint samples of the parameters and measurements produced by the forward and observation models. For biaxial experiments, the results show that low angles have a relatively low information content compared to high angles. The results also show that a smaller number of angles with suitably chosen combinations can result in higher information gains when compared to a larger number of angles which are poorly combined. Finally, it is shown that the proposed framework is consistent with classical approaches, particularly D-optimal design.

Keywords: optimal design; soft tissue mechanics; mutual information; biaxial experiment; inverse problems; information theory

MSC: 62K05; 94A15; 92C10

1. Introduction

Soft tissues exhibit complex biomechanical behaviour, including nonlinearity, anisotropy and heterogeneity [1]. Moreover, the tissues also demonstrate inelastic properties, such as rate-dependence, hysteresis and permanent set. The important link between biomechanics and their physiological function has motivated a large number of ex-vivo studies aimed at characterising their biomechanical properties. Given the complex interplay between the different aspects of their biomechanical properties, the experimental design of ex-vivo soft tissues is extremely challenging and has been a subject of investigation, and a variety of experiments have been proposed [2–6].

Since a variety of soft tissues are thin—e.g., blood vessels, heart valves and skin—biaxial testing is a widely used experimental technique that allows the independent stretching of the tissue in two orthogonal directions and for the corresponding forces to be measured [7,8]. Applying different stretches in two directions allows the characterization of the in-plane anisotropic behavior of a given tissue, while a range of stretches provides us with its nonlinear elastic response. However, even with this relatively simple set of options, the choices of which stretches to apply are unclear. Moreover, it is not obvious upon what these choices will depend.

A variety of hyperelastic models have been developed to describe the anisotropic and nonlinear elastic properties of specific soft tissues [4,9–11]. Biaxial experimental data are commonly fit to these models in order to determine the model parameters. As the unknown parameters depend on specific models, the choice of experimental setup—the problem of optimal design—might depend on the choice of model. However, in practice, a predetermined set of experimental protocols is used.

In the present work, an optimal design problem is defined to find the most suitable protocol in view of estimating the parameters of the material model. A comprehensive overview of the optimal design problem can be found in [12,13], and several criteria for optimal design have been proposed in the literature, often based on the minimisation of the variance of the parameters and sensitivities [14,15]. In the present work, we investigate a criterion based on information theoretic quantities, in the spirit of what has been proposed in [16,17] (from a Bayesian point of view) and [18]. Several works have recently proposed information-based criteria to better define experimental protocols. In [19], the authors proposed the maximisation of the mutual information between the parameters and the observations under the assumption that the model error is a Gaussian process. In [20], the authors proposed a framework based on mutual information maximisation to deal with the design of chemistry experiments. The same criterion is proposed in [21]; the authors maximise the mutual information by using a stochastic gradient ascent method. An application to system biology is investigated in [22]. In [23], the maximisation of the information is exploited in order to choose high-fidelity model resolutions in a multi-fidelity modelling framework.

While mutual information has been used for optimal design in previous studies, the novelty of this work is in the proposal of a combination of information-theoretic quantities of both mutual and conditional mutual information. A further novelty is the application of this framework in the optimal design of soft tissue experiments. Estimating information-theoretic quantities is in general a challenging problem, and this is especially the case in high-dimensional settings. In the present work, a model reduction method is coupled with non-parametric sample-based mutual information estimation in order to provide a pertinent estimation of the information-theoretic quantities involved in the optimal design problem and then apply this to the biaxial testing of soft tissues.

The structure of the work is as follows: in Section 2, the model and information-theoretical aspects of the problem are introduced. In particular, in Section 2.1, we detail the mathematical model of the biaxial experiments for soft tissues: after having introduced the notation and the non-linear elasticity model, in Section 2.1.1, we apply it to the biaxial testing experimental setup. In Section 2.1.2, we introduce the experimental protocol definition; the second part of the section is devoted to the description of the information-theoretic framework used to solve the optimal design problem. In Section 2.2.1, we introduce the problem; in Sections 2.2.2 and 2.2.3, the information-theoretic quantities and their numerical estimation are detailed. We then present the reduce order modeling method used and how to validate the results obtained by the proposed approach. The section ends with an overview of the method. The results and the discussion are presented in Section 3, followed by the conclusion and perspectives on future work.

2. Methods

The methodological aspects are divided into two broad categories: the mathematical model of the biaxial experiments and the information-theoretic optimal design framework.

2.1. Mathematical Model of the Biaxial Experiments

We begin by defining the notation: a material point at its reference position $X \in \mathbb{R}^3$ moves to $x \in \mathbb{R}^3$ after deformation. The elastic behaviour of soft tissues is described using the hyperelastic strain energy density Ψ, which depends on the deformation gradient tensor $\mathbf{F} = \nabla_X x$. The ratio of the volume after deformation to that before deformation is given by $J = \det(\mathbf{F})$. Soft tissues are commonly regarded as incompressible due to their high water content; i.e. J is constrained to be unity.

We consider the hyperelastic model proposed by Gasser et al. [24], which defines the strain energy density as

$$\Psi = \frac{k_1}{2k_2}\left[e^{k_2(\kappa I_1 + (1-3\kappa)I_4 - 1)} - 1\right] + \mu(I_1 - 3), \tag{1}$$

where $I_1 = \operatorname{tr}(\mathbf{F}^\top \mathbf{F})$ is the first invariant of the right Cauchy–Green strain tensor $\mathbf{C} = \mathbf{F}^\top \mathbf{F}$ and $I_4 = \mathbf{M} \cdot \mathbf{C}\mathbf{M}$ is the fourth invariant representing the stretch along fiber direction $\mathbf{M}$. The resulting Cauchy stress is given by

$$\sigma = 2\mathbf{F} \cdot \frac{\partial \Psi}{\partial \mathbf{C}} \cdot \mathbf{F}^\top - p\mathbf{I}, \tag{2}$$

where p acts as the Lagrange multiplier to enforce incompressibility and $\mathbf{I}$ is the identity matrix.

For this model, the set of unknown parameters can be written as $\{k_1, k_2, \kappa, \mu\}$, assuming that the fiber direction $\mathbf{M}$ is known a priori (based on another experiment; e.g., light scattering [6]). κ represents the dispersion of collagen fibers, which is usually measured from optical experiments. Its value lies between 0 (perfectly anisotropic) and $1/3$ (perfectly isotropic). The value of μ corresponds to the shear modulus of the neo-Hookean term in (1), which represents the amorphous and non-fibrous extracellular matrix. Its role in the mechanics of soft tissues is limited to small strains and is largely constant across different tissues. In this paper, in order to simplify the problem, we assume that $\kappa = 0.1$ and $\mu = 1$ kPa are known and fixed. Thus, the aim of an ex-vivo biomechanical experiment is to determine parameters $k_1 \in [5, 100]$ kPa and $k_2 \in [5, 80]$ robustly and with high confidence [25,26]. A commonly used experiment called biaxial testing is described bellow.

2.1.1. Biaxial Experiments for Soft-Tissues

Many of the soft tissue types are planar with a small thickness. In a biaxial experiment, a square-shaped tissue sample is mounted via clamps or rakes and stretched along two orthogonal directions aligned with the sample edges (Figure 1a). If these directions are used as the two coordinate axes and incompressibility is assumed, the stretching results in a diagonal deformation gradient tensor:

$$\mathbf{F} = \operatorname{diag}\left[\lambda_1, \lambda_2, \frac{1}{\lambda_1 \lambda_2}\right], \tag{3}$$

where λ_1 is the stretch along the first in-plane direction and λ_2 is the stretch along the second in-plane direction. The fiber direction $\mathbf{M}$ is generally aligned with the first coordinate axis, which results in only normal stress components. As no force is applied along the thickness of the tissue, $\sigma_{33} = 0$ is used to determine the Lagrange multiplier p. Thus, we obtain

$$\sigma_{11} = 2\frac{\partial \Psi}{\partial I_1}\left[\lambda_1^2 - \frac{1}{\lambda_1^2 \lambda_2^2}\right] + 2\frac{\partial \Psi}{\partial I_4}\lambda_1^2 \tag{4}$$

$$\sigma_{22} = 2\frac{\partial \Psi}{\partial I_1}\left[\lambda_2^2 - \frac{1}{\lambda_1^2 \lambda_2^2}\right]. \tag{5}$$

The applied stresses σ_{11}, σ_{22} are controlled using load cells. The resulting strains, defined as $e_1 := \lambda_1 - 1$ and $e_2 := \lambda_2 - 1$, are measured from the marker positions (although e_1 and e_2 are not the usual strain measures, we use these as our observations). It is important to note that a homogeneous stress and strain state is assumed in the middle of the sample (Figure 1a). Therefore, an implicit assumption is that the material properties and sample thickness are homogeneous. Moreover, these measurement techniques carry an error due to the limitations in measurement tools and/or the deviation from homogeneity, incompressibliity and material direction.

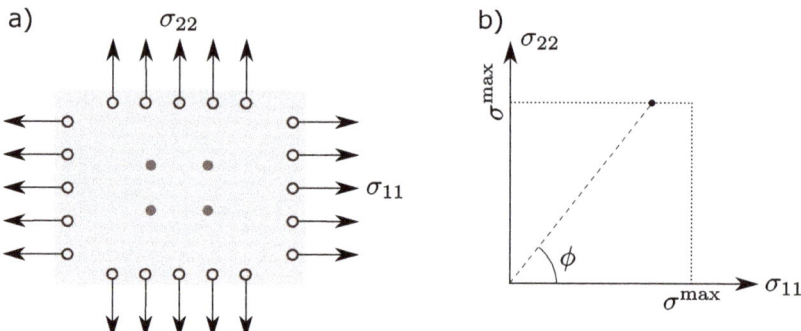

Figure 1. (**a**) A schematic of a biaxial experimental setup in which a thin planar tissue sample (in light gray) is mounted via rakes and two orthogonal forces are applied to induce stresses σ_{11} and σ_{22}, and the resulting strains are measured by tracking the locations of the markers (in dark gray). (**b**) The $\sigma_{11} - \sigma_{22}$ space, where the applied stresses lie on the dotted line with a finite number of protocol angles ϕ used.

2.1.2. Protocol Definition

In practice, there are two approaches to the biaxial experiment: (1) displacement-controlled, where known stretches are imposed and forces are measured; and (2) force-controlled, where known forces are applied and stretches are measured. Generally, the force-controlled approach is used as it is easier to implement. Therefore, in the force-controlled approach, different values of stresses σ_{11} and σ_{22} can be applied.

A single-angle biaxial protocol is defined as a straight line in the σ_{11}-σ_{22} space (Figure 1b). That is, the ratio between the two stresses is kept constant while the applied forces are increased until a maximum value $\sigma^{\max} = 200$ kPa. Thus, for a chosen angle ϕ, we apply

$$\sigma_{11} = \begin{cases} \sigma & \text{if } \phi \leq \dfrac{\pi}{4} \\ \tan(\phi)\sigma & \text{else} \end{cases} \tag{6}$$

$$\sigma_{22} = \begin{cases} \cot(\phi)\sigma & \text{if } \phi \leq \dfrac{\pi}{4} \\ \sigma & \text{else} \end{cases}, \tag{7}$$

where $\sigma \in [0, \sigma^{\max}]$. For σ, 100 linearly spaced observation points between zero and the maximum stress ($\sigma^{\max} = 200$ kPa) are used. The resulting strains are calculated by iteratively solving Equation (5) for $\lambda_{1,2}$ and thereby obtaining $e_{1,2}$. In practice, a combination of angles can be successively tested. We refer to this combination as the experimental protocol that needs to be optimally designed.

For each angle, it is easy to acquire large numbers of points as the sample is continuously stretched. However, to vary between angles, it is essential to restart the experiment at zero applied force, which further requires the "pre-conditioning" of the sample by cyclically applying small stretches. This makes it practically difficult to apply an arbitrarily large number of angles. Therefore, in practice, usually only five angles are tested.

2.2. Information-Theoretic Framework for Optimal Design

The problem of optimal design typically refers to the choice of a design of experiments such that the design is optimal with respect to a pre-determined statistical criterion. We propose that the information-theoretic measures naturally define such statistical criteria.

The central idea is that information gain [27,28] from an experiment or protocol—as quantified by the information-theoretic quantities of mutual information and conditional mutual information—can be directly used as a reasonable statistical criterion for optimal design. These quantities are described next after presenting the framework for optimal design.

2.2.1. Optimal Design Problem

Consider the following general model:

$$\mathbf{y} = \mathcal{M}(\boldsymbol{\theta}), \tag{8}$$

where $\mathcal{M}$ denotes a forward model that takes $\boldsymbol{\theta} \in \mathbb{R}^m$ and outputs $\mathbf{y} \in \mathbb{R}^n$. Note that $\boldsymbol{\theta}$ may contain initial and boundary conditions of the model and that $\mathbf{y}$ may subsume the output at many time-points in the case of a dynamic system. Subsequently, consider that the measurement model is as follows:

$$\mathbf{z} = \mathcal{H}_p(\mathbf{y}, \boldsymbol{\theta}) + \boldsymbol{\varepsilon}, \tag{9}$$

where $\mathcal{H}_p$ represents the observation operator, $\mathbf{z} \in \mathbb{R}^d$ represents the measurement vector, and $\boldsymbol{\varepsilon}$ represents the vector of measurement error/noise. Note that the the observation operator $\mathcal{H}_p$ depends on the design of experiments, which specifies which quantities are measured. Given a set of possible $\mathcal{H}_p = \{\mathcal{H}_1, \mathcal{H}_2, \cdots, \mathcal{H}_h\}$, and a statistical criterion $\mathcal{S}(\mathcal{H}_p)$ to be maximised, the optimal design is given by

$$\hat{\mathcal{H}}_p = \arg\max_{\mathcal{H}_p} \mathcal{S}(\mathcal{H}_p). \tag{10}$$

In the case of the biaxial experiments, the model $\mathcal{M}$ represents the model for the force controlled experiment (Sections 2.1.1 and 2.1.2) and $\mathcal{H}_p$ essentially denotes the experimental protocol (see Section 2.1.2) representing the combination of angles—with each representing a straight line in the σ_{11}–σ_{22} plane—along which the strain measurements of e_1 and e_2 are acquired. With the possible variation of each angle between 0 and $\pi/2$, the set Φ of possible angles ϕ is constructed through a uniform discretisation of the space between 0 and $\pi/2$ into α levels; thus,

$$\Phi = \{\phi_0, \phi_1, \cdots, \phi_\alpha\}. \tag{11}$$

The possible set of protocols is then given by any combination of elements in Φ with the restriction that the number of elements in a protocol must be limited to $\mathcal{C}$. Thus, if $\overline{\Phi} \subset \Phi$ is a subset of angles representing a protocol, our set of protocols is given by

$$\mathcal{H}_p = \{\overline{\Phi} \subset \Phi \,|\, 1 \leq |\overline{\Phi}| \leq \mathcal{C}\}, \tag{12}$$

where $|\cdot|$ represents the number of elements in the set. In other words, we choose at least 1 and up to $\mathcal{C}$ elements from Φ, with the total elements in $\mathcal{H}_p$ being

$$|\mathcal{H}_p| = \binom{\alpha}{1} + \binom{\alpha}{2} \cdots + \binom{\alpha}{\mathcal{C}}. \tag{13}$$

2.2.2. Information-Theoretic Quantities for Optimal Design

In the framework of Section 2.2.1, we propose that information-theoretic quantities of mutual information and conditional mutual information are a natural choice for the statistical criterion $\mathcal{S}$. Denoting the random variables associated with $\boldsymbol{\theta}$ and $\mathbf{z}$ as $\boldsymbol{\Theta}$ and $\mathbf{Z}$, respectively, the mutual information (MI) between the parameters $\boldsymbol{\Theta}$ and the measurements $\mathbf{Z}$ is defined as [27]

$$\mathcal{I}(\boldsymbol{\Theta}; \mathbf{Z}) = \int_{\mathcal{X}_\Theta \times \mathcal{X}_Z} p_{\boldsymbol{\Theta}, \mathbf{Z}}(\boldsymbol{\theta}, \mathbf{z}) \frac{p_{\boldsymbol{\Theta}, \mathbf{Z}}(\boldsymbol{\theta}, \mathbf{z})}{p_{\boldsymbol{\Theta}}(\boldsymbol{\theta}) p_{\mathbf{Z}}(\mathbf{z})} d\boldsymbol{\theta} d\mathbf{z}, \tag{14}$$

where $p_X(x)$ represents the probability density of a random variable X with a realisation $X = x$ and support $\mathcal{X}_X$. The mutual information $\mathcal{I}(\boldsymbol{\Theta}; \mathbf{Z})$ quantifies the amount of information that can be gained on average by one random variable—e.g., $\mathbf{Z}$—knowing about the other—e.g., $\boldsymbol{\Theta}$. Indeed, with this interpretation, MI is a good candidate for the statistical criterion $\mathcal{S}$ for optimal design. For an individual parameter, Θ_i, or indeed for any combination of parameters $\{\Theta_i, \Theta_j\}$, the corresponding information gains can be similarly computed through $\mathcal{I}(\Theta_i; \mathbf{Z})$ and $\mathcal{I}(\{\Theta_i, \Theta_j\}; \mathbf{Z})$, respectively. Thus, while $\mathcal{I}(\Theta_i; \mathbf{Z})$ quantifies the information gain individually for the parameter Θ_i, the quantity $\mathcal{I}(\{\Theta_i, \Theta_j\}; \mathbf{Z})$ quantifies information gain for the pair $\{\Theta_i, \Theta_j\}$ jointly. A measure of correlation between the parameters Θ_i and Θ_j is, however, missing and is provided by conditional mutual information (CMI), defined as

$$\underbrace{\mathcal{I}(\Theta_i; \Theta_j | \mathbf{Z})}_{\text{I}} = \underbrace{\mathcal{I}(\Theta_i; \{\Theta_j, \mathbf{Z}\})}_{\text{II}} - \underbrace{\mathcal{I}(\Theta_i; \mathbf{Z})}_{\text{III}} . \quad (15)$$

The CMI $\mathcal{I}(\Theta_i; \Theta_j | \mathbf{Z})$ represents the additional information gained about the parameter Θ_i when both Θ_j and Z are known (term II) relative to when only the measurements Θ_i alone are known (term III). Note that CMI is symmetrical—i.e., $\mathcal{I}(\Theta_i; \Theta_j | \mathbf{Z}) = \mathcal{I}(\Theta_j; \Theta_i | \mathbf{Z})$—and can be interpreted as a measure of dependence between the parameters given the measurements $\mathbf{Z}$. It should also be noted that both MI and CMI are non-negative.

With the above background, many statistical measures can be constructed. For example:

1. The mutual information for any single parameter may be maximised, giving $\mathcal{S} = \mathcal{I}(\Theta_i; \mathbf{Z})$. This approach only concerns the posterior of the parameter Θ_i and ignores all other parameters;
2. The joint mutual information may be maximised, giving $\mathcal{S} = \mathcal{I}(\boldsymbol{\Theta}; \mathbf{Z})$. In the sense of classical optimal design, this can be interpreted as D-optimal design. This is because D-optimal designs minimise the determinant of the inverse Fisher Information Matrix, and $\mathcal{S} = \mathcal{I}(\boldsymbol{\Theta}; \mathbf{Z})$ measures the information gain in the joint $\boldsymbol{\Theta}$ space;
3. The sum of individual parameter mutual information may be be maximised, giving $\mathcal{S} = \sum_{i=1}^{m} \mathcal{I}(\Theta_i; \mathbf{Z})$. In the sense of classical optimal design, this can be interpreted as A-optimal design. This is because A-optimal design minimises the trace of the inverse Fisher Information Matrix, and $\mathcal{S} = \sum_{i=1}^{m} \mathcal{I}(\Theta_i; \mathbf{Z})$ measures the sum of the information gains for all the parameters;
4. Alternatively, one may seek to maximise individual parameter information gain while minimising pairwise CMI, thus seeking both small posterior variances and minimising pairwise correlations between the parameters. In this case, the statistical criterion is

$$\mathcal{S} = \sum_{i=1}^{m} \mathcal{I}(\Theta_i; \mathbf{Z}) - \tau \sum_{i=1}^{m} \sum_{j=i}^{m} \mathcal{I}(\Theta_i; \Theta_j | \mathbf{Z}), \quad (16)$$

where $\tau > 0$ is a regularisation parameter. Note that high CMI implies that a large amount of information can be gained only about a combination of the two parameters (for instance, their sum or product), but not for each parameter individually. Thus, we seek to minimise the CMI.

Note that the above list is not exhaustive, and based on the interpretations of MI and CMI, other criteria may be constructed based on the desired sense of optimality.

2.2.3. Estimating Mutual Information

In general, the forward model in Equation (8) is non-linear, and thus even if the observation operator is linear (implying linear combinations of the state are measured), the analytical computation of mutual information is intractable. Thus, the information-theoretic quantities of MI and CMI must be estimated. A common method is to generate samples of $\boldsymbol{\Theta}$ through the specification of an appropriate prior probability density $p_{\boldsymbol{\Theta}}(\boldsymbol{\theta})$. Denoting these N_s samples as $\theta^{(i)}, i = \{1, 2, \cdots N_s\}$, each $\theta^{(i)}$ can be propagated through the forward and observation models of Equations (8) and (9) to produce corresponding samples of **Z**, denoted as $\mathbf{z}^{(i)}$. The samples of $\boldsymbol{\theta}^{(i)}$ and $\mathbf{z}^{(i)}$ can subsequently be used on non-parametric estimators of MI and CMI. Such non-parametric estimators can broadly be classified into two categories: kernel density estimators (KDE) [29] and k-nearest neighbour (kNN) estimators [30,31]. For an overview of such methods, we refer to [32]. While the estimator proposed by Kraskov et. al. [30] is widely used and performs very well across a range of scenarios, one of its drawbacks is that it suffers from higher errors when extreme correlations are present between the variables and/or when the the data are effectively in a lower-dimensional manifold. Since we are working with models that specify explicit relationships between the variables through the forward and observation model, this is likely to be true for the data set of $(\boldsymbol{\theta}^{(i)}, \mathbf{z}^{(i)})$. Thus, in this study, we employ the local non-uniformity correction (LNC) proposed in [33], which includes a correction term to the original estimator by Kraskov et al. [30]. This term accounts for strong dependencies between the variables through local principle component analysis [33]. The method of [33] is used for the estimation of all MIs, and CMIs are estimated from the difference of two MIs; see Equation (15).

2.2.4. Dimensionality Reduction for the Biaxial Experiment

One of the main difficulties in estimating information-theoretic quantities is related to the data dimension. Non-parametric estimation is particularly challenging whenever the data are close to manifolds embedded in high-dimensional spaces. This is indeed the case when a physical model relates parameters and observable quantities. One of the possible ways to overcome this difficulty, or at least to mitigate it, is (dimension or) model reduction, which aims at discovering the underlying low-dimensional structure of a set of data (a comprehensive review of the topic can be found in [34–37]). A large spectrum of methods has been proposed in the literature. In the present contribution, we adopt a local reduced-basis method (similar in spirit to the methods proposed in [38,39]). Let the strains computed by the model be $e_{1,2}(\sigma; \phi; k_1, k_2)$, where k_1 and k_2 are the model parameters $(k_1, k_2) \in \Omega_k \subset \mathbb{R}^2$, and $\sigma \in \Omega_\sigma \subset \mathbb{R}$ is the variable defined in Section 2.1.2. Let $n \in \mathbb{N}^*$; thus, we introduce the following approximation:

$$e_{1,2} \approx \sum_{i=1}^{n} \eta_i r_i(\sigma, \phi) s_i(k_1, k_2, \phi), \tag{17}$$

which is well defined by virtue of the Eckart–Young theorem. First, let us observe that a given protocol consists of a set of known angles $\overline{\Phi}$. An efficient way to construct the local reduced basis is therefore to introduce a Proper Orthogonal Decomposition (POD) for each of the angles $\phi_j \in \overline{\Phi}$. This corresponds to the search for an approximation of the form

$$e_{1,2}^{(j)}(\sigma; \phi_j; k_1, k_2) \approx \sum_{i=1}^{n} \eta_i^{(j)} r_i^{(j)}(\sigma) s_i^{(j)}(k_1, k_2), \tag{18}$$

where $\langle r_i^{(j)}, r_k^{(j)} \rangle_{\Omega_\sigma} = \delta_{ik}$ and $\langle s_i^{(j)}, s_k^{(j)} \rangle_{\Omega_k} = \delta_{ik}$ ($\langle \cdot, \cdot \rangle_{\Omega_\sigma, with \Omega_k}$ being the standard L^2 scalar product). The error in the approximation is related to the number n of modes retained:

$$\|e_{1,2}^{(j)} - \sum_{i=1}^{n} \eta_i^{(j)} r_i^{(j)}(\sigma) s_i^{(j)}(k_1, k_2)\|^2_{L^2(\Omega_\sigma \times \Omega_k)} = \sum_{i=n+1}^{\infty} \eta_i^{(j)2} \tag{19}$$

In the present work, a number $n = 4$ of modes proved to be sufficient in order to obtain errors smaller than 10^{-3} in L^2 norm in the solution reconstruction. This means that the set of elements $e_{1,2}(\sigma; \phi_j; k_1, k_2)$ was close to the linear subspace spanned by the first $n = 4$ modes $r_i^{(j)}$. Henceforth, instead of considering the discretised $e_{1,2}$ we consider their coordinates in the subspace given by

$$z_{1,2}^{(j)}|_i = \langle e_{1,2}, r_i^{(j)} \rangle_{\Omega_\sigma} = \eta_i^{(j)} s_i^{(j)}(k_1, k_2). \tag{20}$$

2.2.5. Validation of Results against Existing Methods

Several methods and criteria to define and reach an optimal design of experiments have been proposed [12]. Among them, D–optimality criterion attempts at maximising the determinant of the information matrix. In the present case, this is equivalent to minimize the determinant of the inverse of the average Hessian of the loss function we would introduce in a classical parameter estimation method. In a noisy setting, and, in particular, when the noise is Gaussian, this cost function is equivalent to minus the logarithm of the likelihood function. Let the misfit function be $f(\theta)$ and $\mathbb{E}_\Theta$ denote the expectation operator. The average of the Hessian reads:

$$H = \mathbb{E}_\Theta[\partial^2_\theta f|_{\theta_*}], \tag{21}$$

where θ_* is the value of the parameter minimising the loss function.

2.2.6. Overview of Approach for the Biaxial Experiments

In the context of the biaxial experiments, the parameters are k_1 and k_2, represented as random variables K_1 and K_2, respectively. The variability in these parameters is considered to be uniform (thus imposing a uniform prior distribution) in the following intervals: $k_1 \in [5, 100]$ kPa and $k_2 \in [5, 80]$. For a single value of angle ϕ, the measurements are the strain values e_1 and e_2 and are measured at 100 points along the line defined by the angle ϕ. Here, we consider $\alpha = 16$ discrete values of possible measurement angles ϕ uniformly distributed between, and including, 0° and 90°. For each angle ϕ, separate reduced bases of four modes for e_1 and e_2 are constructed through POD over 400 values of (K_1, K_2) sampled uniformly in the aforementioned parametric space. Thus, for any angle ϕ, the dimensionality reduction approach projects e_1 and e_2 measured at 100 points along the line defined by ϕ to a basis of 4 + 4 modes. For a given protocol consisting of multiple angles, the measurement vector $\mathbf{z}$ (with a corresponding random variable $\mathbf{Z}$) is the collection of all the reduced basis representations of e_1 and e_2 along the angles in the protocol. Lastly, the maximum number of angles in a protocol is restricted to $\mathcal{C} = 5$, giving a total of 6884 unique combinations of the $\alpha = 16$ angles.

For the estimation of MI and CMI, a total of $N = 10,000$ values of (K_1, K_2) are uniformly distributed in the parametric space. For each sample $(k_1^{(i)}, k_2^{(i)})$, the numerical model of the biaxial experiment is run to produce $e_1^{(i)}$ and $e_2^{(i)}$, which are then projected on to the reduced basis, giving $\mathbf{z}^{(i)}$. The N triplets of $(k_1^{(i)}, k_2^{(i)}, \mathbf{z}^{(i)})$ are subsequently used for the estimation of MI and CMI through the LNC estimator (see Section 2.2.3). In Equation (16), we use $\tau = 1$.

3. Results and Discussion

For all the 6884 combinations of angles, three statistical criteria are evaluated: (i) $\mathcal{I}(K_1; \mathbf{Z})$, (ii) $\mathcal{I}(K_2; \mathbf{Z})$ and (iii) $\mathcal{I}(K_1; \mathbf{Z}) + \mathcal{I}(K_2; \mathbf{Z}) - \mathcal{I}(K_1; K_2|\mathbf{Z})$. While the first two criteria aim to maximise the information gain about K_1 and K_2 individually, the third criterion aims to maximise the information gain about K_1 and K_2 simultaneously while minimising the information dependence between them. Figures 2–4 show the variation in these three criteria when grouped by the number of angles in a protocol. In these figures, the values of information criterion when using two approaches to uniformly discretise the angular space within protocols are also presented. Observations from these plots are as follows:

1. Generally, all the three information criteria increase with the increasing number of angles in the protocol. Intuitively, this is expected, as a higher number of angles implies more measurement data and hence a higher potential for the improved estimation of the parameters. This observation is true for the maximum information gain, minimum information gain and the mean information gain;
2. Across all the three criteria, it is observed that the uniform discretisation is not necessarily reflective of the best protocol for estimating the parameters. In fact, in

most cases, the performance of uniform discretisation is close to the mean information gain observed across all the angle combinations;
3. From Figures 2 and 3, it is observed that the angular combinations that maximise information gain for K_1 are not identical—and vary significantly when more than two angles are simultaneously used—to those that maximise information gain for K_2. This further motivates the use of a criterion that balances information gains in both the parameters while minimising their interdependence;
4. Figure 4 shows that the best combinations that maximise a balanced criterion, such as $\mathcal{I}(K_1;\mathbf{Z}) + \mathcal{I}(K_2;\mathbf{Z}) - \mathcal{I}(K_1;K_2|\mathbf{Z})$, are a trade-off between the combinations of angles that maximise $\mathcal{I}(K_1;\mathbf{Z})$ and $\mathcal{I}(K_2;\mathbf{Z})$ individually. For example, when five angles are considered, the angles that maximise $\mathcal{I}(K_1;\mathbf{Z})$ are $\phi_a = [66, 72, 78, 84, 90]$ and those that maximise $\mathcal{I}(K_2;\mathbf{Z})$ are $\phi_b = [30, 36, 42, 48, 54]$, while the combination that maximises $\mathcal{I}(K_1;\mathbf{Z}) + \mathcal{I}(K_2;\mathbf{Z}) - \mathcal{I}(K_1;K_2|\mathbf{Z})$ is $[30, 36, 48, 78, 90]$, which has two angles from ϕ_a and three angles from ϕ_b. It should be noted that such a trade-off between maximising individual parameter gains is still significantly different to a uniform discretisation;
5. Finally, it is observed that the worst combinations are all low angles: $[0, 6, 12, 18, 24]$. This can be related to the fact that, at low angles, the applied stress is largely aligned along the stiff fibers of the tissue, thus resulting in lower strain values. Thus, the lower angles provide a small range of the observations, while the larger angles provide a larger range (Figure 5a), thereby containing more information about the parameters.

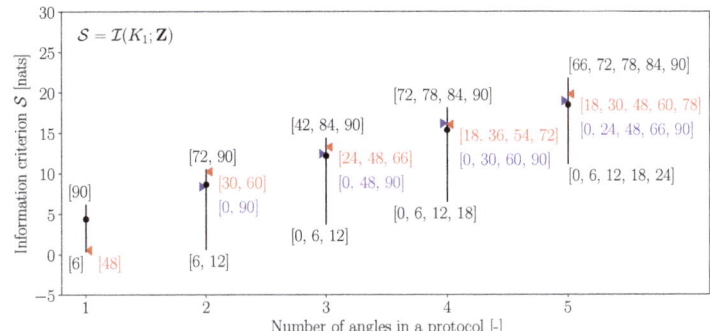

Figure 2. The variation of information criterion $\mathcal{S} = \mathcal{I}(K_1;\mathbf{Z})$ across the 6884 combinations grouped by the number of angles in a protocol. The vertical lines represent the variation around the mean value, which is shown in black circles. Black text shows the combinations that produce maximum and minimum values of $\mathcal{S}$. The red and blue pointers show $\mathcal{S}$ for angle combinations that follow a uniform discretisation of the angular space between 0 and 90 degrees. Red and blue texts show the associated angle combinations.

From this point onward, we present results only for the balanced information criterion $\mathcal{S} = \mathcal{I}(K_1;\mathbf{Z}) + \mathcal{I}(K_2;\mathbf{Z}) - \mathcal{I}(K_1;K_2|\mathbf{Z})$. Figure 6 shows the variation in $\mathcal{S}$ across all the combinations (x-axis and in log-scale to capture the spread) grouped by the number of angles in a protocol and sorted according to the increasing order of $\mathcal{S}$ within each such group. Within each group, observing the minimum and maximum values of $\mathcal{S}$ shows that a better choice of angles can lead to more than a 100% increase in the information gain compared to a poor choice. Furthermore, this shows that good combinations of a lower number of angles can lead to higher information gain compared to a higher number of angles with poor combinations. For example, the maximum $\mathcal{S}$ when only one angle is used is higher than many combinations with two to four angles. This emphasises the utility of optimal design and the proposed framework.

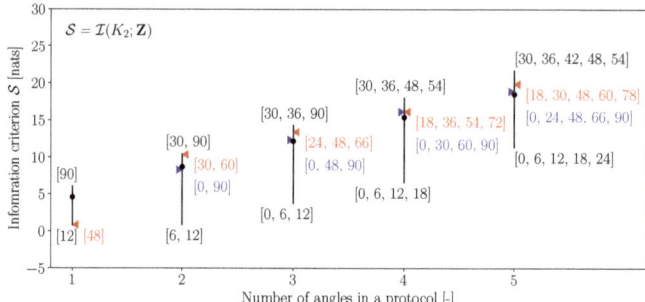

Figure 3. The variation of information criterion $\mathcal{S} = \mathcal{I}(K_2; \mathbf{Z})$ across the 6884 combinations grouped by the number of angles in a protocol. The vertical lines represent the variation around the mean value, which is shown in black circles. Black text shows the combinations that produce maximum and minimum values of $\mathcal{S}$. The red and blue pointers show $\mathcal{S}$ for angle combinations that follow a uniform discretisation of the angular space between 0 and 90 degrees. Red and blue texts show the associated angle combinations.

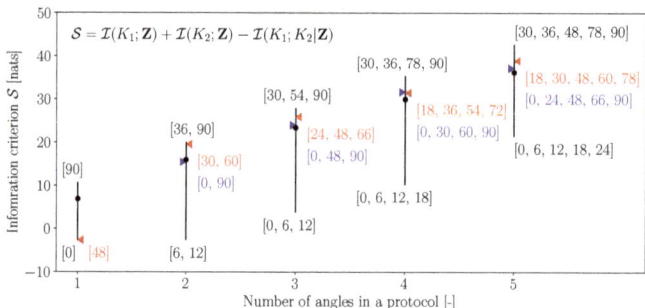

Figure 4. The variation of information criterion $\mathcal{S} = \mathcal{I}(K_1; \mathbf{Z}) + \mathcal{I}(K_2; \mathbf{Z}) - \mathcal{I}(K_1; K_2|\mathbf{Z})$ across the 6884 combinations grouped by the number of angles in a protocol. The vertical lines represent the variation around the mean value, which is shown in black circles. Black text shows the combinations that produce maximum and minimum values of $\mathcal{S}$. The red and blue pointers show $\mathcal{S}$ for angle combinations that follow a uniform discretisation of the angular space between 0 and 90 degrees. Red and blue texts show the associated angle combinations.

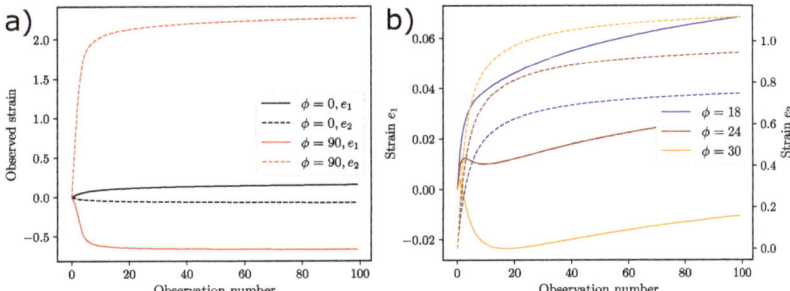

Figure 5. Representative observations from the model with $k_1 = 40$ kPa and $k_2 = 40$. (**a**) The observations using angles $\phi = 0$ and 90 degrees, with the latter covering a significantly larger range. (**b**) The change in observations the angle is changed from 18, 24 to 30 degrees shows a transition in e_1 from positive to negative values, indicating a coupling between the two directions. Note that e_1 is shown here in solid lines (left y-axis) and e_2 is shown in dashed lines (right y-axis).

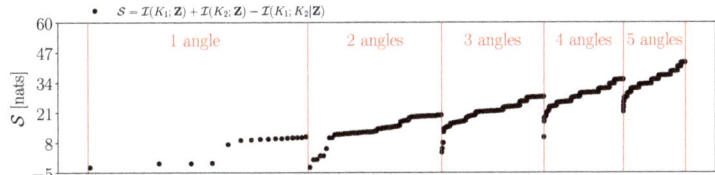

Figure 6. The variation of information criterion $\mathcal{S} = \mathcal{I}(K_2; \mathbf{Z}) + \mathcal{I}(K_2; \mathbf{Z}) - \mathcal{I}(K_1; K_2|\mathbf{Z})$ across the 6884 combinations. The vertical red lines show the groupings with respect to the number of angles in a protocol and $\mathcal{S}$ values are sorted in increasing order within each such grouping. The x-axis is represents the index associated with the protocol and is in logarithmic scale to capture the spread between one angle in a protocol (16 values) vs five angles in a protocol (4368 values).

Figure 7 shows $\mathcal{S}$ for all the 6884 combinations in increasing order of magnitude, and Figure 8 shows a zoomed plot for the first 150 combinations along with the corresponding combinations of angles. Observing the index values of 26 (red) and 28 (blue) in Figure 8 shows that even though four combinations are used in the index 26 protocol, it produces a lower $\mathcal{S}$ compared to when only a single angle is used in the index 28 protocol. Furthermore, since Figure 8 shows the first 150 out of 6884 combinations of Figure 7 (which is sorted in creasing order of $\mathcal{S}$), all combinations here are relatively low $\mathcal{S}$-producing protocols. Observing the high density of angles in the region $\phi < 24°$ is indicative that lower values of angles—in particular, those less than $24°$—are relatively less informative when compared to higher values of angles. This behaviour is also apparent in Figure 9, which shows $\mathcal{S}$ values for protocols that use only one angle, and where a sharp jump can be observed when transitioning from $18°$ to $24°$. This peculiar behaviour may be explained by the physics of the biaxial experiment. Looking at the resulting strains e_1 and e_2 of this transition (Figure 5b), we observe that the e_1 changes from positive to negative values. This behavior captures the important coupling between the two normal stresses and strains and is also related to the fiber dispersion in our constitutive model (Equation (1), [24]). It is remarkable and encouraging that the information-theoretic framework captures the physics of the problem without explicitly considering it in the framework. While for simpler low-dimensional models, the association between physics and optimal design may be relatively easy to see, inferring such behaviour is, in general, not trivial for more complex and higher-dimensional models.

Similarly to Figure 9, the results of the information-theoretic optimal design are further analysed for a higher number of angles. When two angles are considered, the $\mathcal{S}$

values in increasing order of magnitude and the corresponding angle combinations are shown in Figure 10. This figure re-iterates observations made previously: (i) the choice of combinations significantly affects the information gain, where the best combination gives approximately 20 nats more information compared to the worst combination; and (ii) generally speaking, higher angles are more informative compared to lower angles—in particular, angles below 24°. While a similar analysis for more than two angle combinations can be easily performed, the efficient visual representation of such results is cumbersome and avoided in this manuscript.

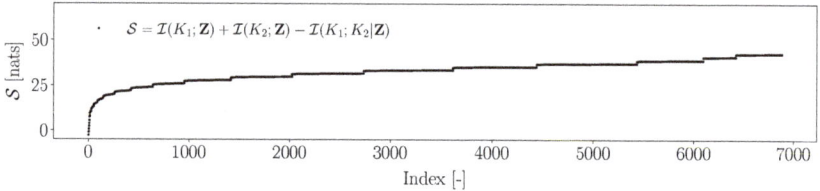

Figure 7. The variation of information criterion $\mathcal{S} = \mathcal{I}(K_1; \mathbf{Z}) + \mathcal{I}(K_2; \mathbf{Z}) - \mathcal{I}(K_1; K_2|\mathbf{Z})$ across the 6884 combinations sorted in increasing order of $\mathcal{S}$.

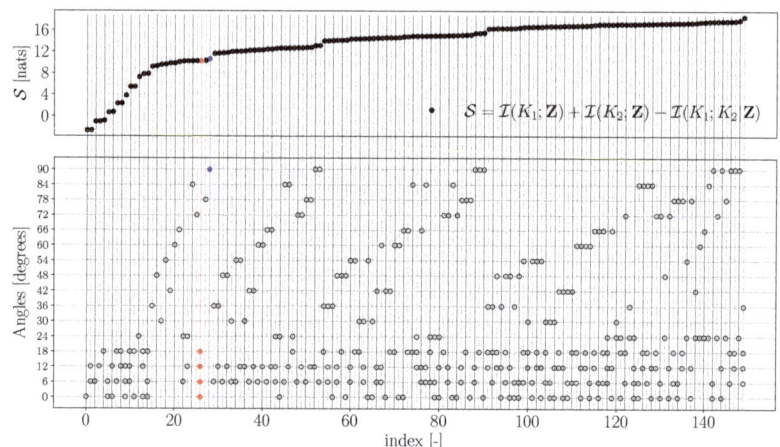

Figure 8. Zoomed view of the first 150 protocols from Figure 7. The upper panel shows $\mathcal{S}$ and the lower panel shows the angles (by circles) in the corresponding protocol. The red points showcase a protocol with four measurement angles which yet produces a lower $\mathcal{S}$, implying a poorer protocol, with respect to the blue points which show a protocol with only one measurement angle.

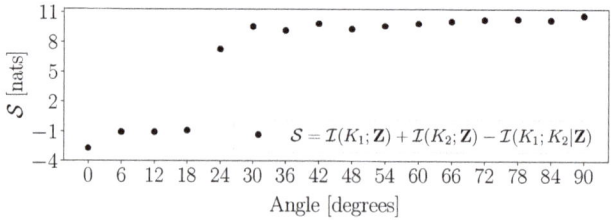

Figure 9. Information criterion against the angle when the protocols are restricted to a maximum of one angle.

To further illustrate the validity of the information-theoretic approach, a comparison with a classical method (see Section 2.2.5) is presented. For one and two angles in a protocol, Figure 11 shows a comparison between $\mathcal{S}$ and the log of the determinant of the inverse Fisher Information Matrix, $\log |H^{-1}|$. It is encouraging that a high correspondence between the two metrics is observed. In particular, increases in $\mathcal{S}$, implying higher information gains, are accompanied by corresponding decreases in $\log |H^{-1}|$, implying a smaller volume of the parameter posteriors. A Pearson correlation coefficient of $r = -0.76$ is observed between $\mathcal{S}$ and $\log |H^{-1}|$, implying a high similarity between the two metrics and validating the information-theoretic approach in part. We note that, when the number of the parameters become large, evaluating the Hessian would imply a non-negligible computational cost. On the contrary, the method used to evaluate the mutual information, as a primarily Monte Carlo-based estimation, is less severely dependent on the number of parameters. Furthermore, the computation of derivatives (either numerically or through adjoint based methods) may be cumbersome for certain types of models. Finally, we note that the effect of noise on information gain, and hence optimal design, can be easily assessed in the proposed framework by adding noise to the samples of $\mathbf{Z}$ (see Equation (9)).

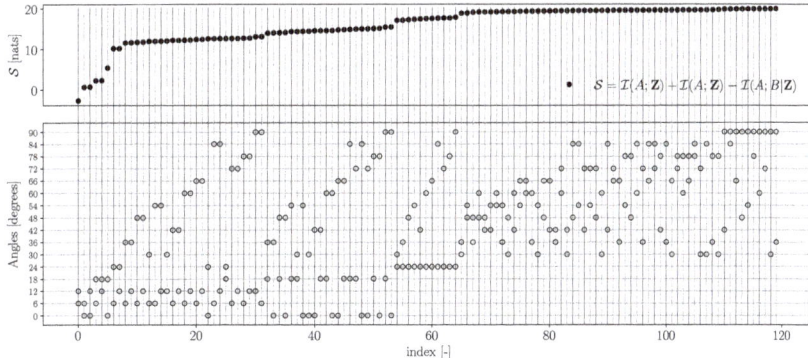

Figure 10. Information criterion against the angles when the protocols are restricted to only two angles. The upper panel shows $\mathcal{S}$ and the lower panel shows the angles (with circles) in the corresponding protocol.

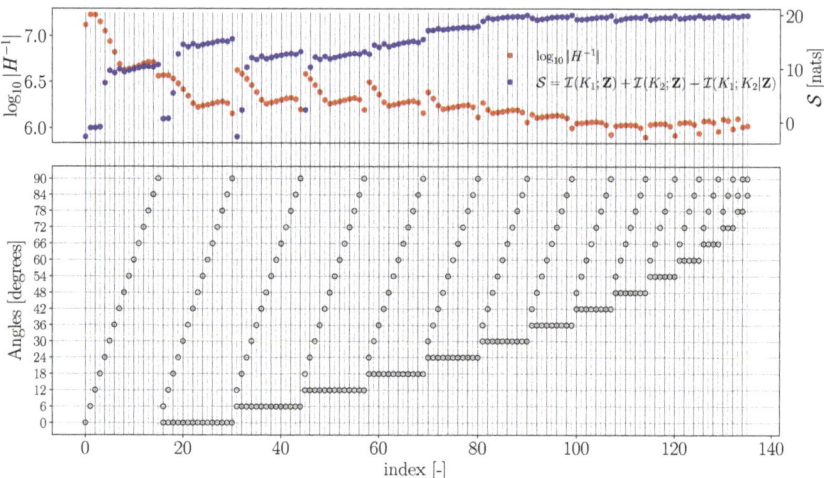

Figure 11. Information criterion $\mathcal{S}$ (in blue) and the log of the determinant of the inverse Fisher Information Matrix $\log_{10}|H^{-1}|$ (in red) against the angles when the protocols are restricted to a maximum of two angles. The upper panel shows $\mathcal{S}$ and $\log_{10}|H^{-1}|$, while the lower panel shows the angles (with circles) in the corresponding protocol.

4. Conclusions

A framework for optimal design based on information-theoretic quantities of mutual information and conditional mutual information is proposed. The framework treats information gain as the central criterion for inverse problems and proposes several information-theoretic frameworks for a desired sense of optimality. The capabilities of this framework are tested on the optimal design problem for biaxial experiments, where the effect of the angle combinations along which the strains are measured is assessed in terms of parameter estimation through information gain. Without including any physics-based reasoning, and purely through the information-theoretic measures, it is found that low angles $\leq 24°$ are not very informative regarding the parameters relative to high angles. These observations are then found to be consistent based on physics-based reasoning, thereby showing the efficacy of the proposed framework. Furthermore, it is demonstrated that measurements for a low total number of angles which are carefully chosen can be more informative compared to the case when measurements along a high number of poorly chosen angles are acquired, thus highlighting both the importance of optimal design for biaxial experiments and the utility of the proposed framework in determining good angle combinations. The application of the proposed framework to classical optimal design is performed, and it is shown that the results produced by the new framework are consistent with classical frameworks.

5. Limitations and Future Work

While the proposed framework is shown to perform well on a two-parameter problem, its performance in higher parameter problems is not assessed. This assessment represents the primary limitation of this work and an area of future assessment. In particular, the problems envisaged are largely related to the performance of the MI and CMI estimators in higher dimensions of both parameters and the measurements. While a dimensionality reduction approach was adopted in this study to minimise the adverse effects of the latter, this may not be possible in many forward and inverse problems. Thus, a large area of future work is related to the development of efficient and robust MI and CMI estimators. Note that several approaches are being proposed by researchers to solve this problem; see for example [40–46]. Lastly, a thorough comparison against classical optimal design methods

(C, E, T and V-optimal designs, etc.) needs to be performed, along with the construction and analysis of corresponding information-theoretic metrics.

Author Contributions: All authors contributed equally to writing, editing, and reviewing the manuscript. S.P. and D.L. conceptualised the information-theoretic framework. A.A. and S.P. conceptualised the application to the biaxial experiments. AA wrote the numerical code for the biaxial experiment. S.P. and D.L. wrote the code for dimensionality reduction and the estimation of information-theoretic measures. All authors contributed equally to the analysis of the results. All authors have read and agreed to the published version of the manuscript.

Funding: This research was funded by the Engineering and Physical Sciences Research Council of the UK (Grant reference EP/R010811/1 to SP and grant reference EP/P018912/1 and EP/P018912/2 to AA).

Data Availability Statement: Not applicable.

Conflicts of Interest: The authors declare no conflict of interest.

References

1. Holzapfel, G.A. *Nonlinear Solid Mechanics*; Wiley: Chichester, UK, 2000; Volume 24.
2. Zhang, W.; Feng, Y.; Lee, C.H.; Billiar, K.L.; Sacks, M.S. A generalized method for the analysis of planar biaxial mechanical data using tethered testing configurations. *J. Biomech. Eng.* **2015**, *137*, 064501. [CrossRef] [PubMed]
3. Labrosse, M.R.; Jafar, R.; Ngu, J.; Boodhwani, M. Planar biaxial testing of heart valve cusp replacement biomaterials: Experiments, theory and material constants. *Acta Biomater.* **2016**, *45*, 303–320. [CrossRef]
4. Humphrey, J.; Yin, F. On constitutive relations and finite deformations of passive cardiac tissue: I. A pseudostrain-energy function. *J. Biomech. Eng.* **1987**, *109*, 298–304. [CrossRef]
5. Laurence, D.; Ross, C.; Jett, S.; Johns, C.; Echols, A.; Baumwart, R.; Towner, R.; Liao, J.; Bajona, P.; Wu, Y.; et al. An investigation of regional variations in the biaxial mechanical properties and stress relaxation behaviors of porcine atrioventricular heart valve leaflets. *J. Biomech.* **2019**, *83*, 16–27. [CrossRef] [PubMed]
6. Jett, S.V.; Hudson, L.T.; Baumwart, R.; Bohnstedt, B.N.; Mir, A.; Burkhart, H.M.; Holzapfel, G.A.; Wu, Y.; Lee, C.H. Integration of polarized spatial frequency domain imaging (pSFDI) with a biaxial mechanical testing system for quantification of load-dependent collagen architecture in soft collagenous tissues. *Acta Biomater.* **2020**, *102*, 149–168. [CrossRef] [PubMed]
7. Billiar, K.L.; Sacks, M.S. Biaxial mechanical properties of the native and glutaraldehyde-treated aortic valve cusp: Part II–a structural constitutive model. *J. Biomech. Eng.* **2000**, *122*, 327–335. [CrossRef] [PubMed]
8. Ross, C.; Laurence, D.; Wu, Y.; Lee, C.H. Biaxial Mechanical Characterizations of Atrioventricular Heart Valves. *J. Vis. Exp. JoVE* **2019**. [CrossRef] [PubMed]
9. Maurel, W.; Thalmann, D.; Wu, Y.; Thalmann, N.M. Constitutive Modeling. In *Biomechanical Models for Soft Tissue Simulation*; Springer: Berlin/Heidelberg, Germany, 1998; pp. 79–120.
10. Holzapfel, G.A.; Gasser, T.C.; Ogden, R.W. A new constitutive framework for arterial wall mechanics and a comparative study of material models. *J. Elast. Phys. Sci. Solids* **2000**, *61*, 1–48.
11. May-Newman, K.; Yin, F.C.P. A constitutive law for mitral valve tissue. *J. Biomech. Eng.* **1998**, *120*, 38–47. [CrossRef]
12. Pukelsheim, F. *Optimal Design of Experiments*; Society for Industrial and Applied Mathematics (SIAM): Philadelphia, PA, USA, 2006.
13. Banks, H.T.; Holm, K.; Kappel, F. Comparison of optimal design methods in inverse problems. *Inverse Probl.* **2011**, *27*, 075002. [CrossRef] [PubMed]
14. Banks, H.T.; Dediu, S.; Ernstberger, S.L.; Kappel, F. Generalized sensitivities and optimal experimental design. *J. Inv. Ill-Posed Problems.* **2010**, *18*, 25–83. [CrossRef]
15. Banks, H.T.; Rubio, D.; Saintier, N.; Troparevsky, M.I. Optimal design techniques for distributed parameter systems. In Proceedings of the 2013 Conference on Control and Its Applications, San Diego, CA, USA, 8–10 July 2013; pp. 83–90.
16. Lindley, D.V. On a measure of the information provided by an experiment. *Ann. Math. Stat.* **1956**, *27*, 986–1005. [CrossRef]
17. Sebastiani, P.; Wynn, H.P. Maximum entropy sampling and optimal Bayesian experimental design. *J. R. Stat. Soc. Ser. (Stat. Methodol.)* **2000**, *62*, 145–157. [CrossRef]
18. Capellari, G.; Chatzi, E.; Mariani, S. Parameter identifiability through information theory. In Proceedings of the 2nd ECCOMAS Thematic Conference on Uncertainty Quantification in Computational Sciences and Engineering (UNCECOMP), Rhodes Island, Greece, 15–17 June 2017; pp. 15–17.
19. Bryant, C.; Terejanu, G. An information-theoretic approach to optimally calibrate approximate models. In Proceedings of the 50th AIAA Aerospace Sciences Meeting including the New Horizons Forum and Aerospace Exposition, Nashville, TN, USA, 9–12 January 2012; p. 153.
20. Terejanu, G.; Upadhyay, R.R.; Miki, K. Bayesian experimental design for the active nitridation of graphite by atomic nitrogen. *Exp. Therm. Fluid Sci.* **2012**, *36*, 178–193. [CrossRef]

21. Huan, X.; Marzouk, Y.M. Simulation-based optimal Bayesian experimental design for nonlinear systems. *J. Comput. Phys.* **2013**, *232*, 288–317. [CrossRef]
22. Liepe, J.; Filippi, S.; Komorowski, M.; Stumpf, M.P. Maximizing the information content of experiments in systems biology. *PLoS Comput. Biol.* **2013**, *9*, e1002888. [CrossRef] [PubMed]
23. Lewis, A.; Smith, R.; Williams, B.; Figueroa, V. An information theoretic approach to use high-fidelity codes to calibrate low-fidelity codes. *J. Comput. Phys.* **2016**, *324*, 24–43. [CrossRef]
24. Gasser, T.C.; Ogden, R.W.; Holzapfel, G.A. Hyperelastic modelling of arterial layers with distributed collagen fibre orientations. *J. R. Soc. Interface* **2006**, *3*, 15–35. [CrossRef] [PubMed]
25. Aggarwal, A. An improved parameter estimation and comparison for soft tissue constitutive models containing an exponential function. *Biomech. Model. Mechanobiol.* **2017**, *16*, 1309–1327. [CrossRef] [PubMed]
26. Aggarwal, A. Effect of Residual and Transformation Choice on Computational Aspects of Biomechanical Parameter Estimation of Soft Tissues. *Bioengineering* **2019**, *6*, 100. [CrossRef] [PubMed]
27. Pant, S.; Lombardi, D. An information-theoretic approach to assess practical identifiability of parametric dynamical systems. *Math. Biosci.* **2015**, *268*, 66–79. [CrossRef]
28. Pant, S. Information sensitivity functions to assess parameter information gain and identifiability of dynamical systems. *J. R. Soc. Interface* **2018**, *15*, 20170871. [CrossRef]
29. Moon, Y.I.; Rajagopalan, B.; Lall, U. Estimation of mutual information using kernel density estimators. *Physical Rev. E* **1995**, *52*, 2318. [CrossRef] [PubMed]
30. Kraskov, A.; Stögbauer, H.; Grassberger, P. Estimating mutual information. *Phys. Rev. E* **2004**, *69*, 066138. [CrossRef] [PubMed]
31. Lombardi, D.; Pant, S. Nonparametric k-nearest-neighbor entropy estimator. *Phys. Rev. E* **2016**, *93*, 013310. [CrossRef] [PubMed]
32. Beirlant, J.; Dudewicz, E.J.; Györfi, L.; Van der Meulen, E.C. Nonparametric entropy estimation: An overview. *Int. J. Math. Stat. Sci.* **1997**, *6*, 17–39.
33. Gao, S.; Ver Steeg, G.; Galstyan, A. Efficient estimation of mutual information for strongly dependent variables. In Proceedings of the Eighteenth International Conference on Artificial Intelligence and Statistics, San Diego, CA, USA, 9–12 May 2015; pp. 277–286.
34. Benner, P.; Gugercin, S.; Willcox, K. A survey of projection-based model reduction methods for parametric dynamical systems. *SIAM Rev.* **2015**, *57*, 483–531. [CrossRef]
35. Benner, P.; Ohlberger, M.; Cohen, A.; Willcox, K. *Model Reduction and Approximation: Theory and Algorithms*; Society for Industrial and Applied Mathematics (SIAM): Philadelphia, PA, USA, 2017.
36. Quarteroni, A.; Rozza, G. *Reduced Order Methods for Modeling and Computational Reduction*; Springer: Berlin/Heidelberg, Germany, 2014; Volume 9.
37. Ma, Y.; Fu, Y. *Manifold Learning Theory and Applications*; CRC Press: Boca Raton, FL, USA, 2011.
38. Amsallem, D.; Haasdonk, B. PEBL-ROM: Projection-error based local reduced-order models. *Adv. Model. Simul. Eng. Sci.* **2016**, *3*, 1–25. [CrossRef]
39. Maday, Y.; Stamm, B. Locally adaptive greedy approximations for anisotropic parameter reduced basis spaces. *SIAM J. Sci. Comput.* **2013**, *35*, A2417–A2441. [CrossRef]
40. Belghazi, M.I.; Baratin, A.; Rajeshwar, S.; Ozair, S.; Bengio, Y.; Courville, A.; Hjelm, D. Mutual information neural estimation. In Proceedings of the Machine Learning Research, Stockholmsmässan, Stockholm Sweden, 10–15 July 2018; pp. 531–540.
41. Singh, S.; Póczos, B. Generalized exponential concentration inequality for Rényi divergence estimation. In Proceedings of the 31st International Conference on Machine Learning, Bejing, China, 22–24 June 2014; pp. 333–341.
42. Kleinegesse, S.; Drovandi, C.; Gutmann, M.U. Sequential Bayesian experimental design for implicit models via mutual information. *Bayesian Anal.* **2021**, *1*, 1–30. [CrossRef]
43. Fukumizu, K. Nonparametric Bayesian inference with kernel mean embedding. In *Modern Methodology and Applications in Spatial-Temporal Modeling*; Springer: Berlin/Heidelberg, Germany, 2015; pp. 1–24.
44. Moon, K.R.; Hero, A.O. Ensemble estimation of multivariate f-divergence. In Proceedings of the 2014 IEEE International Symposium on Information Theory, Honolulu, HI, USA, 29 June–4 July 2014; pp. 356–360.
45. Brodu, N.; Crutchfield, J.P. Discovering Causal Structure with Reproducing-Kernel Hilbert Space ϵ-Machines. *arXiv* **2020**, arXiv:2011.14821.
46. Gökmen, D.E.; Ringel, Z.; Huber, S.D.; Koch-Janusz, M. Phase diagrams with real-space mutual information neural estimation. *arXiv* **2021**, arXiv:2103.16887.

Article

A Fractional-in-Time Prey–Predator Model with Hunting Cooperation: Qualitative Analysis, Stability and Numerical Approximations

Maria Francesca Carfora * and Isabella Torcicollo

Istituto per le Applicazioni del Calcolo "Mauro Picone" CNR, 80131 Napoli, Italy; i.torcicollo@iac.cnr.it
* Correspondence: f.carfora@iac.cnr.it

Abstract: A prey–predator system with logistic growth of prey and hunting cooperation of predators is studied. The introduction of fractional time derivatives and the related persistent memory strongly characterize the model behavior, as many dynamical systems in the applied sciences are well described by such fractional-order models. Mathematical analysis and numerical simulations are performed to highlight the characteristics of the proposed model. The existence, uniqueness and boundedness of solutions is proved; the stability of the coexistence equilibrium and the occurrence of Hopf bifurcation is investigated. Some numerical approximations of the solution are finally considered; the obtained trajectories confirm the theoretical findings. It is observed that the fractional-order derivative has a stabilizing effect and can be useful to control the coexistence between species.

Keywords: Caputo fractional derivative; Allee effect; existence and stability; Hopf bifurcation; implicit schemes

MSC: 34A08; 34D20; 65R10

1. Introduction

Population dynamics are regulated by several factors: availability of resources, predation, diseases, etc. Among these factors, the interaction between prey and predators is probably the most studied one in ecology, tracing back to the works of Lotka and Volterra in the early 20th century. Since then, a number of predator–prey models have been proposed and studied (see the excellent reviews [1,2]), considering different extensions of the original one: the realistic assumption that prey populations are limited by food resources and not just by predation leads to the inclusion of terms representing carrying capacity for the prey population; the characterization of specific behaviors of the predator population results in the introduction of several functional responses for them; the complexity and dishomogeneity in the environment often requires a spatial description [3]. It is well-known that species diffusing at different rates can generate spatial patterns, observed in several biological contexts. Such Turing patterns affecting spatial predator–prey models have been deeply investigated in the literature. In addition, the inclusion of group defense has a significant impact on the dynamics of the predator–prey system. Of course, this movement is conditioned by the abundance or scarcity of the other species, so that spatio-temporal population models can include cross-diffusion terms in addition to self-diffusion ones.

Cooperation is a common behavior in many biological groups and can sensibly affect the growth rate of populations in an ecological community. Limiting our interest to the predator–prey dynamics, many mechanisms have been identified, all capable of facilitating reproduction, breeding, foraging and defense in the prey population. All of them can induce a demographic Allee effect [4]. In predators, hunting cooperation, among other interactions, can also result in Allee effects; different intensities of such a cooperative behavior impact on the survival of both species and modify the stability of the ecological system (see [5] and references therein).

Even if the interest on fractional calculus traces back at least to the 1970s (without considering the seminal suggestions given by Leibniz and Euler more than 300 years ago), in the last decades there has been an explosion of research activities on its application to several areas. Fractional order systems are not only an extension of conventional integer-order systems in mathematics but also have memory and hereditary properties that integer order systems do not have. In the last decades, there has been a huge interest in such models, due to their specific feature of describing memory effects in dynamical systems: many nonlinear models with time fractional derivatives have been considered, not only in population dynamics [6,7], but also in chemistry and biochemistry, medicine, mechanics, engineering, finance, psychology (see, among the recent surveys, [8,9]). In some situations, indeed, fractional-order models, enabling the description of the memory and hereditary properties inherent in various materials, processes and biological systems, seem more consistent with the real phenomena than integer-order models. Models with fractional-order derivatives can take different forms depending on the system considered; among them, fractional differential equations and fractional partial differential equations are greatly applied to describe continuous systems, both deterministic and stochastic [10]. In addition, heterogeneous non-ergodic diffusion processes [11] and the effect of anomalous diffusion on a population survival have been investigated [12].

The authors considered, in previous studies, some interacting population models that explored the mentioned characteristics (limited resources, nonlinear growth, fear effect, cooperative behavior), also in the presence of spatial dishomogeneity [13–17]. In the above-mentioned research, systems of both ordinary and partial differential equations have been studied. Precisely, ODE descriptions for the dynamics of an intraguild predator–prey model [13] and for the spreading of waterborne diseases [15,17] have been considered and the stability of the model solutions discussed. Furthermore, in order to highlight how the spatial diffusion can both play an important role in the population evolution and lead to the formation of spatial patterns, a reaction–diffusion system modeling hunting cooperation [14] and a predator–prey system with fear and group defense [16] have been analyzed. In such researches, the effect of spatial diffusion on the stability of the equilibria has been highlighted and the conditions on the parameters ensuring the formation of patterns (stripes, spots, mixed, etc.) have been found.

In the present paper, the authors generalize the model introduced in [5], by replacing the ordinary time derivative with a fractional one so to investigate how the fractional-in-time derivative impacts the system dynamics. It is worth underlining, for the sake of completeness, that such a model has been already extended in [14] to include spatio-temporal dynamics and the related occurrence of Turing patterns. Here, instead, the authors, to better highlight the fractional derivative impact on the populations dynamics, take into account the original ODE model and consider the corresponding fractional-in-time system.

Such a modeling provides challenges and ideas in many other fields of applied mathematics in which nonlinear mathematical models having a similar structure are considered [18–22]. After reviewing in Section 2 the main prerequisites on fractional calculus, the model is formulated in Section 3 and existence and boundedness of solutions is proved. Section 4 discusses the stability of the coexistence equilibrium, while Section 5 shows some numerical approximations by different schemes. Section 6 concludes the manuscript.

2. Preliminaries

In this section, we recall the fundamental definitions, concepts and results that we will use throughout the paper. For further details, we refer the reader to [23–25].

Definition 1. *The fractional integral of order $\theta > 0$ for a Lebesgue integrable function $f : \mathbb{R}^+ \to \mathbb{R}$ is defined by*

$$I^\theta f(t) = \frac{1}{\Gamma(\theta)} \int_{t_0}^{t} (t-\tau)^{\theta-1} f(\tau) d\tau$$

and the Caputo fractional derivative of order $\theta \in (m-1, m)$ of a sufficiently differentiable function $f(t)$ is defined by

$$D_t^\theta f(t) = \frac{1}{\Gamma(m-\theta)} \int_{t_0}^t (t-\tau)^{m-\theta-1} f^{(m)}(\tau) d\tau,$$

where $\Gamma(\theta)$ is the Euler's Gamma function.

We underline that, under natural conditions on $f(t)$, the Caputo derivative coincides with the classical derivative whenever $\theta \in \mathbb{N}$.

Lemma 1. Let $x(t)$ be a continuous function on $[t_0, \infty)$ satisfying $D^\theta x(t) \leq -Ax(t) + B$, $x(t_0) = x_0$, with $0 < \theta < 1$ and $A, B \in \mathbb{R}$, $A > 0$, $t_0 \geq 0$. Then for $x(t)$ it holds true

$$x(t) \leq \left(x_0 - \frac{B}{A}\right) E_\theta[-A(t-t_0)^\theta] + \frac{B}{A}, \tag{1}$$

where $E_\theta(\cdot)$ is the Mittag–Leffler function defined by

$$E_\theta = \sum_{k=0}^\infty \frac{z^k}{\Gamma(\theta k + 1)}. \tag{2}$$

Lemma 2. Let $D^\theta x(t) = \phi(t, x)$, with $t > t_0$, be the system with the initial condition $x(t_0)$ and $0 < \theta \leq 1$, $\phi : [t_0, \infty) \times \Omega \to \mathbb{R}^m, \Omega \subseteq \mathbb{R}^m$. If $\phi(t, x)$ satisfies the Lipschitz condition with respect to x, then there exists a unique solution of the above system on $[t_0, \infty) \times \Omega$.

Lemma 3. Let $D_t^\theta \mathbf{x} = \phi(\mathbf{x})$, with $\mathbf{x}(t_0) = \mathbf{x}_0$ and $0 < \theta < 1$, be an autonomous nonlinear fractional-order system. A point $\mathbf{x}^*$ is called an equilibrium point of the system if it satisfies $\phi(\mathbf{x}^*) = 0$. This equilibrium point is locally asymptotically stable if all eigenvalues λ_i of the Jacobian matrix $J = \frac{\partial \phi}{\partial x}$ evaluated at $\mathbf{x}^*$ satisfy the Matignon conditions $|\arg(\lambda_i)| > \frac{\theta \pi}{2}$.

3. Fractional Model with Caputo Derivative: Statement of the Problem, Boundedness, Existence and Uniqueness

We present the following fractional-order prey–predator model corresponding to the non-dimensional model introduced in [5]

$$\begin{cases} D_t^\theta n = \sigma n \left(1 - \frac{n}{k}\right) - (1 + \alpha p)np, \\ D_t^\theta p = (1 + \alpha p)np - p, \end{cases} \tag{3}$$

where D_t^θ represents the Caputo fractional derivative, given in Definition 1, with $0 < \theta < 1$ and n, p are the non-dimensional variables corresponding to the prey and predator densities. All the non-dimensional parameters are positive constant. Precisely, k comprises the dimensional carrying capacity, the conversion efficiency as well as the per-capita predator mortality and attack rate, while σ is linked to the per-capita intrinsic growth rate of prey and per-capita mortality rate of predator and α is linked to the predator cooperation in hunting rate, the attack rate and the per-capita predator mortality rate. Details on both the derivation of model and the biological meaning of parameters can be found in [5,14]. To (3) we append the initial conditions

$$n(t_0) = n_0, \quad p(t_0) = p_0. \tag{4}$$

3.1. Boundedness

In this subsection, we prove the positivity and boundedness of the solution of the system (3). Let $\mathbb{R}_+^2 = \{\mathbf{x}(t) \in \mathbb{R}^2 : \mathbf{x}(t) \geq \mathbf{0}\}$ and $\mathbf{x}(t) = \begin{pmatrix} n(t) \\ p(t) \end{pmatrix}$.

Theorem 1. *The solution of the fractional-order system (3) and (4) is bounded in $\mathbb{R}_+^2$. Moreover, the density of the population remains in a nonnegative region.*

Proof. Let us define the function

$$U(t) = n(t) + p(t) \qquad (5)$$

and ξ be a positive constant. Applying the Caputo fractional derivative on both sides in (5) and using (3), it follows that

$$\begin{aligned}
D_t^\theta U(t) + \xi U(t) &= D_t^\theta n(t) + D_t^\theta p(t) + \xi n(t) + \xi p(t) \\
&= \sigma n\left(1 - \frac{n}{k}\right) - p + \xi n + \xi p \\
&= -\frac{\sigma n^2}{k} + (\sigma + \xi)n + (\xi - 1)p \\
&\leq -\frac{\sigma}{k}\left(n - \frac{k(\sigma + \xi)}{2\sigma}\right)^2 + \frac{k(\sigma + \xi)^2}{2\sigma} \\
&\leq \frac{k(\sigma + \xi)^2}{2\sigma},
\end{aligned}$$

where $\xi < 1$. Using Lemma 1, it follows that

$$\begin{aligned}
U(t) &\leq \left(U(t_0) - \frac{k(\sigma + \xi)^2}{2\sigma\xi}\right)E_\theta[-\xi t^\theta] + \frac{k(\sigma + \xi)^2}{2\sigma\xi} \\
&\leq U(t_0)E_\theta[-\xi t^\theta] + \frac{k(\sigma + \xi)^2}{2\sigma\xi}\left[1 - E_\theta[-\xi t^\theta]\right].
\end{aligned} \qquad (6)$$

Then, for $t \to \infty$ it follows that $U(t) \leq \frac{k(\sigma + \xi)^2}{2\sigma\xi}$. Hence, all the solutions to (3) starting from $\mathbb{R}_+^2$ are confined in the following domain $\mathcal{A} \subseteq \mathbb{R}_+^2$

$$\mathcal{A} = \{(n(t), p(t)) \in \mathbb{R}_+^2 : n(t) + p(t) \leq \frac{k(\sigma + \xi)^2}{2\sigma\xi} + \varepsilon, \forall \varepsilon > 0\}$$

which is positively invariant. □

3.2. Existence and Uniqueness

In this subsection, we find the conditions for the existence and uniqueness of the solution to fractional-order prey–predator model (3) in the region $\Theta \times (t_0, T]$, $T < \infty$, where

$$\Theta = \{(n, p) \in \mathbb{R}^2 : \max(|n|, |p|) \leq K\},$$

by using the fixed point technique. The existence of K is guaranteed by the boundedness of the solution. The model can be reformulated in the fractional integral form which gives

$$\begin{aligned}
n(t) - n(t_0) &= \frac{1}{\Gamma(\theta)}\int_{t_0}^t (t - \tau)^{\theta-1}\left(\sigma n(\tau)\left(1 - \frac{n(\tau)}{k}\right) - (1 + \alpha p(\tau))n(\tau)p(\tau)\right)d\tau, \\
p(t) - p(t_0) &= \frac{1}{\Gamma(\theta)}\int_{t_0}^t (t - \tau)^{\theta-1}((1 + \alpha p(\tau))n(\tau)p(\tau) - p(\tau))d\tau.
\end{aligned} \qquad (7)$$

Denoting by F_1 and F_2 the following kernels

$$\begin{aligned}
F_1(t, n(t), p(t)) &= \sigma n(t)\left(1 - \frac{n(t)}{k}\right) - [1 + \alpha p(t)]n(t)p(t), \\
F_2(t, n(t), p(t)) &= [1 + \alpha p(t)]n(t)p(t) - p(t),
\end{aligned} \qquad (8)$$

the following theorem holds.

Theorem 2. Let $M_1 = \sigma + (1 + \alpha K)K + 2\frac{\sigma}{k}K$ and $M_2 = |K - 1 + 2\alpha K^2|$. If $0 < M_i < 1$, $(i = 1, 2)$, then the kernels $F_i(t, n, p)$, $(i = 1, 2)$ agree with the contraction and Lipschitz conditions in the region $\Theta \times (t_0, T]$.

Proof. Let $n(t_1) = n_1$ and $n(t_2) = n_2$ be two functions for the kernel F_1 and $p(t_1) = p_1$ and $p(t_2) = p_2$ be two functions for the kernel F_2. Then we have

$$\begin{aligned} &\|F_1(t, n_1, p) - F_1(t, n_2, p)\| \\ &= \left\|\sigma\left(1 - \frac{n_1}{k}\right)n_1 - (1+\alpha p)pn_1 - \sigma\left(1 - \frac{n_2}{k}\right)n_2 + (1+\alpha p)pn_2\right\| \\ &= \left\|[\sigma - (1+\alpha p)p]n_1 - [\sigma - (1+\alpha p)p]n_2 - \frac{\sigma}{k}(n_1^2 - n_2^2)\right\| \\ &= \left\|[\sigma - (1+\alpha p)p](n_1 - n_2) - \frac{\sigma}{k}(n_1 - n_2)(n_1 + n_2)\right\| \\ &\leq \left(\sigma + (1+\alpha K)K + 2\frac{\sigma}{k}K\right)\|n_1 - n_2\| \\ &\leq M_1\|n_1 - n_2\| \end{aligned} \tag{9}$$

and

$$\begin{aligned} &\|F_2(t, n, p_1) - F_2(t, n, p_2)\| \\ &= \|(1 + \alpha p_1)np_1 - p_1 - (1 + \alpha p_2)np_2 + p_2\| \\ &= \|(n-1)(p_1 - p_2) + \alpha n(p_1^2 - p_2^2)\| \\ &= \|[(n-1) + \alpha n(p_1 + p_2)](p_1 - p_2)\| \\ &\leq |K - 1 + 2\alpha K^2|\|p_1 - p_2\| \\ &\leq M_2\|p_1 - p_2\|, \end{aligned} \tag{10}$$

where $M_1 = \sigma + (1 + \alpha K)K + 2\frac{\sigma}{k}K$ and $M_2 = |K - 1 + 2\alpha K^2|$. Therefore, the Lipschitz conditions are satisfied for kernels F_1 and F_2, and if $0 < M_i < 1$, $(i = 1, 2)$ then M_1 and M_2 are also contractions for F_1 and F_2, respectively. Assume that the conditions (9) and (10) hold and let us consider the kernels F_1 and F_2. Then (7) can be written

$$\begin{aligned} n(t) &= n(t_0) + \frac{1}{\Gamma(\theta)}\int_{t_0}^{t}(t-\tau)^{\theta-1}F_1(\tau, n, p)d\tau, \\ p(t) &= p(t_0) + \frac{1}{\Gamma(\theta)}\int_{t_0}^{t}(t-\tau)^{\theta-1}F_2(\tau, n, p)d\tau. \end{aligned} \tag{11}$$

The initial conditions and the recurrence form of the model (11) are, respectively

$$n_0(t) = n(t_0), \quad p_0(t) = p(t_0) \tag{12}$$

and

$$\begin{aligned} n_m(t) &= n(t_0) + \frac{1}{\Gamma(\theta)}\int_{t_0}^{t}(t-\tau)^{\theta-1}F_1(\tau, n_{m-1}, p)d\tau, \\ p_m(t) &= p(t_0) + \frac{1}{\Gamma(\theta)}\int_{t_0}^{t}(t-\tau)^{\theta-1}F_2(\tau, n, p_{m-1})d\tau. \end{aligned} \tag{13}$$

The successive difference between the terms is defined as

$$\Phi_{1m}(t) = n_m(t) - n_{m-1}(t) = \frac{1}{\Gamma(\theta)} \int_{t_0}^{t} (t-\tau)^{\theta-1} [F_1(\tau, n_{m-1}, p) - F_1(\tau, n_{m-2}, p)] d\tau,$$
$$\Phi_{2m}(t) = p_m(t) - p_{m-1}(t) = \frac{1}{\Gamma(\theta)} \int_{t_0}^{t} (t-\tau)^{\theta-1} [F_2(\tau, n, p_{m-1}) - F_2(\tau, n, p_{m-2})] d\tau, \tag{14}$$

where

$$n_m(t) = \sum_{i=1}^{m} \Phi_{1i}(t), \quad p_m(t) = \sum_{i=1}^{m} \Phi_{2i}(t). \tag{15}$$

Taking the norm of (14), it follows from the conditions (9)–(10) that

$$\|\Phi_{1m}(t)\| = \|n_m(t) - n_{m-1}(t)\| \leq \frac{M_1}{\Gamma(\theta)} \int_{t_0}^{t} (t-\tau)^{\theta-1} \|\Phi_{1(m-1)}(\tau)\| d\tau,$$
$$\|\Phi_{2m}(t)\| = \|p_m(t) - p_{m-1}(t)\| \leq \frac{M_2}{\Gamma(\theta)} \int_{t_0}^{t} (t-\tau)^{\theta-1} \|\Phi_{2(m-1)}(\tau)\| d\tau. \tag{16}$$

□

The following theorem holds.

Theorem 3. *Assume that the conditions (9)–(10) hold. If*

$$\frac{M_i T^\theta}{\Gamma(\theta+1)} < 1, \quad i=1,2, \tag{17}$$

then the solution of the fractional model given in (3)–(4) exists and is unique.

Proof. Let us consider $n(t), p(t)$ bounded functions and the kernels F_1 and F_2 satisfy the Lipschitz condition. From (15) and (16), it follows that

$$\|\Phi_{1m}(t)\| \leq \|n_0(t)\| \left\{ \frac{M_1 T^\theta}{\Gamma(\theta+1)} \right\}^m,$$
$$\|\Phi_{2m}(t)\| \leq \|p_0(t)\| \left\{ \frac{M_2 T^\theta}{\Gamma(\theta+1)} \right\}^m. \tag{18}$$

This shows the existence for the solutions. Moreover, in order to prove that (18) are solutions to (3) and (4), we consider

$$n(t) - n(t_0) = n_m(t) - r_{1m}(t),$$
$$p(t) - p(t_0) = p_m(t) - r_{2m}(t), \tag{19}$$

where r_{1m}, r_{2m} are the remaining terms. We show that the terms in (19) satisfy $\|r_{1\infty}(t)\| \to 0$ and $\|r_{2\infty}(t)\| \to 0$. Now, we consider the conditions

$$\|r_{1m}(t)\| \leq \left\| \frac{1}{\Gamma(\theta)} \int_{t_0}^{t} (t-\tau)^{\theta-1} |F_1(\tau, n, p) - F_1(\tau, n_{m-1}, p)| d\tau \right\|$$
$$\leq \frac{1}{\Gamma(\theta)} \int_{t_0}^{t} (t-\tau)^{\theta-1} \|F_1(\tau, n, p) - F_1(\tau, n_{m-1}, p)\| d\tau$$
$$\leq \frac{M_1 T^\theta}{\Gamma(\theta+1)} \|n - n_{m-1}\|. \tag{20}$$

On using recursive techniques, we get

$$\|r_{1m}(t)\| \leq \left\{ \frac{T^\theta}{\Gamma(\theta+1)} \right\}^{m+1} M_1^{m+1}.$$

For $m \to \infty$, it follows that $r_{1m}(t) \to 0$. In a similar way, we conclude that $r_{2m}(t) \to 0$. In order to prove the uniqueness of the solution to the model (3) and (4), let us suppose there exists another solution of the system $\bar{n}(t)$ and $\bar{p}(t)$. Then

$$\|n(t) - \bar{n}(t)\| \leq \frac{1}{\Gamma(\theta)} \int_{t_0}^{t} (t-\tau)^{\theta-1} \|F_1(\tau, n, p) - F_1(\tau, \bar{n}, p)\| d\tau,$$
$$\|p(t) - \bar{p}(t)\| \leq \frac{1}{\Gamma(\theta)} \int_{t_0}^{t} (t-\tau)^{\theta-1} \|F_2(\tau, n, p) - F_2(\tau, n, \bar{p})\| d\tau. \tag{21}$$

In view of the Lipschitz condition, it follows that

$$\|n(t) - \bar{n}(t)\| \leq \frac{M_1 t^\theta}{\Gamma(\theta+1)} \|n(t) - \bar{n}(t)\|,$$
$$\|p(t) - \bar{p}(t)\| \leq \frac{M_2 t^\theta}{\Gamma(\theta+1)} \|p(t) - \bar{p}(t)\|, \tag{22}$$

and hence

$$\|n(t) - \bar{n}(t)\| \left(1 - \frac{M_1 t^\theta}{\Gamma(\theta+1)}\right) \leq 0,$$
$$\|p(t) - \bar{p}(t)\| \left(1 - \frac{M_2 t^\theta}{\Gamma(\theta+1)}\right) \leq 0. \tag{23}$$

In view of (17), it follows that $\|n(t) - \bar{n}(t)\| = 0$ and $\|p(t) - \bar{p}(t)\| = 0$. □

4. Stability Analysis

The equilibrium points of system (3) are obtained by considering

$$D_t^\theta n(t) = 0, \quad D_t^\theta p(t) = 0.$$

According to [5,14], besides the trivial equilibrium $E_0 \equiv (0,0)$, the system admits the predator-free equilibrium $E_b \equiv (k, 0)$ and the coexistence equilibrium $E^* = (n^*, p^*) \equiv \left(\frac{1}{(1+\alpha p^*)}, p^*\right)$ where p^* satisfies $\alpha^2 k p^3 + 2\alpha k p^2 + k(1 - \alpha\sigma)p + \sigma(1-k) = 0$. As shown in [14], if $k > 1$, E^* is unique, while if $\frac{-1 + \sqrt{1+3\sigma\alpha}}{\alpha\sigma} < k \leq 1$ and $\sigma > \frac{1}{\alpha}$ there exist two coexistence equilibria. In all other cases, no coexistence equilibria are admissible.

In the following, we will discuss the stability properties of the coexistence equilibrium by using the linearization method.

The Jacobian matrix of system (3) evaluated at the equilibrium point $E^* = (n^*, p^*)$ is given by

$$J(E^*) = \begin{pmatrix} -\frac{\sigma}{k(1+\alpha p^*)} & -\frac{1+2\alpha p^*}{1+\alpha p^*} \\ (1+\alpha p^*)p^* & \frac{\alpha p^*}{1+\alpha p^*} \end{pmatrix}. \tag{24}$$

Now, the characteristic equation of the Jacobian matrix is $\lambda^2 - I\lambda + A = 0$ and the roots of the characteristic equation are

$$\lambda_{1,2} = \frac{I \pm \sqrt{I^2 - 4A}}{2}, \tag{25}$$

where I and A are the trace and determinant of the Jacobian matrix $J(E^*)$ given by

$$I = \frac{k\alpha p^* - \sigma}{k(1+\alpha p^*)},$$
$$A = \frac{p^*}{k(1+\alpha p^*)^2}[k(1+\alpha p^*)^2(1+2\alpha p^*) - \alpha\sigma].$$

The stability analysis of E^* will be divided in the following four cases:

(1) If $I^2 - 4A \geq 0$ and $I < 0$, then $\lambda_i (i = 1,2)$ are negative real numbers. Hence, $|\arg(\lambda_i)| = \pi > \theta\frac{\pi}{2}$. The coexistence equilibrium is asymptotically stable.

(2) If $I^2 - 4A \geq 0$ and $I \geq 0$ then at least one of λ_1, λ_2 will be positive. Therefore, at least one of λ_i will be such that $|\arg \lambda_i| = 0 < \theta\frac{\pi}{2}$ and hence the coexistence point is unstable.

(3) When $I^2 - 4A < 0$, then the eigenvalues are a pair of complex conjugate $\lambda_{1,2} = \frac{I}{2} \pm i\frac{\sqrt{4A - I^2}}{2}$.

   (a) If $I > 0$ then $Re(\lambda_1) = Re(\lambda_2) > 0$ and $Im(\lambda_1) = -Im(\lambda_2) = \frac{\sqrt{4A - I^2}}{2} > 0$.
   Then, when $\frac{\sqrt{4A - I^2}}{I} > \tan\theta\frac{\pi}{2}$, the coexistence equilibrium is asymptotically stable, otherwise it is unstable.

   (b) If $I < 0$, then $Re(\lambda_1) = Re(\lambda_2) < 0$. Then, $|\arg(\lambda_i)| > \frac{\pi}{2}$ and the equilibrium is asymptotically stable.

(4) When $I = 0$ and $I^2 - 4A = -4A < 0$, then $Re(\lambda_1) = Re(\lambda_2) = 0$. Hence, $|\arg(\lambda_i)| = \frac{\pi}{2} > \theta\frac{\pi}{2}$. In this case, the equilibrium is asymptotically stable.

A Hopf bifurcation occurs when a pair of complex eigenvalues of the Jacobian matrix at an equilibrium point exists and the stability of the equilibrium changes from stable to unstable when a bifurcation parameter crosses a critical value. We can choose the order of derivation θ to be the bifurcation parameter and by using the following well-known result we find the conditions for the Hopf bifurcation to appear.

Lemma 4. *[26] When bifurcation parameter θ passes through the critical value $\theta^* \in (0,1)$, fractional-order system (3) undergoes a Hopf bifurcation at the equilibrium point, if the following conditions hold*

1. the corresponding characteristic equation of system (3) has a pair of complex conjugate roots $\lambda_{1,2} = \gamma \pm i\omega$ with $\gamma > 0$;
2. $m(\theta^*) = \theta^*\frac{\pi}{2} - \min_{1 \leq i \leq 2}|\arg \lambda_i| = 0$;
3. $\frac{dm(\theta)}{d\theta}|_{\theta=\theta^*} \neq 0$.

We find the conditions for the model (3) tp undergo a Hopf bifurcation at equilibrium point $E^* \equiv (n^*, p^*)$ when the order of derivation passes through the critical value $\theta^* = \frac{2}{\pi}\arctan[\frac{\sqrt{|I^2 - 4A|}}{I}]$.

Let us assume that $I^2 - 4A < 0$ and $I > 0$; let the critical value $\theta^* = \frac{2}{\pi}\arctan[\frac{\sqrt{|I^2 - 4A|}}{I}]$.

Let us define $\gamma = \frac{I}{2}$ and $\omega = \frac{\sqrt{|I^2 - 4A|}}{2}$. Then, it follows that the eigenvalues are a pair of complex conjugate $\lambda_{1,2} = \gamma \pm i\omega$ with $\gamma > 0$. In addition, let $m(\theta) = \theta\frac{\pi}{2} - \min_{1 \leq i \leq 2}|\arg \lambda_i|$, then, it follows that

$$m(\theta^*) = \theta^*\frac{\pi}{2} - \min_{1 \leq i \leq 2}|\arg \lambda_i| = \theta^*\frac{\pi}{2} - \arctan\frac{\omega}{\gamma}$$
$$= \arctan\frac{\omega}{\gamma} - \arctan\frac{\omega}{\gamma} = 0.$$

Finally,

$$\frac{dm(\theta)}{d\theta}|_{\theta=\theta^*} = \frac{\pi}{2} \neq 0.$$

Therefore, we can conclude that a Hopf bifurcation of (3) will appear at E^*.

5. Numerical Solution

Even if the interest on fractional calculus traces back at least to the 1970s (without considering the seminal suggestions given by Leibniz and Euler more than 300 years ago), in the last decades there has been an explosion of research activities on its application to several areas. Such a growing interest for fractional-order models had led to the development of specific algorithms devoted to the numerical approximation of the solution to fractional-order differential problems. Several solvers have been proposed in the last years [27–29], all trying to balance efficiency and accuracy while guaranteeing reliability of the numerical approximation. It is well-known, indeed, that the persistent memory of fractional-order operators reflects on the numerical evaluation of the solution: as a natural consequence, at each new timestep all the past history of the solution is to be considered. The number of involved steps increases with time, and so does the computational burden.

In a sense, numerical methods for fractional-order differential systems are then naturally multi-step. They generalize some of the ODE methods, but with significant differences in both complexity and accuracy. We follow here the excellent survey [30] and consider some of the schemes reported there. The simplest schemes are derived by approximating the integrand function $\mathbf{x}(t) = (n(t), p(t))$ in (7) by a piecewise polynomial and proceeding to its exact integration. This leads to "rectangular" (explicit or implicit) product integration formulas when a zero-order approximation of the integrand function in each subinterval is assumed, or "trapezoidal" formulas where first-order approximation is chosen. Of course, implicit formulas are more stable, but they require solving the nonlinear equations that generally arise. To avoid this additional burden, a predictor–corrector approach can be useful.

The convergence order of the rectangular formulas is one: the distance between the numerical approximation and the exact solution in any time point t_n decreases linearly with the timestep, and this result replicates the well-known one for the ODE formula. Unfortunately, this is not always true for the trapezoidal rule: its convergence order is limited by the minimum between $1 + \theta$ and 2 [28], so that for $0 < \theta < 1$ the rate of the corresponding ODE method cannot be reached. Similarly, higher order formulas for fractional systems do not lead to significant improvements in accuracy and convergence rate, and they are not worthy to be considered. As an alternative to product integration formulas, Fractional Linear Multistep methods have been proposed to generalize the standard multi-step methods for ODEs. They are very robust and can reach higher convergence order, but their convolution weights are not known explicitly in advance and have to be evaluated. This computation can however be performed very efficiently by FFT derived algorithms.

Numerical approximations of the solution of (3) can then be obtained by applying several different schemes. Even developed from the same ODE methods, they present distinguishing characteristics that deserve to be investigated. For this reason, we considered and compared the following schemes:

- Implicit Rectangular Product Integration rule (PI1)

$$x_n = x_0 + h^\theta \sum_{i=1}^{n} b_{n-i}^{(\theta)} F(t_i, x_i), \qquad b_n^{(\theta)} = \frac{(n+1)^\theta - n^\theta}{\Gamma(\theta + 1)} ;$$

- Implicit Trapezoidal Product Integration rule (PI2)

$$x_n = x_0 + h^\theta \frac{1}{\Gamma(\theta + 2)} \left(\tilde{a}_n^{(\theta)} F(t_0, x_0) + \sum_{i=1}^{n} a_{n-i}^{(\theta)} F(t_i, x_i) \right),$$

with $a_0^{(\theta)} = 1$ and

$$\tilde{a}_n^{(\theta)} = (n-1)^{\theta+1} - n^\theta (n - \theta - 1), \qquad a_n^{(\theta)} = (n-1)^{\theta+1} - 2n^{\theta+1} + (n+1)^{\theta+1} ;$$

- Predictor–Corrector Product Integration rule (PI12)

$$x_n^P = x_0 + h^\theta \sum_{i=0}^{n-1} b_{n-i-1}^{(\theta)} F(t_i, x_i),$$

$$x_n = x_0 + h^\theta \frac{1}{\Gamma(\theta+2)} \left(\tilde{a}_n^{(\theta)} F(t_0, x_0) + \sum_{i=1}^{n-1} a_{n-i}^{(\theta)} F(t_i, x_i) + a_0^{(\theta)} F(t_n, x_n^P) \right);$$

- Fractional Backward Differentiation Formula (FLMM2)

$$x_n = x_0 + h^\theta \sum_{i=0}^n (-1)^{n-i} \binom{-\theta}{n-i} F(t_i, x_i), \quad \text{where} \quad \binom{-\theta}{n-i} = \frac{\Gamma(1-\theta)}{\Gamma(n+1)\Gamma(-\theta-n+1)}.$$

All these numerical schemes have been implemented in Matlab routines as given in [30,31]. For the problem at hand, the performance of these schemes can be evaluated only qualitatively, due to the lack of an exact (closed form) solution, by comparing their behavior for decreasing timesteps; however, we refer the reader to the above cited literature for an exhaustive assessment of their results on several test cases.

5.1. Preliminary Assessment

As a preliminary test, we checked the results of all methods in reproducing trajectories when $\theta = 1$, because these results can be directly compared with the reference solutions given by classical ODE solvers. Even if the considered methods are all devised for fractional-order systems, they are expected to reproduce a good numerical approximation of the solution also for the integer order case. When the parameter settings ($\sigma = 3$, $\alpha = 0.3$, $k = 5$) lead to a stable coexistence equilibrium, all methods are able to reproduce the expected trajectories: the left panel of Figure 1 shows the convergence of all numerical approximations of $n(t)$ towards the equilibrium value $n^* \approx 0.66$. On the other hand, when Hopf instability occurs ($\sigma = 3$, $\alpha = 10$, $k = 0.8$), the simplest PI1 method shows a sensible damping of the oscillations around the equilibrium $n^* \approx 0.188$, as shown in the right panel of the same Figure, while the other methods' trajectories are practically indistinguishable from the reference one (as reported in [14]).

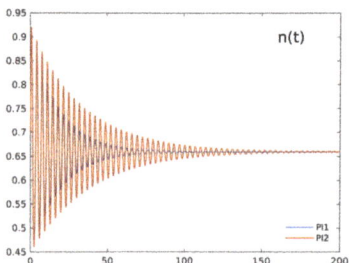
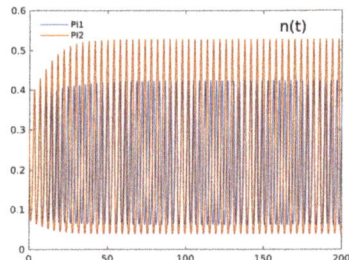

Figure 1. Trajectories for the numerical approximations PI1 and PI2 of $n(t)$ solution of system (3) for $\theta = 1$ in case of a stable equilibrium (**left panel**) and in the presence of Hopf instability (**right panel**). The lowest order approximation PI1 clearly shows damped oscillations, while the trajectories of the higher order approximations are in excellent agreement with the ones obtained by classical ODE solvers.

5.2. Stabilizing Effect of the Persistent Memory

As a second test case, we consider a parameter setting for which the corresponding ODE system shows instability: if we choose as before $\sigma = 3$, $\alpha = 10$, $k = 0.8$, simply considering a slightly lower value for θ ($\theta = 0.9$) has a powerful stabilizing effect. As Figure 2 shows, all the numerical approximations agree in a fast damping of the oscillations for both the variables. Again, the damping is more evident in the lower order method PI1, while the other considered schemes agree in accurately reconstructing the system trajectories. To further analyze the different performance of the considered numerical methods, we present in the following Figure 3 a detail of the first part of the same trajectories as reconstructed by any of the numerical schemes for decreasing time steps $h_1 = 2^{-4}, h_2 = 2^{-5}, h_3 = 2^{-6}, h_4 = 2^{-7}$. The less accurate results of PI1 with respect to the other schemes can be clearly seen. Of course, lower values for the fractional-order θ result in an even faster damping of the oscillations in the trajectories, confirming the strong stabilizing effect of the persistent memory in the model. For the same test case, Figure 4 compares the numerical approximations by PI2 of the populations' trajectories for different values of the order θ (0.95, 0.9, 0.85, 0.8), confirming the strong stabilizing effect of the fractional derivative.

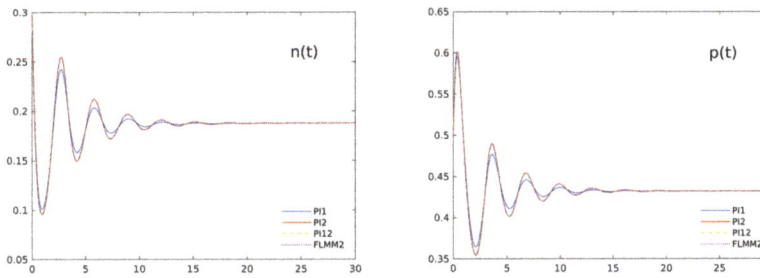

Figure 2. Trajectories for all the numerical approximations of the solutions of system (3) in case $\theta = 0.9$. While the corresponding ODE systems show instability (see the right panel of Figure 1), the trajectories of the fractional system rapidly stabilizes for both variables ($n(t)$ shown in the left panel, $p(t)$ in the right one). The lowest order approximation PI1 clearly shows more damped oscillations, while the higher order approximations are all in excellent agreement.

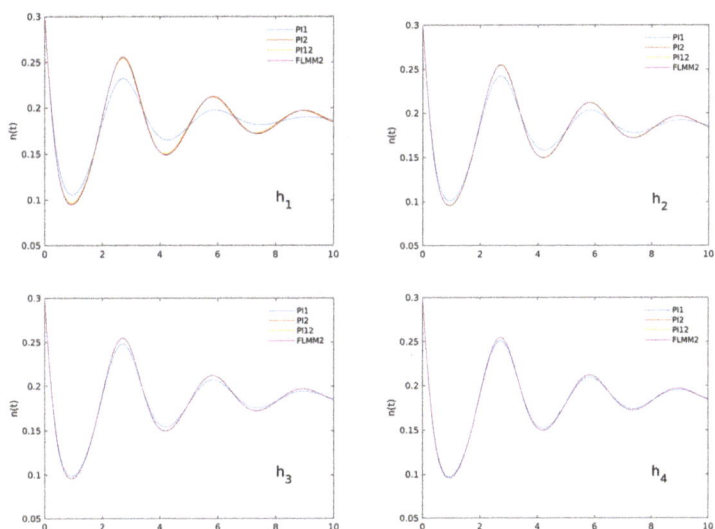

Figure 3. Details of the initial part of the trajectories shown in the left panel of Figure 2 as reconstructed by all the numerical methods for decreasing timesteps h_1, h_2, h_3, h_4. The lowest order approximation PI1 clearly shows more damped oscillations, and only for the smallest timestep it agrees with the other methods, whose results are practically identical.

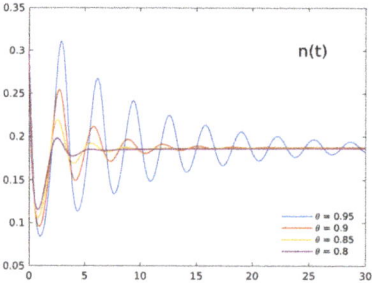
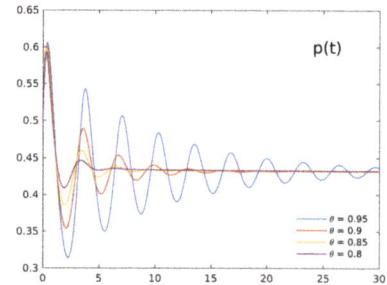

Figure 4. Trajectories for the numerical approximations PI2 of the solutions of system (3) for different θ values (0.95, 0.9, 0.85, 0.8) for both populations ($n(t)$ shown in the left panel, $p(t)$ in the right one). The stabilizing effect of the fractional operators can be clearly seen.

5.3. Hopf Instability for the Fractional System

As a third example, we show how the instability can still occur for the fractional system with a suitable choice of the parameters. We start by considering a parameter setting for which, as stated in Section 4, the conditions for the Hopf instability to appear are met: we set $\sigma = 3$, $\alpha = 3.5$, $k = 5$. In this case, it is $I > 0$ and $I^2 - 4A < 0$. The eigenvalues of the Jacobian are a couple of complex conjugate numbers with a positive real part so that for $\theta > \theta^* = \arctan 2/\pi \sqrt{4A - I^2}/I \approx 0.89$ the coexistence equilibrium $E^* \approx (0.27, 0.77)$ loses its stability. The following Figure 5 shows the trajectories of $(n(t), p(t))$ when $\theta = 0.92$ as reconstructed by the more accurate numerical algorithms and the corresponding phase plan portrait, clearly showing the appearance of a limit cycle.

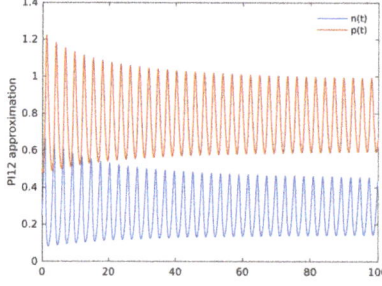
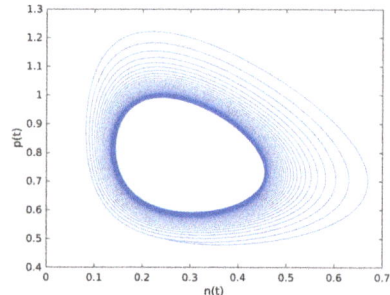

Figure 5. Numerical approximation PI12 of the solutions of system (3) confirming the Hopf instability predicted by the theory in case $\theta = 0.92$ with $\sigma = 3$, $\alpha = 3.5$, $k = 5$. In the left panel, the trajectories for both variables are reported as functions of time; the right panel shows the limit cycle in the phase plan.

6. Conclusions

Fractional calculus, implying non-locality and memory effects, allows the description of numerous phenomena in a wide variety of scientific domains. Fractional-order operators have been proved to be a very powerful modeling instrument to represent a variety of processes and biological systems. In this framework, we studied in this paper the dynamic behavior of a fractional-in-time prey–predator model with hunting cooperation. The existence and uniqueness of a non-negative solution has been proved and the local stability of the coexistence equilibrium point has been analyzed by using the linearization technique. The conditions for the occurrence of a Hopf bifurcation have been found. Finally, numerical simulations have been presented to confirm by some selected examples how the order of derivation θ affects the dynamical behavior of prey and predator density. The findings of the present study, in our opinion, can provide hints for the investigation of the dynamics of other processes describing real-world phenomena. As a final consideration, we outline that several authors investigated pattern formation in the related spatial fractional-order systems. Recently, Yin and Wen [32] have shown that fractional-derivative systems can result in persistent spatial patterns even though their ODE counterpart does not induce any steady pattern. Such mechanisms of pattern formation along with the ones induced by anomalous diffusion suggest further directions of research in spatial generalizations of the considered model.

Author Contributions: Conceptualization, M.F.C. and I.T.; formal analysis, I.T.; numerical simulations, M.F.C.; writing (original draft preparation, review and editing), M.F.C. and I.T.; funding acquisition, M.F.C. and I.T. All authors have read and agreed to the published version of the manuscript.

Funding: This research was partially supported by Regione Campania Projects REMIAM, ADVISE and MEDIA.

Acknowledgments: This paper has been performed under the auspices of the G.N.F.M. and G.N.C.S. of INdAM.

Conflicts of Interest: The authors declare no conflict of interest.

References

1. Abrams, P.A. The Evolution of Predator-Prey Interactions: Theory and Evidence. *Annu. Rev. Ecol. Syst.* **2000**, *31*, 79–105. [CrossRef]
2. Berryman, A.A. The Origins and Evolution of Predator-Prey Theory. *Ecology* **1992**, *73*, 1530–1535. [CrossRef]
3. Briggs, C.J.; Hoopes, M.F. Stabilizing effects in spatial parasitoid–host and predator–prey models: A review. *Theor. Popul. Biol.* **2004**, *65*, 299–315. [CrossRef] [PubMed]
4. Courchamp, F.; Berec, L.; Gascoigne, J. *Allee Effects in Ecology and Conservation*; Oxford University Press: Oxford, UK, 2008. [CrossRef]
5. Alves, M.T.; Hilker, F. Hunting cooperation and Allee effects in predators. *J. Theor. Biol.* **2017**, *419*, 13–22. [CrossRef]
6. Tang, B. Dynamics for a fractional-order predator-prey model with group defense. *Sci. Rep.* **2020**, *10*, 4906. [CrossRef]
7. Amirian, M.M.; Towers, I.; Jovanoski, Z.; Irwin, A.J. Memory and mutualism in species sustainability: A time-fractional Lotka-Volterra model with harvesting. *Heliyon* **2020**, *6*, e04816. [CrossRef]
8. Patnaik, S.; Hollkamp, J.P.; Semperlotti, F. Applications of variable-order fractional operators: a review. *Proc. R. Soc. Math. Phys. Eng. Sci.* **2020**, *476*, 20190498. [CrossRef] [PubMed]
9. Sun, H.; Zhang, Y.; Baleanu, D.; Chen, W.; Chen, Y. A new collection of real world applications of fractional calculus in science and engineering. *Commun. Nonlinear Sci. Numer. Simul.* **2018**, *64*, 213–231. [CrossRef]
10. Abundo, M.; Ascione, G.; Carfora, M.F.; Pirozzi, E. A fractional PDE for first passage time of time-changed Brownian motion and its numerical solution. *Appl. Numer. Math.* **2020**, *155*, 103–118. [CrossRef]
11. Cherstvy, A.G.; Chechkin, A.V.; Metzler, R. Ageing and confinement in non-ergodic heterogeneous diffusion processes. *J. Phys. Math. Theor.* **2014**, *47*, 485002. [CrossRef]
12. Wang, W.; Cherstvy, A.G.; Liu, X.; Metzler, R. Anomalous diffusion and nonergodicity for heterogeneous diffusion processes with fractional Gaussian noise. *Phys. Rev. E* **2020**, *102*, 012146. [CrossRef]
13. Capone, F.; Carfora, M.F.; De Luca, R.; Torcicollo, I. On the dynamics of an intraguild predator–prey model. *Math. Comput. Simul.* **2018**, *149*, 17–31. [CrossRef]
14. Capone, F.; Carfora, M.F.; De Luca, R.; Torcicollo, I. Turing patterns in a reaction-diffusion system modeling hunting cooperation. *Math. Comput. Simul.* **2019**, *165*, 172–180. [CrossRef]
15. Capone, F.; Carfora, M.F.; De Luca, R.; Torcicollo, I. Analysis of a model for waterborne diseases with Allee effect on bacteria. *Nonlinear Anal. Model. Control* **2020**, *25*, 1035–1058. [CrossRef]
16. Carfora, M.F.; Torcicollo, I. Cross-diffusion-driven instability in a predator-prey system with fear and group defense. *Mathematics* **2020**, *8*, 1244. [CrossRef]
17. Carfora, M.F.; Torcicollo, I. Identification of epidemiological models: the case study of Yemen cholera outbreak. *Appl. Anal.* **2020**. [CrossRef]
18. Torcicollo, I. On the nonlinear stability of a continuous duopoly model with constant conjectural variation. *Int. J. Non-Linear Mech.* **2016**, *81*, 268–273. [CrossRef]
19. Rionero, S.; Torcicollo, I. On the dynamics of a nonlinear reaction-diffusion duopoly model. *Int. J. Non-Linear Mech.* **2018**, *99*, 105–111. [CrossRef]
20. Rionero, S.; Torcicollo, I. Stability of a Continuous Reaction-Diffusion Cournot-Kopel Duopoly Game Model. *Acta Appl. Math.* **2014**, *132*, 505–513. [CrossRef]
21. De Angelis, F.; De Angelis, M. On solutions to a FitzHugh-Rinzel type model. *Ric. Mat.* **2020**. [CrossRef]

22. Capone, F.; De Luca, R.; Rionero, S. On the stability of non-autonomous perturbed Lotka-Volterra models. *Appl. Math. Comput.* **2013**, *219*, 6868–6881. [CrossRef]
23. Podlubny, I. *Fractional Differential Equations. An Introduction to Fractional Derivatives, Fractional Differential Equations, Some Methods of Their Solution and Some of Their Applications*; Academic Press: San Diego, CA, USA; London, UK, 1999.
24. Diethelm, K. *The Analysis of Fractional Differential Equations: An Application-Oriented Exposition Using Differential Operators of Caputo Type*; Springer: Berlin/Heidelberg, Germany, 2010. [CrossRef]
25. Kilbas, A.A.; Srivastava, H.M.; Trujillo, J.J. *Theory and Applications of Fractional Differential Equations, Volume 204 (North-Holland Mathematics Studies)*; Elsevier Science Inc.: Amsterdam, The Netherlands, 2006.
26. Li, X.; Wu, R. Hopf bifurcation analysis of a new commensurate fractional-order hyperchaotic system. *Nonlinear Dyn.* **2014**, *78*, 279–288. [CrossRef]
27. Diethelm, K.; Ford, N.J.; Freed, A.D.; Luchko, Y. Algorithms for the fractional calculus: A selection of numerical methods. *Comput. Methods Appl. Mech. Eng.* **2005**, *194*, 743–773. [CrossRef]
28. Garrappa, R. Trapezoidal methods for fractional differential equations: Theoretical and computational aspects. *Math. Comput. Simul.* **2015**, *110*, 96–112. [CrossRef]
29. Cai, M.; Li, C. Numerical Approaches to Fractional Integrals and Derivatives: A Review. *Mathematics* **2020**, *8*, 43. [CrossRef]
30. Garrappa, R. Numerical Solution of Fractional Differential Equations: A Survey and a Software Tutorial. *Mathematics* **2018**, *6*, 16. [CrossRef]
31. Garrappa, R. FLMM2. MATLAB Central File Exchange. 2021. Available online: https://www.mathworks.com/matlabcentral/fileexchange/47081-flmm2 (accessed on 26 March 2021).
32. Yin, H.; Wen, X. Pattern Formation through Temporal Fractional Derivatives. *Sci. Rep.* **2018**, *8*, 5070. [CrossRef] [PubMed]

Article

A Quadratic Mean Field Games Model for the Langevin Equation

Fabio Camilli

Dipartimento di Scienze di Base e Applicate per l'Ingegneria, Sapienza Università di Roma, Via Scarpa 16, 00161 Roma, Italy; fabio.camilli@uniroma1.it

Abstract: We consider a Mean Field Games model where the dynamics of the agents is given by a controlled Langevin equation and the cost is quadratic. An appropriate change of variables transforms the Mean Field Games system into a system of two coupled kinetic Fokker–Planck equations. We prove an existence result for the latter system, obtaining consequently existence of a solution for the Mean Field Games system.

Keywords: langevin equation; Mean Field Games system; kinetic Fokker–Planck equation; hypoelliptic operators

MSC: 35K40; 91A16

Citation: Fabio Camilli A Quadratic Mean Field Games Model for the Langevin Equation. *Axioms* **2021**, *10*, 68. https://doi.org/10.3390/axioms10020068

Academic Editor: Gabriella Bretti

Received: 10 January 2021
Accepted: 16 April 2021
Published: 19 April 2021

Publisher's Note: MDPI stays neutral with regard to jurisdictional claims in published maps and institutional affiliations.

Copyright: © 2021 by the author. Licensee MDPI, Basel, Switzerland. This article is an open access article distributed under the terms and conditions of the Creative Commons Attribution (CC BY) license (https://creativecommons.org/licenses/by/4.0/).

1. Introduction

The Mean Field Games (MFG in short) theory concerns the study of differential games with a large number of rational, indistinguishable agents and the characterization of the corresponding Nash equilibria. In the original model introduced in [1,2], an agent can typically act on its velocity (or other first order dynamical quantities) via a control variable. Mean Field Games where agents control the acceleration have been recently proposed in [3–5].

A prototype of stochastic process involving acceleration is given by the Langevin diffusion process, which can be formally defined as

$$\ddot{X}(t) = -b(X(t)) + \sigma \dot{B}(t), \qquad (1)$$

where $\ddot{X}$ is the second time derivative of the stochastic process X, B a Brownian motion and σ a positive parameter. The solution of (1) can be rewritten as a Markov process (X, V) solving

$$\begin{cases} \dot{X}(t) = V(t), \\ \dot{V}(t) = -b(X(t)) + \sigma \dot{B}(t). \end{cases}$$

The probability density function of the previous process satisfies the kinetic Fokker–Planck equation

$$\partial_t p - \frac{\sigma^2}{2} \Delta_v p - b(x) \cdot D_v p + v \cdot D_x p = 0 \quad \text{in} \quad (0, \infty) \times \mathbb{R}^d \times \mathbb{R}^d.$$

The previous equation, in the case $b \equiv 0$, was first studied by Kolmogorov [6] who provided an explicit formula for its fundamental solution. Then considered by Hörmander [7] as motivating example for the general theory of the hypoelliptic operators (see also [8–10]).

We consider a Mean Field Games model where the dynamics of the single agent is given by a controlled Langevin diffusion process, i.e.,

$$\begin{cases} \dot{X}(s) = V(s), & s \geq t \\ \dot{V}(s) = -b(X(s)) + \alpha(s) + \sigma \dot{B}(s) & s \geq t \\ X(t) = x, \ V(t) = v \end{cases} \quad (2)$$

for $(t, x, v) \in [0, T] \times \mathbb{R}^d \times \mathbb{R}^d$. In (2), the control law $\alpha : [t, T] \to \mathbb{R}^d$, which is a progressively measurable process with respect to a fixed filtered probability space such that $\mathbb{E}[\int_t^T |\alpha(t)|^2 dt] < +\infty$, is chosen to *maximize* the functional

$$J(t, x, v; \alpha) = \mathbb{E}_{t,(x,v)} \left\{ \int_t^T \left[f(X(s), V(s), m(s)) - \frac{1}{2} |\alpha(s)|^2 \right] ds + u_T(X(T), V(T)) \right\},$$

where $m(s)$ is the distribution of the agents at time s. Let u the value function associated with the previous control problem, i.e.,

$$u(t, x, v) = \sup_{\alpha \in \mathcal{A}_t} \{ J(t, x, v; \alpha) \}$$

where $\mathcal{A}_t$ is the the set of the control laws. Formally, the couple (u, m) satisfies the MFG system (see Section 4.1 in [3] for more details)

$$\begin{cases} \partial_t u + \frac{\sigma^2}{2} \Delta_v u - b(x) \cdot D_v u + v \cdot D_x u + \frac{1}{2} |D_v u|^2 = -f(x, v, m) \\ \partial_t m - \frac{\sigma^2}{2} \Delta_v m - b(x) \cdot D_v m + v \cdot D_x m + \text{div}_v(m D_v u) = 0 \\ m(0, x, v) = m_0(x, v), \quad u(T, x, v) = u_T(x, v). \end{cases} \quad (3)$$

for $(t, x, v) \in (0, T) \times \mathbb{R}^d \times \mathbb{R}^d$. The first equation is a backward Hamilton–Jacobi–Bellman equation, degenerate in the x-variable and with a quadratic Hamiltonian in the v variable, and the second equation is forward kinetic Fokker–Planck equation. In the standard setting, MFG systems with quadratic Hamiltonians has been extensively considered in literature both as a reference model for the general theory and also since, thanks to the Hopf-Cole change of variable, the nonlinear Hamilton-Jacobi-Bellman equation can be transformed into a linear equation, allowing to use all the tools developed for this type of problem (see for example [2,11–15]). Recently, a similar procedure has been used for ergodic hypoelliptic MFG with quadratic cost in [16] and for a flocking model involving kinetic equations in Section 4.7.3 of [17].

We study (3) by means of a change of variable introduced in [11,14] for the standard case. By defining the new unknowns $\phi = e^{u/\sigma^2}$ and $\psi = m e^{-u/\sigma^2}$, the system (3) is transformed into a system of two kinetic Fokker–Planck equations

$$\begin{cases} \partial_t \phi + \frac{\sigma^2}{2} \Delta_v \phi - b(x) \cdot D_v \phi + v \cdot D_x \phi = -\frac{1}{\sigma^2} f(x, v, \psi \phi) \phi \\ \partial_t \psi - \frac{\sigma^2}{2} \Delta_v \psi - b(x) \cdot D_v \psi + v \cdot D_x \psi = \frac{1}{\sigma^2} f(x, v, \psi \phi) \psi \\ \psi(0, x, v) = \frac{m_0(x,v)}{\phi(0,x,v)}, \quad \phi(T, x, v) = e^{\frac{u_T(x,v)}{\sigma^2}}. \end{cases} \quad (4)$$

for $(t, x, v) \in (0, T) \times \mathbb{R}^d \times \mathbb{R}^d$. In the previous problem, the coupling between the two equations is only in the source terms. Following [14], we prove existence of a weak solution to (4) by showing the convergence of an iterative scheme defined, starting from $\psi^{(0)} \equiv 0$, by solving alternatively the backward problem

$$\begin{cases} \partial_t \phi^{(k+\frac{1}{2})} + \frac{\sigma^2}{2}\Delta_v \phi^{(k+\frac{1}{2})} - b(x) \cdot D_v \phi^{(k+\frac{1}{2})} + v \cdot D_x \phi^{(k+\frac{1}{2})} \\ \qquad\qquad\qquad\qquad = -\frac{1}{\sigma^2} f(\psi^{(k)} \phi^{(k+\frac{1}{2})}) \phi^{(k+\frac{1}{2})} \\ \phi^{(k+\frac{1}{2})}(T,x,v) = e^{\frac{u_T(x,v)}{\sigma^2}}, \end{cases} \quad (5)$$

and the forward one

$$\begin{cases} \partial_t \psi^{(k+1)} - \frac{\sigma^2}{2}\Delta_v \psi^{(k+1)} - b(x) \cdot D_v \psi^{(k+1)} + v \cdot D_x \psi^{(k+1)} \\ \qquad\qquad\qquad\qquad = \frac{1}{\sigma^2} f(\psi^{(k+1)} \phi^{(k+\frac{1}{2})}) \psi^{(k+1)} \\ \psi^{(k+1)}(0,x,v) = \frac{m_0(x,v)}{\phi^{(k+\frac{1}{2})}(0,x,v)}. \end{cases} \quad (6)$$

We show that the resulting sequence $(\phi^{(k+\frac{1}{2})}, \psi^{(k+1)})$, $k \in \mathbb{N}$, monotonically converges to the solution of (4). Hence, by the inverse change of variable (see again [11,14] for details)

$$u = \frac{\ln(\phi)}{\sigma^2}, \qquad m = \phi \psi, \quad (7)$$

we obtain a solution of the original problem (3). We have

Theorem 1. *The sequence $(\phi^{(k+\frac{1}{2})}, \psi^{(k+1)})$ defined by (5) and (6) converges in $L^2([0,T] \times \mathbb{R}^d \times \mathbb{R}^d)$ and a.e. to a weak solution (ϕ, ψ) of (4). Moreover, the couple (u,m) defined by (7) is a weak solution to (3).*

The main difficulty in the study of problems (3) and (4) is due both in the degeneracy of the second order operator with respect to x and in the unbounded dependence of the coefficients of the first order terms with respect to v. To overcome the previous difficulties we rely on the results for linear kinetic Fokker–Planck equations developed in [18]. We mention that existence of weak solutions for the standard MFG problem, possibly degenerate, has been studied in [19], but the results in this paper do not cover the present setting. The previous iterative procedure also suggests a monotone numerical method for the approximation of (4), hence for (3). Indeed, by approximating (5) and (6) by finite differences and solving alternatively the resulting discrete equations, we obtain an approximation of the sequence $(\phi^{(k+\frac{1}{2})}, \psi^{(k+1)})$. A corresponding procedure for the standard quadratic MFG system was studied in [14], where the convergence of the method is proved. We plan to study the properties of the previous numerical procedure in a future work.

2. Well Posedness of the Kinetic Fokker–Planck System

In this section, we study the existence of a solution to system (4). The proof of the result follows the strategy implemented in Section 2 of [14] for the case of a standard MFG system with quadratic Hamiltonian and relies on the results for linear kinetic Fokker–Planck equations in Appendix A of [18]. We remark the model here studied does not fit exactly the problem treated in [18] because of the presence of a zero order term in the Fokker-Planck equation. Hence some technical aspects should be analyzed in more detail, however the present paper is mainly intended to give some idea on the change of variabile for the kinetic MGF.

We fix the assumptions we will assume in the whole paper. The vector field $b: \mathbb{R}^d \to \mathbb{R}^d$ and the coupling cost $f: \mathbb{R}^d \times \mathbb{R}^d \times \mathbb{R} \to \mathbb{R}$ are assumed to satisfy

$$b \in L^\infty(\mathbb{R}^d),$$
$$f \in L^\infty(\mathbb{R}^d \times \mathbb{R}^d \times \mathbb{R}), f \leq 0 \text{ and } f(x,v,\cdot) \text{ strictly decreasing}.$$

Moreover, the diffusion coefficient σ is positive and the initial and terminal data satisfy

$$m_0 \in L^\infty(\mathbb{R}^d \times \mathbb{R}^d), \, m_0 \geq 0, \, \iint m_0(x,v)dxdv = 1, \tag{8}$$

$$\text{and } \exists R_0 > 0 \text{ s.t. supp}\{m_0\} \subset \mathbb{R}^d \times B(0, R_0)$$

and

$$u_T \in C^0(\mathbb{R}^d \times \mathbb{R}^d) \text{ and } \exists C_0, C_1 > 0 \text{ s.t. } \forall (x,v) \in \mathbb{R}^d \times \mathbb{R}^d$$
$$-C_0(|v|^2 + |x|) - C_0 \leq u_T(x,v) \leq -C_1(|v|^2 + |x|) + C_1. \tag{9}$$

Note that (9) implies that $e^{u_T/\sigma^2} \in L^\infty(\mathbb{R}^d \times \mathbb{R}^d) \cap L^2(\mathbb{R}^d \times \mathbb{R}^d)$. We denote with $(\cdot,\cdot)$ the scalar product in $L^2([0,T] \times \mathbb{R}^d \times \mathbb{R}^d)$ and with $\langle\cdot,\cdot\rangle$ the pairing between $\mathcal{X} = L^2([0,T] \times \mathbb{R}_x^d; H^1(\mathbb{R}_v^d))$ and its dual $\mathcal{X}' = L^2([0,T] \times \mathbb{R}_x^d; H^{-1}(\mathbb{R}_v^d))$. We define the following functional space

$$\mathcal{Y} = \left\{ g \in L^2([0,T] \times \mathbb{R}_x^d, H^1(\mathbb{R}_v^d)), \partial_t g + v \cdot D_x g \in L^2([0,T] \times \mathbb{R}_x^d, H^{-1}(\mathbb{R}_v^d)) \right\}$$

and we set $\mathcal{Y}_0 = \{g \in \mathcal{Y} : g \geq 0\}$. If $g \in \mathcal{Y}$, then it admits (continuous) trace values $g(0,x,v), g(T,x,v) \in L^2(\mathbb{R}^d \times \mathbb{R}^d)$ (see [18, Lemma A.1]) and therefore the initial/terminal conditions for (4) are well defined in L^2 sense. We first prove the well posedness of problems (5) and (6).

Proposition 2. *We have*

(i) *For any $\psi \in \mathcal{Y}_0$, there exists a unique solution $\phi \in \mathcal{Y}_0$ to*

$$\begin{cases} \partial_t \phi + \frac{\sigma^2}{2}\Delta_v \phi - b(x) \cdot D_v \phi + v \cdot D_x \phi = -\frac{1}{\sigma^2}f(x,v,\psi\phi)\phi \\ \phi(T,x,v) = e^{\frac{u_T(x,v)}{\sigma^2}}. \end{cases} \tag{10}$$

Moreover, $\phi \in L^\infty([0,T] \times \mathbb{R}^d \times \mathbb{R}^d)$ and, for any $R > 0$, there exist $\delta_R \in \mathbb{R}$ and $\rho > 0$ such that

$$\phi(t,x,v) \geq C_R := e^{\frac{1}{\sigma^2}(\delta_R - \rho T)} \quad \forall t \in [0,T], \, (x,v) \in B(0,R) \subset \mathbb{R}^d \times \mathbb{R}^d. \tag{11}$$

(ii) *Let $\Phi : \mathcal{Y}_0 \to \mathcal{Y}_0$ be the map which associates to ψ the unique solution of (10). Then, if $\psi_2 \leq \psi_1$, we have $\Phi(\psi_2) \geq \Phi(\psi_1)$.*

Proof. We first prove existence of a solution to the nonlinear problem (10) by a fixed point argument exploiting the results for the corresponding linear problem proved in [18]. Fixed $\psi \in \mathcal{Y}_0$, consider the map $F = F(\varphi)$ from $L^2([0,T] \times \mathbb{R}^d \times \mathbb{R}^d)$ into itself that associates with φ the weak solution $\phi \in L^2([0,T] \times \mathbb{R}^d \times \mathbb{R}^d)$ of the linear problem

$$\begin{cases} \partial_t \phi + \frac{\sigma^2}{2}\Delta_v \phi - b(x) \cdot D_v \phi + v \cdot D_x \phi = -\frac{1}{\sigma^2}f(\psi\varphi)\phi \\ \phi(T,x,v) = e^{\frac{u_T(x,v)}{\sigma^2}}. \end{cases} \tag{12}$$

By Prop. A.2 of [18], ϕ belongs to $\mathcal{Y}$ and it coincides with the unique solution of (12) in this space. Moreover, the following estimate

$$\|\phi\|_{L^2([0,T] \times \mathbb{R}_x^d; H^1(\mathbb{R}_v^d))} + \|\partial_t \phi + v \cdot D_x \phi\|_{L^2([0,T] \times \mathbb{R}_x^d; H^{-1}(\mathbb{R}_v^d))} \leq C \tag{13}$$

holds for some constant C which depends only on $\|e^{u_T/\sigma^2}\|_{L^2}, \|f\|_{L^\infty}$ and σ. Hence F maps B_C, the closed ball of radius C of $L^2([0,T] \times \mathbb{R}^d \times \mathbb{R}^d)$, into itself.

To show that the map F is continuous on B_C, consider $\{\varphi_n\}_{n\in\mathbb{N}}, \varphi \in L^2([0,T] \times \mathbb{R}^d \times \mathbb{R}^d)$ such that $\|\varphi_n - \varphi\|_{L^2} \to 0$ and set $\phi_n = F(\varphi_n)$. Then $\phi_n \in \mathcal{Y}$, and, by the estimate (13), we get that, up to a subsequence, there exists $\overline{\phi} \in \mathcal{Y}$ such that $\phi_n \to \overline{\phi}, D_v\phi_n \to D_v\overline{\phi}$

in $L^2([0,T] \times \mathbb{R}^d \times \mathbb{R}^d)$, $\partial_t \phi_n + v \cdot D_x \phi_n \to \partial_t \overline{\phi}_n + v \cdot D_x \overline{\phi}_n$ in $L^2([0,T] \times \mathbb{R}^d_x; H^{-1}(\mathbb{R}^d_v))$. Moreover, $\phi_n \to \phi$ almost everywhere. By the definition of weak solution to (12), we have that

$$\langle \partial_t \phi_n + v \cdot D_x \phi_n, w \rangle - \frac{\sigma^2}{2}(D_v \phi_n, D_v w) - (b \cdot D_v \phi_n, w) = (-\frac{1}{\sigma^2}\phi_n f(\phi_n \psi), w), \quad (14)$$

for any $w \in \mathcal{D}([0,T] \times \mathbb{R}^d \times \mathbb{R}^d)$, the space of infinite differentiable functions with compact support in $[0,T] \times \mathbb{R}^d \times \mathbb{R}^d$. Employing weak convergence for left hand side of (14) and the Dominated Convergence Theorem for the right hand one, we get for $n \to \infty$

$$\langle \partial_t \overline{\phi} + v \cdot D_x \overline{\phi}, w \rangle - \frac{\sigma^2}{2}(D_v \overline{\phi}, D_v w) - (b \cdot D_v \overline{\phi}, w) = (-\overline{\phi} f(\phi \psi), w)$$

for any $w \in \mathcal{D}([0,T] \times \mathbb{R}^d \times \mathbb{R}^d)$. Hence $\overline{\phi} = F(\phi)$ and $F(\phi_n) \to F(\phi)$ for $n \to \infty$ in $L^2([0,T] \times \mathbb{R}^d \times \mathbb{R}^d)$. The compactness of the map F in $L^2([0,T] \times \mathbb{R}^d \times \mathbb{R}^d)$ follows by the compactness of the set of the solutions to (12), see Theorem 1.2 of [20]. We conclude, by Schauder's Theorem, that there exists a fixed-point of the map F in L^2, hence in $\mathcal{Y}$, and therefore a solution to the nonlinear parabolic Equation (10).

Observe that, if ϕ is a solution of (10), then $\tilde{\phi} = e^{\lambda t}\phi$ is a solution of

$$\partial_t \tilde{\phi} + \frac{\sigma^2}{2}\Delta_v \tilde{\phi} - b(x) \cdot D_v \tilde{\phi} + v \cdot D_x \tilde{\phi} - \lambda \tilde{\phi} = -\frac{1}{\sigma^2} f(e^{-\lambda t}\psi \tilde{\phi})\tilde{\phi} \quad (15)$$

with the corresponding final condition. In the following, we assume that $\lambda > 0$. To show that ϕ is non-negative, we will exploit the following property (see Lemma A.3 of [18]): given $\phi \in \mathcal{Y}$ and defined $\phi^\pm = \max(\pm\phi, 0)$, then $\phi^\pm \in \mathcal{X}$ and

$$\langle \partial_t \phi + v \cdot D_x \phi, \phi^- \rangle = \frac{1}{2}\left(\iint |\phi(0,x,v)^-|^2 dx dv - \iint |\phi(T,x,v)^-|^2 dx dv\right). \quad (16)$$

Let ϕ be a solution of (15), multiply the equation by ϕ^- and integrate. Then, since $\phi(T,x,v)$ is non-negative, by (16) we get

$$-\frac{1}{\sigma^2}(\phi f(e^{\lambda t}\phi\psi), \phi^-) = \langle \partial_t \phi + v \cdot D_x \phi, \phi^- \rangle -$$
$$\frac{\sigma^2}{2}(D_v \phi, D_v \phi^-) - (b \cdot D_v \phi, \phi^-) - \lambda(\phi, \phi^-) =$$
$$\frac{1}{2}\iint |\phi(0,x,v)^-|^2 dx dv + \frac{\sigma^2}{2}(D_v \phi^-, D_v \phi^-) + \lambda(\phi^-, \phi^-) \geq$$
$$\lambda(\phi^-, \phi^-),$$

where it has been exploited that, by integration by parts, $(b \cdot D_v \phi, \phi^-) = 0$. Since $f \leq 0$ and therefore

$$-(\phi f(e^{\lambda t}\phi\psi), \phi^-) = (\phi^- f(e^{\lambda t}\phi\psi), \phi^-) \leq 0,$$

we get $(\phi^-, \phi^-) \equiv 0$, hence $\phi \geq 0$.

To prove the uniqueness of the solution to (10), consider two solutions ϕ_1, ϕ_2 of (15) and set $\overline{\phi} = \phi_1 - \phi_2$. Multiplying the equation for $\overline{\phi}$ by $\overline{\phi}$, integrating and using $\overline{\phi}(x,v,T) = 0$, we get

$$-\frac{1}{\sigma^2}(f(e^{-\lambda t}\psi\phi_1)\phi_1 - f(e^{-\lambda t}\psi\phi_2)\phi_2, \phi_1 - \phi_2) = \langle \partial_t \overline{\phi} + v \cdot D_x \overline{\phi}, \overline{\phi} \rangle -$$
$$\frac{\sigma^2}{2}(D_v \overline{\phi}, D_v \overline{\phi}) - (b \cdot D_v \overline{\phi}, \overline{\phi}) - \lambda(\overline{\phi}, \overline{\phi}) = \quad (17)$$
$$-\frac{1}{2}\iint |\overline{\phi}(x,v,0)|^2 dx dv - \frac{\sigma^2}{2}(D_v \overline{\phi}, D_v \overline{\phi}) - \lambda(\overline{\phi}, \overline{\phi}) \leq -\lambda(\phi_1 - \phi_2, \phi_1 - \phi_2)$$

and, by the strict monotonicity of f, we conclude that $\phi_1 = \phi_2$.

To prove that ϕ is bounded from above, we observe that the function $\overline{\phi}(t,x,v) = e^{C_1 + (T-t)\|f\|_\infty / \sigma^2}$, where C_1 as in (9), is a supersolution of the linear problem (12) for any $\varphi \in L^2([0,T] \times \mathbb{R}^d \times \mathbb{R}^d)$, i.e., $\phi(T,x,v) \geq e^{u_T(x,v)/\sigma^2}$ and

$$\partial_t \overline{\phi} + \frac{\sigma^2}{2} \Delta_v \overline{\phi} - b(x) \cdot D_v \overline{\phi} + v \cdot D_x \overline{\phi} \leq -\frac{1}{\sigma^2} f(\psi \varphi) \overline{\phi}.$$

By the Maximum Principle (see Prop. A.3 (i) in [18]), we get that $\overline{\phi} \geq \phi$, where ϕ is the solution of (12). Since the previous property holds for any $\varphi \in L^2([0,T] \times \mathbb{R}^d \times \mathbb{R}^d)$, we conclude that $\overline{\phi} \geq \phi$, where ϕ is the solution of the nonlinear problem (10).

A similar argument show that $\underline{\phi}(x,v,t) = e^{(-C_0(|v|^2 + |x| + 1) - \rho(T-t))/\sigma^2}$, where C_0 as in (9) and ρ sufficiently large, is a subsolution of (12) for any $\varphi \in L^2([0,T] \times \mathbb{R}^d \times \mathbb{R}^d)$. Indeed, replacing $\underline{\phi}$ in the equation, we get that the inequality

$$\partial_t \underline{\phi} + \frac{\sigma^2}{2} \Delta_v \underline{\phi} - b(x) \cdot D_v \underline{\phi} + v \cdot D_x \underline{\phi} =$$
$$= \frac{\underline{\phi}}{\sigma^2} \left(\rho - C_0 d\sigma^2 + 2C_0^2 \sigma^2 |v|^2 + 2C_0 b(x) \cdot v - C_0 v \cdot \frac{x}{|x|} \right) \geq$$
$$- \frac{1}{\sigma^2} f(\psi \varphi) \underline{\phi}$$

is satisfied for ρ large enough and, moreover, $\underline{\phi} \leq e^{u_T(x,v)/\sigma^2}$. Hence $\underline{\phi} \leq \phi$, where ϕ is the solution of the nonlinear problem (10), and, from this estimate, we deduce (11).

We finally prove the monotonicity of the map Φ. Set $\phi_i = \Phi(\psi_i)$, $i = 1, 2$, and consider the equation satisfied by $\overline{\phi} = e^{\lambda t} \phi_1 - e^{\lambda t} \phi_2$, multiply it by $\overline{\phi}^+$ and integrate. Performing a computation similar to (17), we get

$$-\frac{1}{\sigma^2}(f(\phi_1 \psi_1)\phi_1 - f(\phi_2 \psi_2)\phi_2, \overline{\phi}^+) \leq -\lambda(\overline{\phi}^+, \overline{\phi}^+).$$

Since, by monotonicity of f and non-negativity of ϕ_i, we have

$$-(f(\phi_1 \psi_1)\phi_1 - f(\phi_2 \psi_2)\phi_2, \overline{\phi}^+) = -(f(\phi_1 \psi_1)(\phi_1 - \phi_2), \overline{\phi}^+) -$$
$$((f(\phi_1 \psi_1) - f(\phi_2 \psi_2))\phi_2, \overline{\phi}^+) \geq 0,$$

we get $(\overline{\phi}^+, \overline{\phi}^+) = 0$ and therefore $\phi_1 \leq \phi_2$. □

We set
$$\mathcal{Y}_R = \{\phi \in \mathcal{Y}_0 : \phi \geq C_R \quad \forall (x,v) \in B(0,R), \, t \in [0,T]\},$$
where C_R is defined as in (11).

Proposition 3. *Given $R > R_0$, where R_0 as in (8), we have*

(i) *For any $\phi \in \mathcal{Y}_R$, there exists a unique solution $\psi \in \mathcal{Y}_0$ to*

$$\begin{cases} \partial_t \psi - \frac{\sigma^2}{2} \Delta_v \psi - b(x) \cdot D_v \psi + v \cdot D_x \psi = \frac{1}{\sigma^2} f(x,v,\psi\phi)\psi \\ \psi(0,x,v) = \frac{m_0(x,v)}{\phi(0,x,v)}. \end{cases} \quad (18)$$

Moreover

$$\psi(x,v,t) \leq \frac{\|m_0\|_{L^\infty}}{C_R} \quad \forall t \in [0,T], \, (x,v) \in \mathbb{R}^d \times \mathbb{R}^d, \quad (19)$$

where C_R as in (11).

(ii) Let $\Psi : \mathcal{Y}_R \to \mathcal{Y}_0$ be the map which associates with $\phi \in \mathcal{Y}_R$ the unique solution of (18). Then, if $\phi_2 \leq \phi_1$, we have $\Psi(\phi_2) \geq \Psi(\phi_1)$.

Proof. First observe that, since $R > R_0$, then $\psi(0, x, v)$ is well defined for $\phi \in \mathcal{Y}_R$. The proof of the first part of (i) is very similar to the one of the corresponding result in Proposition 2, hence we only prove the bound (19). If ψ is a solution of (18), then $\tilde{\psi} = e^{-\lambda t}\psi$ is a solution of

$$\partial_t \tilde{\psi} - \frac{\sigma^2}{2}\Delta_v \tilde{\psi} - b(x) \cdot D_v \tilde{\psi} + v \cdot D_x \psi + \lambda \tilde{\psi} = \frac{1}{\sigma^2} f(x, v, e^{\lambda t}\tilde{\psi}\phi)\psi. \tag{20}$$

Let ψ be a solution of (20), set $\bar{\psi} = \psi - e^{-\lambda t}\|m_0\|_{L^\infty}/C_R$ and observe that $\bar{\psi}(0) \leq 0$. Multiply the equation for $\bar{\psi}$ by $\bar{\psi}^+$ and integrate to obtain

$$(\psi f(e^{\lambda t}\psi\phi), \bar{\psi}^+) =$$
$$\langle \partial_t \bar{\psi} + v \cdot D_x \bar{\psi}, \bar{\psi}^+\rangle + \frac{1}{\sigma^2}(D_v \bar{\psi}, D_v \bar{\psi}^+) - (b(x)D_v \bar{\psi}, \bar{\psi}^+) + \lambda(\bar{\psi}, \bar{\psi}^+) \geq$$
$$\iint |\bar{\psi}^+(x, v, T)|^2 dx dv + \lambda(\bar{\psi}^+, \bar{\psi}^+) \geq \lambda(\bar{\psi}^+, \bar{\psi}^+).$$

Since $\psi \geq 0$ and $f \leq 0$, we have

$$(\psi f(e^{\lambda t}\psi\phi), \bar{\psi}^+) \leq 0$$

and therefore $\bar{\psi}^+ \equiv 0$. Hence the upper bound (19).

Now we prove (ii). Set $\psi_i = \Psi(\phi_i)$, $i = 1, 2$, and $\bar{\psi} = e^{-\lambda t}\psi_1 - e^{-\lambda t}\psi_2$. Multiply the equation satisfied by $\bar{\psi}$ by $\bar{\psi}^+$ and integrate. Since, by monotonicity and negativity of f, we have

$$(f(e^{\lambda t}\phi_1\psi_1)\psi_1 - f(e^{\lambda t}\phi_2\psi_2)\psi_2, \bar{\psi}^+) = (f(e^{\lambda t}\phi_1\psi_1)(\psi_1 - \psi_2), \bar{\psi}^+) +$$
$$(\psi_2(f(e^{-\lambda t}\phi_1\psi_1) - f(e^{-\lambda t}\phi_2\psi_2)), \bar{\psi}^+) \leq 0.$$

Then

$$0 \geq \langle \partial_t \bar{\psi} + v \cdot D_x \bar{\psi}, \bar{\psi}^+\rangle + \frac{1}{\sigma^2}(D_v \bar{\psi}, D_v \bar{\psi}^+) - (b(x)D_v \bar{\psi}, \bar{\psi}^+) + \lambda(\bar{\psi}, \bar{\psi}^+) \geq$$
$$\iint |\bar{\psi}^+(x, v, T)|^2 dx dv + \lambda(\bar{\psi}^+, \bar{\psi}^+) \geq \lambda(\bar{\psi}^+, \bar{\psi}^+).$$

Hence $\bar{\psi}^+ \equiv 0$ and therefore $\psi_1 \leq \psi_2$. □

Proof of Theorem 1. Given $\psi^{(0)} \equiv 0$, consider the sequence $(\phi^{(k+\frac{1}{2})}, \psi^{(k+1)})$, $k \in \mathbb{N}$, defined in (5) and (6). It can rewritten as

$$\begin{cases} \phi^{(k+\frac{1}{2})} = \Phi(\psi^{(k)}) \\ \psi^{(k+1)} = \Psi(\phi^{(k+\frac{1}{2})}) \end{cases} \tag{21}$$

where the maps Φ, Ψ are as in Propositions 2 and, respectively 3. Observe that, by (11), we have $\phi^{(k+\frac{1}{2})} \in \mathcal{Y}_R$ for $R > R_0$ and $\psi^{(k+1)} \geq 0$ for any k. Hence the sequence $(\phi^{(k+\frac{1}{2})}, \psi^{(k+1)})$ is well defined. We first prove by induction the monotonicity of the components of $(\phi^{(k+\frac{1}{2})}, \psi^{(k+1)})$. By non-negativity of solutions to (18), we have $\psi^{(1)} = \Phi(\phi^{(\frac{1}{2})}) \geq 0$ and therefore $\psi^{(1)} \geq \psi^{(0)}$. Moreover, by the monotonicity of Φ, $\phi^{(\frac{3}{2})} = \Phi(\psi^{(1)}) \leq \Phi(\psi^{(0)}) = \phi^{(\frac{1}{2})}$. Now assume that $\psi^{(k+1)} \geq \psi^{(k)}$. Then

$$\phi^{(k+\frac{3}{2})} = \Phi(\psi^{(k+1)}) \leq \Phi(\psi^{(k)}) = \phi^{(k+\frac{1}{2})}$$

and
$$\psi^{(k+2)} = \Psi(\phi^{(k+\frac{3}{2})}) \geq \Psi(\phi^{(k+\frac{1}{2})}) = \psi^{(k+1)},$$
therefore the monotonicity of two sequences.

Since $\phi^{(k+\frac{1}{2})} \geq 0$ and, by (19), for $k \to \infty$, the sequence $\psi^{(k+1)} \leq \|m_0\|_{L^\infty}/C_R$, $(\phi^{(k+\frac{1}{2})}, \psi^{(k+1)})$ converges a.e. and in $L^2([0,T] \times \mathbb{R}^d \times \mathbb{R}^d)$ to a couple (ϕ, ψ). Taking into account the estimate (13), the a.e. convergence of the two sequences and repeating an argument similar to the one employed for the continuity of the map F in Proposition 2, we get that the couple (ϕ, ψ) satisfies, in weak sense, the first two equations in (4). The terminal condition for ϕ is obviously satisfied, while the initial condition for ψ, in L^2 sense, follows by convergence of $\phi^{(k+\frac{1}{2})}(0)$ to $\phi(0)$.

We now consider the couple (u, m) given by the change of variable in (7). We first observe that, by Theorem 1.5 of [10], we have $\partial_t \phi + v \cdot D_x \phi$, $D_v \phi$, $\Delta_v \phi \in L^2([0,T] \times \mathbb{R}^d \times \mathbb{R}^d)$ and a corresponding regularity for ψ. Taking into account the boundedness of ϕ and the estimate in (11), we have that $u, \partial_t u + v \cdot D_x u, D_v u, \Delta_v u \in L^2_{loc}([0,T] \times \mathbb{R}^d \times \mathbb{R}^d)$. Hence we can write the equation for u in weak form, i.e.,

$$(\partial_t u + v \cdot D_x u, w) - \frac{\sigma^2}{2}(D_v u, D_v w) - (b \cdot D_v u, w) + \frac{1}{2}(|D_v u|^2, w) = -(f(m), w),$$

for any $w \in \mathcal{D}([0,T] \times \mathbb{R}^d \times \mathbb{R}^d)$, with final datum in trace sense. In a similar way, since $m, \partial_t m + v \cdot D_x m, D_v m, \Delta_v m \in L^2_{loc}([0,T] \times \mathbb{R}^d \times \mathbb{R}^d)$ and m is locally bounded, we can rewrite also the equation for m in weak form, i.e.,

$$(\partial_t m + v \cdot D_x m, w) + \frac{\sigma^2}{2}(D_v m, D_v w) - (b \cdot D_v m, w) - (m D_v u, Dw) = 0,$$

for any $w \in \mathcal{D}([0,T] \times \mathbb{R}^d \times \mathbb{R}^d)$ with the initial datum in trace sense. □

Funding: This research received no external funding.

Acknowledgments: The author wishes to thank Alessandro Goffi (Univ. di Padova) and Sergio Polidoro (Univ. di Modena e Reggio Emilia) for useful discussions.

Conflicts of Interest: The author declares no conflict of interest.

References

1. Huang, M.; Caines, P.E.; Malhame, R.P. Large-population cost-coupled LQG problems with non uniform agents: Individual-mass behaviour and decentralized ε-Nash equilibria. *IEEE Trans. Autom. Control* **2007**, *52*, 1560–1571. [CrossRef]
2. Lasry, J.-M.; Lions, P.-L. Mean field games. *Jpn. J. Math.* **2007**, *2*, 229–260. [CrossRef]
3. Achdou, Y.; Mannucci, P.; Marchi, C.; Tchou, N. Deterministic mean field games with control on the acceleration. *Nodea Nonlinear Differ. Eq. Appl.* **2020**, *27*, 33. [CrossRef]
4. Bardi, M.; Cardaliaguet, P. Convergence of some Mean Field Games systems to aggregation and flocking models. *arXiv* **2004**, arXiv:2004.04403.
5. Cannarsa, P.; Mendico, C. Mild and weak solutions of Mean Field Games problem for linear control systems. *Minimax Theory Appl.* **2020**, *5*, 221–250.
6. Kolmogoroff, A. Zufällige Bewegungen (zur Theorie der Brownschen Bewegung). *Ann. Math.* **1934**, *35*, 116–117. [CrossRef]
7. Hörmander, L. Hypoelliptic second order differential equations. *Acta Math.* **1967**, *119*, 147–171. [CrossRef]
8. Lanconelli, E.; Polidoro, S. On a class of hypoelliptic evolution operators. *Rend. Sem. Mat. Univ. Politec. Torino* **1994**, *52*, 29–63
9. Armstrong, S.; Mourrat, J.-C. Variational methods for the kinetic Fokker-Planck equation. *arXiv* **1902**, arXiv:1902.04037.
10. Bouchut, F. Hypoelliptic regularity in kinetic equations. *J. Math. Pures Appl.* **2002**, *81*, 1135–1159. [CrossRef]
11. Guéant, O.; Lasry, J.; Lions, P. Mean field games and applications. In *Paris-Princeton Lectures on Mathematical Finance 2010*; Lecture Notes in Math; Springer: Berlin, Germany, 2011; Volume 2003; pp. 205–266.
12. Gomes, D.A.; Mitake, H. Existence for stationary mean-field games with congestion and quadratic Hamiltonians. *Nodea Nonlinear Differ. Eq. Appl.* **2015**, *22*, 1897–1910. [CrossRef]
13. Gomes, D.A.; Pimentel, E.A.; Voskanyan, V. *Regularity Theory for Mean-Field Game Systems*; Springer Briefs in Mathematics; Springer: Berlin, Germany, 2016.

14. Guéant, O. Mean field games equations with quadratic Hamiltonian: A specific approach. *Math. Models Methods Appl. Sci.* **2012**, *22*, 37. [CrossRef]
15. Ullmo, D.; Swiecicki, I.; Gobron, T. Quadratic mean field games. *Phys. Rep.* **2019**, *799*, 1–35. [CrossRef]
16. Feleqi, E.; Gomes, D.; Tada, T. Hypoelliptic mean field games—A case study. *Minimax Theory Appl.* **2020**, *5*, 305–326.
17. Carmona, R.; Delarue, F. *Probabilistic Theory of Mean Field Games with Applications. I Mean Field FBSDEs, Control, and Games*; Probability Theory and Stochastic Modelling, 83; Springer: Cham, Switzerland, 2018.
18. Degond, P. Global existence of smooth solutions for the Vlasov-Fokker-Planck equation in 1 and 2 space dimensions. *Ann. Sci. École Norm. Sup.* **1986**, *19*, 519–542. [CrossRef]
19. Cardaliaguet, P.; Graber, P.J.; Porretta, A.; Tonon, D. Second order mean field games with degenerate diffusion and local coupling. *Nodea Nonlinear Differ. Eq. Appl.* **2015**, *22*, 1287–1317. [CrossRef]
20. Camellini, F.; Eleuteri, M.; Polidoro, S. A compactness result for the Sobolev embedding via potential theory. *arXiv* **1806**, arXiv:1806.03606.

Article

Non-Standard Discrete RothC Models for Soil Carbon Dynamics

Fasma Diele [1],*, Carmela Marangi [1] and Angela Martiradonna [1,2]

[1] Istituto per Applicazioni del Calcolo Mauro Picone, CNR, via Amendola 122/D, 70126 Bari, Italy; c.marangi@ba.iac.cnr.it (C.M.); a.martiradonna@ba.iac.cnr.it (A.M.)
[2] Department of Mathematics, University of Bari, via Orabona 4, 70125 Bari, Italy
* Correspondence: f.diele@ba.iac.cnr.it

Abstract: Soil Organic Carbon (SOC) is one of the key indicators of land degradation. SOC positively affects soil functions with regard to habitats, biological diversity and soil fertility; therefore, a reduction in the SOC stock of soil results in degradation, and it may also have potential negative effects on soil-derived ecosystem services. Dynamical models, such as the Rothamsted Carbon (RothC) model, may predict the long-term behaviour of soil carbon content and may suggest optimal land use patterns suitable for the achievement of land degradation neutrality as measured in terms of the SOC indicator. In this paper, we compared continuous and discrete versions of the RothC model, especially to achieve long-term solutions. The original discrete formulation of the RothC model was then compared with a novel non-standard integrator that represents an alternative to the exponential Rosenbrock–Euler approach in the literature.

Keywords: soil organic carbon; RothC; non-standard integrators; Exponential Rosenbrock–Euler

MSC: 34C60; 65L05; 65D30

Citation: Diele, F.; Marangi, C.; Martiradonna, A. Non-Standard Discrete RothC Models for Soil Carbon Dynamics. *Axioms* **2021**, *10*, 56. https://doi.org/10.3390/axioms10020056

Academic Editors: Ioannis Dassios and Clemente Cesarano

Received: 5 February 2021
Accepted: 17 March 2021
Published: 8 April 2021

Publisher's Note: MDPI stays neutral with regard to jurisdictional claims in published maps and institutional affiliations.

Copyright: © 2021 by the authors. Licensee MDPI, Basel, Switzerland. This article is an open access article distributed under the terms and conditions of the Creative Commons Attribution (CC BY) license (https://creativecommons.org/licenses/by/4.0/).

1. Introduction

The United Nations Convention to Combat Desertification (UNCCD) is an international agreement, established in 1994, that links the environment and development with sustainable land management. The first objective indicated in the UNCCD 2018–2030 Strategic Framework is to improve the conditions of affected ecosystems, combat desertification/land degradation and promote sustainable land management [1]. For each country, the commitment is to achieve no net loss of land-based natural capital by 2030 [2]. No net loss means that the quantity and quality of land-based natural capital are maintained or increased, despite the impacts of global environmental change, whether due to human or natural causes. Land degradation is monitored through the changes of the values of a specific set of consistently measured indicators from their baseline quantities, conventionally identified as their initial values. The deviations from the baseline values of these indicators are the basis for monitoring land degradation.

Soil Organic Carbon (SOC) is one key indicator of land degradation [3]. Monitoring the SOC stocks and the loss of soil organic carbon due to land use changes is fundamental for maintaining the physical, chemical and biological quality of soil [4]. Soil organic carbon positively affects soil functions with regard to habitats, biological diversity, soil fertility, crop production potential, erosion control and water retention. A high SOC content improves the processes of soil formation, nutrient storage, water holding capacity and the absorption of organic or inorganic pollutants. Thus, a reduction of the SOC stock not only indicates soil degradation, but may also have potential negative effects on soil-derived ecosystem services.

Starting from the SOC baseline, predictive spatial modelling can simulate the carbon dynamics, estimate carbon sequestration under the actual land use and evaluate the deviation from the baseline average value of total carbon [5]. Moreover, a dynamical model may determine the optimal potential land use pattern that is suitable to achieve land degradation neutrality in terms of SOC indicator values. Well-validated models, such as Rothamsted Carbon (RothC) [6], CENTURY [7] and MOMOS [8], which take into account the interactions among the climate, pedology, cropping systems and soil and crop management, can be used to predict SOC changes under different management practices and climatic conditions. These models are essentially compartmental, meaning that they represent soil organic matter as a few discrete compartments (generally two to five) characterized by different chemical characteristics of the soil's degradation. The decomposition rates, applied to each compartment, are governed by kinetic and stoichiometric laws and are mainly ruled by the environmental conditions (e.g., soil moisture level, aeration and soil temperature). These models are used in a variety of ways and often for long-term studies. Indeed, being able to compute and predict long-term solutions is extremely valuable for various reasons: it gives a synthetic view of the system in the given agro-climatic conditions; it makes it possible to test if a studied soil has reached equilibrium or not; and it allows envisioning what would be the consequences of specific events on a given soil [9]. In this paper, we compared continuous and discrete versions of the RothC model, especially regarding long-term solutions. Moreover, since the discrete RothC version can be interpreted as a first-order approximation of the continuous model, we introduced a non-standard discrete approximation that can be interpreted as a novel discrete model of soil carbon dynamics. The original discrete formulation of the RothC model was then compared to the novel non-standard integrator, which represents a different approach with respect to the Exponential Rosenbrock–Euler (ERE) discretization [10].

2. Rothamsted Carbon Model—Continuous Formulation

The SOC indicator is considered to be the result of the equilibrium between the inputs and outputs of the soil system. SOC contained in the organic matter is constantly built up and decomposed and is then released into the atmosphere as CO_2 and recaptured through photosynthesis. Inputs in the soil organic matter decomposition model consist of two major components: living organisms' biomass (mainly plant roots and microbial biomass) and plant and animal residues at various stages of decomposition. Outputs result from the heterotrophic respiration processes when soil organic carbon is used as an energy source by soil organisms and returned to the atmosphere as CO_2 fluxes. The Rothamsted Carbon model [6,11] (RothC) is a model of carbon turnover in non-waterlogged soils [12]. Initially developed for arable soils, it was later expanded to grasslands and forests. It takes into account the effects of temperature, moisture content and soil type. The RothC model divides the soil carbon into four active compartments and one inactive, characterized by different chemical decomposition rates of degradability (see Figure 1). The Inert Organic Matter (IOM) represents the inactive pool, resistant to decomposition, which does not receive carbon (C) inputs. At each time step, incoming plant residues are split between easily Decomposable Plant Material (DPM) and Resistant Plant Material (RPM), depending on the ratio $\frac{\gamma}{1-\gamma}$, which estimates the decomposability of the particular plant material inputs, which in turn depends on the specific cultivation being considered. The fraction 2η of the input of Farmyard Manure (FYM), if any, is equally split between the DPM and RPM compartments; the remaining part $1 - 2\eta$ enters in the system directly as Humified organic matter (HUM). Both DPM and RPM decompose to form CO_2, microbial Biomass (BIO) and more HUM. The fraction $\alpha + \beta$ of metabolised C incorporated into the sum of compartments BIO+HUM is determined by the clay content of the soil, while the remaining part $1 - \alpha - \beta$ is released as CO_2 and lost by the system. The BIO+HUM carbon content is then split into $\frac{\alpha}{\alpha + \beta}$ percent BIO and $\frac{\beta}{\alpha + \beta}$ percent HUM. Finally, both BIO and HUM decompose to form more CO_2, BIO and HUM.

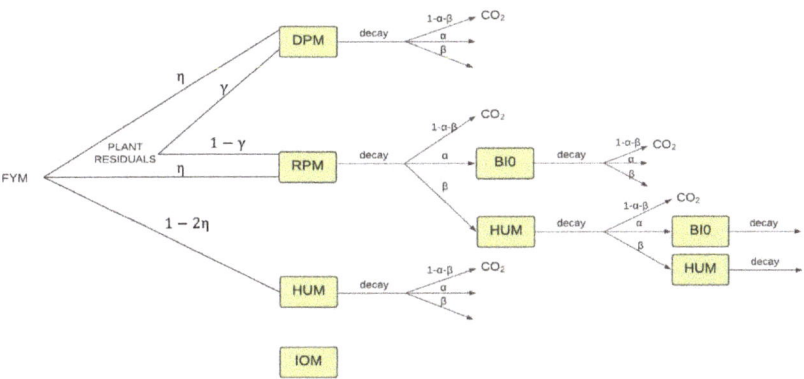

Figure 1. Flow diagram of the Rothamsted Carbon (RothC) model. FYM, Farmyard Manure; DPM, Decomposable Plant Material; RPM, Resistant Plant Material; BIO, microbial Biomass; HUM, Humified organic matter; IOM, Inert Organic Matter.

The four active compartments undergo decomposition as a function of different rate constants, which correspond to the entries of the vector $\mathbf{k} = [k_{dpm}, k_{rpm}, k_{bio}, k_{hum}]$ and of the rate modifier $\rho(t)$, which depends on the clay content of the soil, on climatic variables (rainfall, temperature, open pan evaporation) and land cover.

In real soil systems, processes involved in the RothC model are continuous in time, and thus, in [10], the author proposed the following continuous formulation:

$$\dot{\mathbf{c}} = \rho(t)\, A\, \mathbf{c} + \mathbf{b}(t), \qquad \mathbf{c}(t_0) = \mathbf{c}_0 \geq 0 \qquad (1)$$

where $\mathbf{c}(t) = [c_{dpm}(t), c_{rpm}(t), c_{bio}(t), c_{hum}(t)]^T$ and $\mathbf{c}_0 \geq 0$ denotes the vector of the initial concentrations. The matrix A is given by:

$$A = \begin{pmatrix} -k_{dpm} & 0 & 0 & 0 \\ 0 & -k_{rpm} & 0 & 0 \\ \alpha\, k_{dpm} & \alpha\, k_{rpm} & (\alpha-1)\, k_{bio} & \alpha\, k_{hum} \\ \beta\, k_{dpm} & \beta\, k_{rpm} & \beta\, k_{bio} & (\beta-1)\, k_{hum} \end{pmatrix}.$$

The vector $\mathbf{b}(t)$ represents the carbon amount entering the system at time t. It considers both the input of plant residues $g(t)\,\mathbf{a}^{(g)}$ and the input of FYM $f(t)\,\mathbf{a}^{(f)}$, so that:

$$\mathbf{b}(t) := g(t)\,\mathbf{a}^{(g)} + f(t)\,\mathbf{a}^{(f)}.$$

The entries of vectors $\mathbf{a}^{(g)} := [\gamma, 1-\gamma, 0, 0]^T$ and $\mathbf{a}^{(f)} := [\eta, \eta, 0, 1-2\eta]^T$ are the fraction inputs $0 \leq \gamma \leq 1$, $0 \leq \eta \leq 1/2$, which sum up to one; the carbon of plant residues enters the soil only through the DPM and RPM compartments, while the carbon amount of FYM enters through the DPM, RPM and HUM compartments.

It can be shown that, under the hypothesis $0 < \alpha + \beta \leq 1$ and $\rho(t) > 0$, the solution of the homogeneous part of the RothC system (1) with non-negative initial data verifies the more general assumptions stated in [13,14] for biochemical systems, which ensure the well-posedness and positivity of the solution. Comparison theorems guarantee the positivity of the solution of the complete RothC system when $g(t)$, $f(t)$ are assumed positive.

Definition 1. *We define as the SOC indicator of the continuous RothC model (1) the function $SOC(t) = c_{iom} + c_{dpm}(t) + c_{rpm}(t) + c_{bio}(t) + c_{hum}(t)$ for $t \geq t_0$, where c_{iom} denotes the constant carbon content in the compartment IOM.*

Theorem 1. *Let $\alpha, \beta > 0$ and $\delta := 1 - \alpha - \beta > 0$; suppose $0 < g(t) + f(t) \leq B$ and $\rho(t)$ are uniformly bounded from below by a constant $\mu > 0$. The set:*

$$\Omega = \left\{ (c_{dpm}, c_{rpm}, c_{bio}, c_{hum}) \in R_+^4 : 0 \leq (c_{dpm} + c_{rpm} + c_{bio} + c_{hum}) \leq \frac{B}{\mu \delta k_{min}} \right\}$$

with $k_{min} = \min_i \mathbf{k}(i)$ is positively invariant and globally attractive for Model (1).

Proof. By denoting $\omega(t) = SOC(t) - c_{iom}$, from System (1), and recalling that $\mathbf{e}^T \mathbf{a}^{(g)} = \mathbf{e}^T \mathbf{a}^{(f)} = 1$, we can show that:

$$\dot{\omega} = \mathbf{e}^T \dot{\mathbf{c}} = \rho(t) \mathbf{e}^T A \mathbf{c} + g \mathbf{e}^T \mathbf{a}^{(g)} + f \mathbf{e}^T \mathbf{a}^{(f)} = g(t) + f(t) - \rho(t) \delta \mathbf{k}^T \mathbf{c} \quad (2)$$

where $\mathbf{e} = [1, 1, 1, 1]^T$. Consequently,

$$\dot{\omega} \leq g(t) + f(t) - \rho(t) \delta k_{min} \omega \leq B - \mu \delta k_{min} \omega$$

so that:

$$\omega(t) \leq \frac{B}{\mu \delta k_{min}} - \left[\frac{B}{\mu \delta k_{min}} - \omega(t_0) \right] e^{-\mu \delta k_{min}(t-t_0)}.$$

Therefore, if $\omega(t_0) \leq \frac{B}{\mu \delta k_{min}}$, then $\omega(t) \leq \frac{B}{\mu \delta k_{min}}$, with Ω, resulting in a positively invariant set for model (1). Moreover, if $\omega(t_0) \geq \frac{B}{\mu \delta k_{min}}$, then $\lim_{t \to \infty} \omega(t) \leq \frac{B}{\mu \delta k_{min}}$ and Ω is globally attractive for Model (1). □

Under the hypothesis of Theorem 1, the set:

$$\Omega_{SOC} = \left\{ SOC \in R_+, \quad c_{iom} \leq SOC \leq c_{iom} + \frac{B}{\mu \delta k_{min}} \right\}$$

is globally attractive for the SOC model:

$$\dot{SOC} = \mathbf{e}^T \dot{\mathbf{c}} = \rho(t) \mathbf{e}^T A \mathbf{c} + g(t) + f(t). \quad (3)$$

Finally, to complete the understanding of the mathematical features of the RothC model, we provide the following theorem:

Theorem 2. *Suppose $g(t)$ and $f(t)$ are integrable on every finite subinterval of $[t_0, +\infty)$. If $\alpha + \beta = 1$, then $SOC(t) = SOC(t_0) + \int_{t_0}^t (g(s) + f(s))\, ds$.*

Proof. Under the assumption $\alpha + \beta = 1$, the unit vector $\mathbf{e} \in \ker(A^T)$. We have $\dot{SOC} = \mathbf{e}^T \dot{\mathbf{c}} = \rho \mathbf{e}^T A \mathbf{c} + g \mathbf{e}^T \mathbf{a}^{(g)} + f \mathbf{e}^T \mathbf{a}^{(f)} = g + f$. □

Long-Term Solutions

When the RothC model is used in real application, the first step is to run the model to equilibrium to calculate the required carbon inputs needed to match the initial SOC content measured. Hence, being able to compute and predict the long-term solution are extremely valuable in order to avoid long-run simulations, which can lead to numerical artefacts if numerical tools are not properly used [15].

We firstly consider the (unrealistic) case when no CO_2 release (i.e., $\alpha + \beta = 1$) is considered.

Theorem 2 indicates that:

- if there is no external carbon input ($g = f = 0$), then the SOC indicator is a linear invariant for the SOC model (3);
- if the carbon input $g + f$ is replaced by its average value $\hat{g} + \hat{f}$, then $SOC(t)$ grows linearly in time;
- if $g(t)$ and $f(t)$ are integrable on every finite subinterval of $[t_0, +\infty)$ and the improper integral $\int_{t_0}^{+\infty}(g(s) + f(s))\,ds$ converges to its value C_∞, the indicator $SOC_R(t)$ increases in time tending to the finite amount $SOC^* := SOC(t_0) + C_\infty$ (steady-state solution);
- if both $g(t)$ and $f(t)$ are integrable on every finite subinterval of $[t_0, +\infty)$ and $\int_{t_0}^{t}(g(s) + f(s))\,ds$ is a bounded periodic function of period T, then $SOC(t)$ oscillates indefinitely with period T (periodic solution).

Theorem 3. *Suppose $\alpha, \beta, \delta > 0$ with $\alpha + \beta < 1$ and $\mathbf{k}(i) > 0$, $i = 1, \ldots 4$. For $\rho, g, f,$ not varying with time, the RothC model admits a unique positive globally stable equilibrium, which has the following expression:*

$$\mathbf{c}^*_{cont} = \frac{1}{\rho}\left[\frac{g\gamma + f\eta}{k_{dpm}}, \frac{g(1-\gamma) + f\eta}{k_{rpm}}, \frac{(g+f)\alpha}{k_{bio}\,\delta}, \frac{f(1-\alpha)(1-2\eta) + \beta(g+2f\eta)}{k_{hum}\,\delta}\right]^\mathsf{T}. \quad (4)$$

Consequently, the SOC indicator has $SOC^*_{cont} = c_{iom} + \mathbf{e}^\mathsf{T}\mathbf{c}_{cont}$ as the equilibrium, which satisfies $SOC^*_{cont} \leq c_{iom} + \dfrac{f+g}{\rho\,k_{min}\,\delta}$.

Proof. From the assumptions, it follows that $0 < \alpha, \beta < 1$. The eigenvalues of the matrix A are given by $\lambda_1 = -k_{dpm} < 0$, $\lambda_2 = -k_{rpm} < 0$, and $\lambda_{3,4}$ are the roots of the second0order polynomial:

$$\lambda^2 + \lambda\,(k_{bio}(1-\alpha) + k_{hum}(1-\beta)) + \delta\,k_{bio}\,k_{hum}$$

with discriminant $\Delta(\lambda) = (\alpha - 1)^2 k_{bio}^2 + (\beta - 1)^2 k_{hum}^2 + 2k_{bio}\,k_{hum}(\alpha(\beta + 1) + \beta - 1) > 0$. Consequently, from Descartes' rule of sign, $\lambda_3, \lambda_4 < 0$. The matrix A turns out to be negative definite and admits an inverse. It is trivial to check that $\det(A) = \delta \prod_{i=1}^{4}\mathbf{k}(i) > 0$.

For constant (positive) values of $\rho, g, f,$ the equilibrium is achieved by solving the linear system $\rho A\mathbf{c} = -\mathbf{b}$, i.e.,

$$\mathbf{c}^*_{cont} = -\frac{1}{\rho} A^{-1}\mathbf{b}, \quad (5)$$

with:

$$-A^{-1} = \frac{1}{\det(A)}\begin{pmatrix} \delta\,k_{2,3,4} & 0 & 0 & 0 \\ 0 & \delta\,k_{1,3,4} & 0 & 0 \\ \alpha\,k_{1,2,4} & \alpha\,k_{1,2,4} & (1-\beta)\,k_{1,2,4} & \alpha\,k_{1,2,4} \\ \beta\,k_{1,2,3} & \beta\,k_{1,2,3} & \beta\,k_{1,2,3} & (1-\alpha)\,k_{1,2,3} \end{pmatrix}$$

where we adopted the convention that $k_{i,j,m} := \mathbf{k}(i)\,\mathbf{k}(j)\,\mathbf{k}(m)$ with $i, j, m \in \{1, 2, 3, 4\}$.

Finally, the stability of the equilibrium is guaranteed as A is negative definite.

As concerns the equilibrium of the SOC indicator, first notice that, from (2), it results that $\mathbf{k}^\mathsf{T}\mathbf{c}^*_{cont} = \dfrac{f+g}{\rho\,\delta}$, then,

$$SOC^*_{cont} = c_{iom} + \mathbf{e}^\mathsf{T}\mathbf{c}^*_{cont} \leq c_{iom} + \frac{1}{k_{min}}\mathbf{k}^\mathsf{T}\mathbf{c}^*_{cont} = c_{iom} + \frac{f+g}{k_{min}\,\rho\,\delta}.$$

□

When climatic and agricultural variables are considered, the functions $\rho(t)$, $g(t)$ and $f(t)$ are chosen to be time varying on a periodical basis, and the system defined by $\mathbf{c}(t)$ is expected to tend toward an oscillatory state as $t \to +\infty$. If we introduce $\xi(t) := \int_{t_0}^{t} \rho(s)\,ds$ for all $t \geq t_0$, then the solution of (1) is given by:

$$\mathbf{c}(t) = e^{\xi(t)A}\mathbf{c}_0 + e^{\xi(t)A}\int_{t_0}^{t} e^{-\xi(s)A}\mathbf{b}(s)\,ds. \tag{6}$$

The study of the eigenvalues of $\xi(t)A$ enables characterizing the solution behaviour. We can first observe that if the eigenvalues of $A\xi(t)$ are negative, $\mathbf{c}(t)$ as $t \to +\infty$ does not depend on initial conditions $\mathbf{c}_0$. Secondly, if there exists a periodic solution $\mathbf{c}(t)$ with period T, then $\mathbf{c}(t_0 + T) = \mathbf{c}_0$, and the following theorem holds:

Theorem 4. *Assume that $\rho(t)$, $g(t)$ and $f(t)$ are periodic with period T. If $\alpha + \beta \neq 1$ and $\mathbf{k}(i) \neq 0$ for all $i \in [dpm, rpm, bio, hum]$, then $I - e^{\xi(t_0+T)A}$ is not singular. Starting from:*

$$\mathbf{c}_0 := (I - e^{\xi(t_0+T)A})^{-1} e^{\xi(t_0+T)A} \int_{t_0}^{t_0+T} e^{-\xi(s)A}\mathbf{b}(s)\,ds$$

the RothC model admits a (unique) periodic solution.

Proof. From the periodicity of $\rho(t)$, it follows that:

$$\xi(t+T) = \int_{t_0}^{t+T} \rho(s)ds = \int_{t_0}^{t_0+T} \rho(s)ds + \int_{t_0+T}^{t+T} \rho(s)ds = \xi(t_0+T) + \xi(t).$$

By imposing in (6) that $\mathbf{c}(t_0 + T) = \mathbf{c}_0$, it follows that:

$$\mathbf{c}_0 = e^{\xi(t_0+T)A}\left(\mathbf{c}_0 + \int_{t_0}^{t_0+T} e^{-\xi(s)A}\mathbf{b}(s)\,ds\right). \tag{7}$$

Under the assumed hypothesis, A is not singular, and consequently, the matrix $I - e^{\xi(t_0+T)A}$ has the inverse. Exploiting the periodicity of $\mathbf{b}(t)$ inherited by the periodicity of both $g(t)$ and $f(t)$, we have that:

$$\begin{aligned}
\mathbf{c}(t+T) &= e^{\xi(t+T)A}\left(\mathbf{c}_0 + \int_{t_0}^{t+T} e^{-\xi(s)A}\mathbf{b}(s)\,ds\right) \\
&= e^{\xi(t)A}e^{\xi(t_0+T)A}\left(\mathbf{c}_0 + \int_{t_0}^{t_0+T} e^{-\xi(s)A}\mathbf{b}(s)\,ds + \int_{t_0+T}^{t+T} e^{-\xi(s)A}\mathbf{b}(s)\,ds\right) \\
&= e^{\xi(t)A}\mathbf{c}_0 + e^{\xi(t)A}e^{\xi(t_0+T)A}\int_{t_0}^{t} e^{-\xi(s+T)A}\mathbf{b}(s)\,ds \\
&= e^{\xi(t)A}\mathbf{c}_0 + e^{\xi(t)A}\int_{t_0}^{t} e^{-\xi(s)A}\mathbf{b}(s)\,ds \\
&= \mathbf{c}(t).
\end{aligned}$$

This proves the periodicity of $\mathbf{c}(t)$. □

The condition $\alpha + \beta < 1$ is always true for the RothC model due to the definition of α and β (see [6]), and thus, the stock of carbon in each compartment tends towards a periodic solution in large times whatever the input values are.

Although the continuous approach (1) gives a simple explicit solution, the function of both the input variables and the parameters of the model, in real applications, first-order discretized versions are applied. We analyse the original discrete formulation of the RothC model, the Exponential Rosenbrock–Euler (ERE) version and a novel first-order non-standard procedure, closer to the classical discrete RothC procedure than the ERE model.

3. Non-Standard RothC Discrete Models
3.1. Original Discrete RothC

By denoting with I the identity matrix and with

$$\Lambda = \begin{pmatrix} 0 & 0 & 0 & 0 \\ 0 & 0 & 0 & 0 \\ \alpha & \alpha & \alpha & \alpha \\ \beta & \beta & \beta & \beta \end{pmatrix}, \quad D = \begin{pmatrix} k_{dpm} & 0 & 0 & 0 \\ 0 & k_{rpm} & 0 & 0 \\ 0 & 0 & k_{bio} & 0 \\ 0 & 0 & 0 & k_{hum} \end{pmatrix}$$

the discrete (monthly) formulation of the RothC Version 26.3 [11] in vectorial form is given by:

$$\mathbf{c}_{n+1} = \left(\Lambda + (I - \Lambda) e^{-\Delta t \rho(t_n) D}\right) \mathbf{c}_n + \Delta t \, \mathbf{b}(t_n) \tag{8}$$

where $\mathbf{c}_n \approx \mathbf{c}(t_n)$ and the discrete temporal grid $t_{n+1} = t_0 + n \Delta t$ advances with stepsize $\Delta t = 1$.

Discrete RothC has been applied using data from long-term experiments across several ecosystems, climate conditions and Land Use (LU) classes. It has been extensively applied in Europe for SOC modelling, and applications of RothC to a long-term experiment in semi-arid conditions in Italy can be found in [16,17].

In the case when $g(t)$, $f(t)$ and $\rho(t)$ do not depend on time t, then $\mathbf{b} = g \mathbf{a}^{(g)} + f \mathbf{a}^{(f)}$; by setting $F(\Delta t) = \Lambda + (I - \Lambda) e^{-\Delta t \rho D}$, one can demonstrate that, whenever $\mathbf{k}(i) \neq 0$, $(I - F(\Delta t))$ has an inverse, and the system yields a steady-state solution.

Theorem 5. *The steady-state solution $\mathbf{c}^*_{cont}$ in (5) of the continuous model (1) and the steady-state solution $\mathbf{c}^*_{RothC}(\Delta t)$ of the discrete RothC model (8) satisfy the following relation:*

$$\mathbf{c}^*_{cont} = \varphi(-\Delta t \rho D) \, \mathbf{c}^*_{RothC}(\Delta t)$$

where $\varphi(z) = z^{-1}(e^z - 1)$.

Proof. Firstly, evaluate $I - F(\Delta t) = (\Lambda - I)(e^{-\Delta t \rho D} - I)$. From $\mathbf{c}^*_{RothC} = F(\Delta t) \mathbf{c}^*_{RothC} + \Delta t \, \mathbf{b}$, one can write:

$$\begin{aligned} \mathbf{c}^*_{RothC}(\Delta t) &= \Delta t \, (I - F(\Delta t))^{-1} \, \mathbf{b} \\ &= \Delta t \, (e^{-\Delta t \rho D} - I)^{-1} (\Lambda - I)^{-1} \, \mathbf{b} \end{aligned} \tag{9}$$

From the definition of $\mathbf{c}^*_{cont}$ in (5) and by noticing that the matrix A that defines the continuous problem (1) verifies the relation $A = (\Lambda - I) D$, it follows that:

$$\mathbf{c}^*_{RothC}(\Delta t) = \varphi(-\Delta t \rho D)^{-1} \mathbf{c}^*_{cont}$$

□

As the original discrete RothC model is applied with $\Delta t = 1$, we can estimate the deviation of the equilibrium of the continuous model with respect to the equilibrium $\mathbf{c}^*_{RothC}(1)$ of the original discrete RothC model. Given the matrix norm $\|\cdot\|$ induced by a vector norm, as $-\rho D$ is negative definite, it results that $\|\varphi(-\rho D)\| \leq \|I\| = 1$ and $\|\mathbf{c}^*_{cont}\| \leq \|\varphi(-\rho D) \, \mathbf{c}^*_{RothC}(1)\| \leq \|\mathbf{c}^*_{RothC}(1)\|$. This indicates that the equilibrium $\mathbf{c}^*_{RothC}(1)$ is, in norm, an overestimation of the theoretical equilibrium $\mathbf{c}^*_{cont}$. The relative error depends on the stepsize Δt according to:

$$\frac{\|\mathbf{c}^*_{RothC}(\Delta t) - \mathbf{c}^*_{cont}\|}{\|\mathbf{c}^*_{cont}\|} \leq \|\varphi(-\Delta t \rho D)^{-1} - I\|. \tag{10}$$

As concerns the equilibrium of the SOC indicator, it results that:

$$SOC^*_{RothC}(\Delta t) = c_{iom} + \mathbf{e}^T \mathbf{c}^*_{RothC}(\Delta t) = c_{iom} + \mathbf{e}^T \varphi(-\Delta t \rho D)^{-1} \mathbf{c}^*_{cont},$$

and consequently,

$$SOC^*_{RothC}(\Delta t) - SOC^*_{cont} = \mathbf{e}^T \left(\varphi(-\Delta t \rho D)^{-1} - I\right) \mathbf{c}^*_{cont} = \mathbf{w}^T(\Delta t) \mathbf{c}^*_{cont}$$

where:

$$\mathbf{w}(\Delta t) = \left[\frac{1 - \varphi(-\Delta t \rho k_{dpm})}{\varphi(-\Delta t \rho k_{dpm})}, \frac{1 - \varphi(-\Delta t \rho k_{rpm})}{\varphi(-\Delta t \rho k_{rpm})}, \frac{1 - \varphi(-\Delta t \rho k_{bio})}{\varphi(-\Delta t \rho k_{bio})}, \frac{1 - \varphi(-\Delta t \rho k_{hum})}{\varphi(-\Delta t \rho k_{hum})} \right] \quad (11)$$

a vector with positive entries that verifies $\lim_{\Delta t \to 0} \mathbf{w}(\Delta t) = \mathbf{0}$.

Since the original discrete RothC model is applied with $\Delta t = 1$, the deviation of the SOC^*_{cont} of the continuous model with respect to the equilibrium $SOC^*_{RothC}(1)$ of the original discrete RothC model is given by $SOC^*_{RothC}(1) = SOC^*_{cont} + \mathbf{w}^T(1) \mathbf{c}^*_{cont}$, thus indicating that the evaluation of soil organic carbon by means of the discrete RothC model is an overestimation of the theoretical value SOC^*_{cont}.

Usually, ρ, g and f vary through time, but it can be assumed that they have a periodic behaviour. Typically, if the agricultural practices are cyclic and if the weather conditions can be considered periodic, then $\rho = \rho(t)$, $g = g(t)$ and $f = f(t)$ will also behave periodically. Assuming that the periodicity of these variables is $T = N\Delta t$, one looks for a solution of $\mathbf{c}$ such that $\mathbf{c}_0 = \mathbf{c}_N$. Then, we can write:

$$\mathbf{c}_{n+1} = \left(\Lambda + (I - \Lambda) e^{-\Delta t \rho(t_n) D}\right) \mathbf{c}_n + \Delta t \, \mathbf{b}(t_n)$$

for $n = 0, \ldots N - 2$ and then impose the periodic condition $\mathbf{c}_N = \mathbf{c}_0$:

$$\mathbf{c}_0 = \left(\Lambda + (I - \Lambda) e^{-\Delta t \rho(t_{N-1}) D}\right) \mathbf{c}_{N-1} + \Delta t \, \mathbf{b}(t_{N-1}).$$

Setting $F_n := F_n(\Delta t) := \Lambda + (I - \Lambda) e^{-\Delta t \rho(t_n) D}$, the above relations can be reformulated as:

$$\begin{pmatrix} 0 & I & 0 & \cdots & 0 \\ \vdots & \ddots & \ddots & \ddots & \vdots \\ \vdots & & \ddots & \ddots & 0 \\ 0 & \cdots & & 0 & I \\ I & 0 & \cdots & 0 & 0 \end{pmatrix} \begin{pmatrix} \mathbf{c}_0 \\ \mathbf{c}_1 \\ \vdots \\ \mathbf{c}_{N-2} \\ \mathbf{c}_{N-1} \end{pmatrix} = \begin{pmatrix} F_0 & 0 & \cdots & & 0 \\ 0 & F_1 & \ddots & & \\ \vdots & & \ddots & \ddots & \vdots \\ & & & \ddots & 0 \\ 0 & & \cdots & 0 & F_{N-1} \end{pmatrix} \begin{pmatrix} \mathbf{c}_0 \\ \mathbf{c}_1 \\ \vdots \\ \mathbf{c}_{N-2} \\ \mathbf{c}_{N-1} \end{pmatrix} + \Delta t \begin{pmatrix} \mathbf{b}_0 \\ \mathbf{b}_1 \\ \vdots \\ \mathbf{b}_{N-2} \\ \mathbf{b}_{N-1} \end{pmatrix}$$

which yields:

$$\begin{pmatrix} \mathbf{c}_0 \\ \mathbf{c}_1 \\ \vdots \\ \mathbf{c}_{N-2} \\ \mathbf{c}_{N-1} \end{pmatrix} = -\Delta t \begin{pmatrix} F_0 & -I & \cdots & & 0 \\ 0 & F_1 & -I & & \cdots \\ \vdots & & \ddots & \ddots & \vdots \\ \vdots & & & \ddots & -I \\ -I & \cdots & & 0 & F_{N-1} \end{pmatrix}^{-1} \begin{pmatrix} \mathbf{b}_0 \\ \mathbf{b}_1 \\ \vdots \\ \mathbf{b}_{N-2} \\ \mathbf{b}_{N-1} \end{pmatrix}, \quad (12)$$

where $\mathbf{b}_n := \mathbf{b}(t_n)$, for $n = 0, \ldots, N-1$. Equation (12) provides a vector of dimension $4 \times N$ and a sequence of states $\mathbf{c}_n = [c_n^{(dpm)}, c_n^{(rpm)}, c_n^{(bio)}, c_n^{(hum)}]^T$, $n = 0, \ldots N-1$, which characterizes the oscillatory state of the carbon stock in each compartment, keeping track of the temporal variability of the forcing variables over the period.

3.2. Exponential Rosenbrock–Euler Model

If we regard the stepping procedure in (8), which defines the original discrete RothC model as a first-order approximation of the solution of the continuous model (1), then different discrete RothC models can be formulated. As an example, the discretization of Equation (1) from $t_n = t_0 + n\Delta t$ to $t_{n+1} = t_n + \Delta t$ by means of the exponential Rosenbrock–Euler model (for non-autonomous systems) [10,18,19] leads to:

$$\begin{aligned} \mathbf{c}_{n+1} &= e^{\Delta t \rho(t_n) A} \, \mathbf{c}_n + \Delta t \, \varphi(\Delta t \, \rho(t_n) \, A) \, \mathbf{b}(t_n) \\ &= \mathbf{c}_n + \Delta t \, \varphi(\Delta t \, \rho(t_n) \, A) \, \mathbf{f}(t_n; \mathbf{c}_n) \end{aligned} \qquad (13)$$

where $\mathbf{f}(t; \mathbf{c}) := \rho(t) \, A \, \mathbf{c}(t) + \mathbf{b}(t)$. Notice that A is negative definite, and for $z < 0$, it results in $0 < \varphi(z) < 1$. In [10], it was shown that $A = (\Lambda - I) D$ and:

$$F_n(\Delta t) = \Lambda + (I - \Lambda) e^{-\Delta t D \rho(t_n)} \approx e^{\Delta t \rho(t_n) A}, \quad \Delta t \, \varphi(\Delta t \, \rho(t_n) \, A) = \mathcal{O}(diag(\Delta t)).$$

Of course, the approximated solutions via the Exponential Rosenbrock–Euler (ERE) method (13) differ from the values given by the original discrete RothC model (8); the major consequence is that the constant discrete steady-state solution, which for the ERE method coincides with the continuous equilibrium $\mathbf{c}^*_{cont}$, does not depend on Δt. As concerns its stability, given A negative definite, it is enough to notice that the eigenvalues of $e^{\Delta t \rho(t_n) A}$ are positive, but all less than one.

To find periodic solutions via the ERE method, we can generalize the approach followed for the discrete original RothC model as follows. Suppose that $T = N \Delta t$, and let us impose that $\mathbf{c}_0 = \mathbf{c}_N$:

$$\mathbf{c}_{n+1} = e^{\Delta t A \rho(t_n)} \mathbf{c}_n + \Delta t \, \varphi(\Delta t \, A \, \rho(t_n)) \, \mathbf{b}(t_n)$$

for $n = 0, \ldots N - 2$ and:

$$\mathbf{c}_0 = e^{\Delta t A \rho(t_{N-1})} \mathbf{c}_{N-1} + \Delta t \, \varphi(\Delta t \, A \, \rho(t_{N-1})) \, \mathbf{b}(t_{N-1})$$

which yields:

$$\begin{pmatrix} \mathbf{c}_0 \\ \mathbf{c}_1 \\ \vdots \\ \mathbf{c}_{N-2} \\ \mathbf{c}_{N-1} \end{pmatrix} = -\Delta t \begin{pmatrix} e^{\Delta t A \rho(t_0)} & -I & \cdots & & 0 \\ 0 & e^{\Delta t A \rho(t_1)} & -I & & \cdots \\ \vdots & & \ddots & \ddots & \vdots \\ & & & \ddots & -I \\ -I & \cdots & & 0 & e^{\Delta t A \rho(t_{N-1})} \end{pmatrix}^{-1} \begin{pmatrix} \tilde{\mathbf{b}}_0 \\ \tilde{\mathbf{b}}_1 \\ \vdots \\ \tilde{\mathbf{b}}_{N-2} \\ \tilde{\mathbf{b}}_{N-1} \end{pmatrix}, \qquad (14)$$

with $\tilde{\mathbf{b}}_n := \varphi(\Delta t \, \rho(t_n) \, A) \, \mathbf{b}_n$, for $n = 0, \ldots, N - 1$.

The sequence of states $\mathbf{c}_n$, $n = 0, \ldots N - 1$ characterizes the oscillatory state of the carbon stock in each compartment. Of course, the solution differs from the periodic solution provided in (12).

3.3. Novel Non-Standard Discrete RothC Model

The ERE procedure described in (13) belongs to the more general class of non-standard finite difference schemes [14,20]. Different first-order non-standard approximations can be used, depending on the function of the stepsize used to advance the first-order procedure. Each of them can be considered as discrete RothC models, alternative to the ERE model.

A discrete model, closer to the original formulation of the RothC model, can be introduced as follows. Consider the discrete formulation of RothC (8). Simple evaluations lead to the equivalent expressions:

$$\begin{aligned} \mathbf{c}_{n+1} &= F_n(\Delta t) \, \mathbf{c}_n + \Delta t \, \mathbf{b}(t_n) \\ &= \mathbf{c}_n + \Delta t \, \varphi(\Delta t \, \rho(t_n) \, \tilde{A}) \, \rho(t_n) \, A \, \mathbf{c}_n + \Delta t \, \mathbf{b}(t_n) \end{aligned} \qquad (15)$$

where the eigenvalues of the matrix $\widetilde{A} = -A(I-\Lambda)^{-1} = -(I-\Lambda)D(I-\Lambda)^{-1}$ are the entries of the vector $-\mathbf{k}$, and the columns of:

$$(I-\Lambda)^{-1} = \frac{1}{\delta}\begin{pmatrix} \delta & 0 & 0 & 0 \\ 0 & \delta & 0 & 0 \\ \alpha & \alpha & \beta-1 & \alpha \\ \beta & \beta & \beta & \alpha-1 \end{pmatrix} \quad (16)$$

are the corresponding eigenvectors.

A novel Non-Standard (NS) discrete RothC model is given by:

$$\begin{aligned}\mathbf{c}_{n+1} &= \mathbf{c}_n + \Delta t \varphi(\Delta t \rho(t_n)\widetilde{A})\,\mathbf{f}(\mathbf{c}_n;t_n) \\ &= F_n(\Delta t)\,\mathbf{c}_n + \Delta t\,\varphi(\Delta t\rho(t_n)\widetilde{A})\,\mathbf{b}(t_n)\end{aligned} \quad (17)$$

Indeed, we can prove that:

Theorem 6. *The NS model (17) is a non-standard first-order approximation of the continuous model (1), i.e.,*

$$\Delta t\,\varphi(\Delta t\rho(t_n)\widetilde{A}) = \mathcal{O}(diag(\Delta t)), \quad \text{for } \Delta t \to 0.$$

Proof. From the definition of the exponential matrix, we have that:

$$\begin{aligned}\Delta t \varphi(\Delta t \rho(t_n)\widetilde{A}) &= \frac{1}{\rho(t_n)}\widetilde{A}^{-1}\left(e^{\Delta t \rho(t_n)\widetilde{A}} - I\right) \\ &= \frac{1}{\rho(t_n)}\widetilde{A}^{-1}\sum_{j=1}^{\infty}\frac{1}{j!}\left(\Delta t\rho(t_n)\widetilde{A}\right)^j \\ &= diag(\Delta t) + \frac{\Delta t^2}{2}\rho(t_n)\widetilde{A} + \sum_{j=2}^{\infty}\frac{\Delta t^{j+1}}{(j+1)!}\left(\rho(t_n)\widetilde{A}\right)^j\end{aligned}$$

It follows that $\Delta t\varphi(\Delta t\rho(t_n)\widetilde{A}) = \mathcal{O}(diag(\Delta t))$, for $\Delta t \to 0$. □

By comparing the ERE flow (13) with the above non-standard formulation (17), we notice that the main difference is in the replacement of the matrix A with the matrix $\widetilde{A}$, which has a simple representation in Jordan form. This simplifies the evaluation of the matrix function $\varphi(\Delta t\rho(t_n)\widetilde{A})$ because it can be evaluated on the known eigenvalues $-\mathbf{k}(i)$ and eigenvectors in (16).

Moreover, the comparison of the original discrete formulation of the RothC model (8) with the novel NS procedure (17) allows us to notice that, different from the ERE method (13), the decomposition dynamic is now treated in the same way as the original discrete RothC model. The time updating procedure related to the carbon amount entering the system at time $t + \Delta t$ is now evaluated up to a factor given by $\varphi(\Delta t\rho(t_n)\widetilde{A})$, different from the ERE approximation, which advances evaluating $\varphi(\Delta t\rho(t_n)A)$.

As the ERE method, in the case of constant functions ρ, g, f, the novel method NS has $\mathbf{c}^*_{cont}$ as the steady-state solution.

Theorem 7. *Under the hypothesis $0 < \alpha + \beta < 1$, the equilibrium $\mathbf{c}^*_{cont}$ of the NS method (17) is globally stable.*

Proof. It is enough to prove that the eigenvalues of $F_n(\Delta t) = \Lambda + (I - \Lambda) e^{-\Delta t \rho(t_n) D}$ are in modulus less than one. It is easy to see that two eigenvalues are given by $e^{-\Delta t \rho(t_n) k_{dpm}}$ and $e^{-\Delta t \rho(t_n) k_{rpm}}$. The others are the eigenvalues of the sub-matrix:

$$F_{3,4} = \begin{pmatrix} \alpha - e^{-\Delta t \rho(t_n) k_{bio}} (\alpha - 1) & \alpha - \alpha e^{-\Delta t \rho(t_n) k_{hum}} \\ \beta - \beta e^{-\Delta t \rho(t_n) k_{bio}} & \beta - e^{-\Delta t \rho(t_n) k_{hum}} (\beta - 1) \end{pmatrix}$$

The eigenvalues lie in the union of the two Gershgorin sets:

$$K_1 = \left\{ z \in \mathbb{C} : |z - \alpha + e^{-\Delta t \rho(t_n) k_{bio}} (\alpha - 1)| \leq |\beta - \beta e^{-\Delta t \rho(t_n) k_{bio}}| \right\},$$

$$K_2 = \left\{ z \in \mathbb{C} : |z - \alpha + \alpha e^{-\Delta t \rho(t_n) k_{hum}}| \leq |\beta - e^{-\Delta t \rho(t_n) k_{hum}} (\beta - 1)| \right\}.$$

Notice that $F_{3,4}$ has positive entries. Set

$$M = \max(F_{3,4}(1,1) + F_{3,4}(2,1), F_{3,4}(1,2) + F_{3,4}(2,2)); \text{ if } z \in K_1 \cup K_2, \text{ then } |z| \leq M.$$

evaluate:

$$F_{3,4}(1,1) + F_{3,4}(2,1) = (\alpha + \beta)\left(1 - e^{-\Delta t \rho(t_n) k_{bio}}\right) + e^{-\Delta t \rho(t_n) k_{bio}} < 1$$

Similarly,

$$F_{3,4}(1,2) + F_{3,4}(2,2) = (\alpha + \beta)\left(1 - e^{-\Delta t \rho(t_n) k_{hum}}\right) + e^{-\Delta t \rho(t_n) k_{hum}} < 1$$

Hence it results that the eigenvalue of $F_{3,4}$ are less than one in modulus. □

To search for periodic solutions via the novel non-standard model, we need to solve the linear system:

$$-\begin{pmatrix} F_0 & -I & \cdots & & 0 \\ 0 & F_1 & -I & & \\ \vdots & & \ddots & \ddots & \vdots \\ \vdots & & & \ddots & -I \\ -I & \cdots & & 0 & F_{N-1} \end{pmatrix} \begin{pmatrix} \mathbf{c}_0 \\ \mathbf{c}_1 \\ \vdots \\ \mathbf{c}_{N-2} \\ \mathbf{c}_{N-1} \end{pmatrix} = \Delta t \begin{pmatrix} \hat{\mathbf{b}}_0 \\ \hat{\mathbf{b}}_1 \\ \vdots \\ \hat{\mathbf{b}}_{N-2} \\ \hat{\mathbf{b}}_{N-1} \end{pmatrix},$$

where $\hat{\mathbf{b}}_n := \varphi(\Delta t \rho(t_n) \widetilde{A}) \mathbf{b}_n$, for $n = 0, \ldots, N-1$. As already observed, we have the same coefficient matrix of the system (12), while the knowledge of the explicit Jordan form of the matrix $\widetilde{A}$ simplifies the evaluation of the coefficients $\hat{\mathbf{b}}_n$.

In the next two sections, we compare the behaviour of the different discrete models on the evaluation of steady-state and long-term periodic solutions. Firstly, the comparison is made on a theoretical basis comparing the accuracy of the methods in approximating the solutions of the continuous model; secondly, we tested both the ERE and the novel non-standard model with respect to the original discrete RothC model on a classical monthly time-scale experiment where real measurements were available.

A freely accessible MATLAB routine named NSRothC [21] was implemented to replicate all the simulations presented in the next sections. The package includes two versions. The first one (contNSRothC) allowed us to run the model with different stepsizes, when the continuous periodic input functions $\rho(t)$ and $\mathbf{b}(t)$ have the particular form presented in Section 4. In the second version (monNSRothC), the stepsize was fixed to one month, and the discrete monthly values of input residuals and FYM were required, as well as the monthly values of weather variables (temperature, rainfall and moisture). As an example, data from the Hoosfield spring barley experiment [6] were used.

4. Numerical Tests

Let us compare long-term solutions obtained by using the original discrete RothC model, the exponential Rosenbrock–Euler method and the novel non-standard first-order scheme. We set parameters $\alpha = 0.1$, $\beta = 0.12$, $\gamma = 0.59$ and $\eta = 0.49$. The vector of the decomposition rates $\mathbf{k} = [0.8333, 0.0250, 0.0550, 0.0017]$ and functions $\rho(t)$, $g(t)$ and $f(t)$ are supposedly expressed on a monthly scale. They vary with the same period $T = 12$ and have Gaussian distributions according to:

$$\rho(t) = a_\rho + \frac{G(t, \mu_\rho^{(1)}, \sigma_\rho^{(1)})}{\int_{t_0}^{t_0+T} G(s, \mu_\rho^{(1)}, \sigma_\rho^{(1)}) \, ds} b_\rho^{(1)} + \frac{G(t, \mu_\rho^{(2)}, \sigma_\rho^{(2)})}{\int_{t_0}^{t_0+T} G(s, \mu_\rho^{(2)}, \sigma_\rho^{(2)}) \, ds} b_\rho^{(2)}$$

$$g(t) = a_g + \frac{G(t, \mu_g, \sigma_g)}{\int_{t_0}^{t_0+T} G(s, \mu_g, \sigma_g) \, ds} b_g, \quad f(t) = a_f + \frac{G(t, \mu_f, \sigma_f)}{\int_{t_0}^{t_0+T} G(s, \mu_f, \sigma_f) \, ds} b_f$$

where $G(t, \mu, \sigma) := \dfrac{1}{\sigma\sqrt{2\pi}} e^{-\dfrac{\left(t - t_0 - \mu - \left\lfloor \frac{t-t_0}{T} \right\rfloor T\right)^2}{2\sigma^2}}$ for $t \geq t_0$.

Values $a_\rho = 0.3561$, $\mu_\rho^{(1)} = \frac{T}{2}$, $\mu_\rho^{(2)} = \frac{5T}{6}$ were set, amplitudes $b_\rho^{(1)} = 0.9596$ and $b_\rho^{(2)} = 1.4996$ were chosen, while dispersion coefficients were given by $\sigma_\rho^{(1)} = \sigma_\rho^{(2)} = 0.6738$. The function $\rho(t)$ in a time-span of three years is shown in Figure 2 on the left.

Values $a_g = 0$, $\mu_g = \frac{T}{2}$, $b_g = 2.7996$, $\sigma_g = 0.9974$ define the function $g(t)$, while values $a_f = 0$, $\mu_f = \frac{T}{6}$, $b_f = 1.5$, $\sigma_f = 0.1995$ define the function $f(t)$. The four components of the input function $\mathbf{b}(t) := g(t)\,\mathbf{a}^{(g)} + f(t)\,\mathbf{a}^{(f)}$ are depicted in Figure 2, on the right.

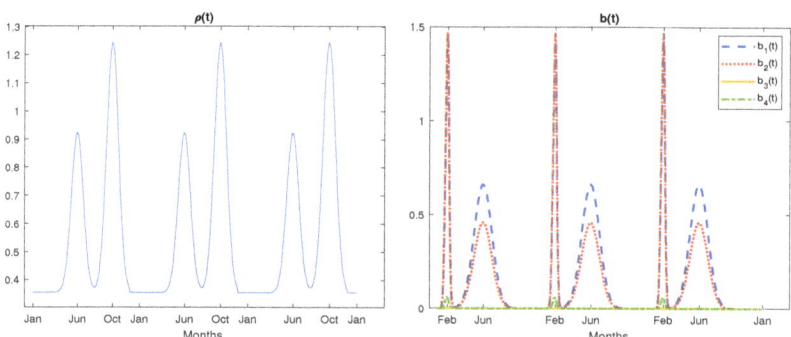

Figure 2. The function $\rho(t)$ (on the left) and the vectorial function $\mathbf{b}(t)$ (on the right) in a three-year time-span.

4.1. Steady-State Solution

To compare long-time periodic and steady-state solutions, we consider the average values of $\rho(t)$ and $\mathbf{b}(t)$ over one period T:

$$\hat{\rho} = a_\rho + \frac{b_\rho^{(1)}}{T} + \frac{b_\rho^{(2)}}{T} = 0.5610,$$

$$\hat{\mathbf{b}} = \left(a_g + \frac{b_g}{T}\right)\mathbf{a}^{(g)} + \left(a_f + \frac{b_f}{T}\right)\mathbf{a}^{(f)} = 0.2333\,\mathbf{a}^{(g)} + 0.125\,\mathbf{a}^{(f)} = [0.1989, 0.1569, 0, 0.025]^T,$$

and we evaluate the steady-state solution (4) of the continuous RothC model (1):

$$\mathbf{c}^*_{cont} = [0.4254, 11.1867, 1.4887, 61.6253]^T. \qquad (18)$$

We recall that both the ERE and the NS method have $\mathbf{c}^*_{cont}$ as the steady-state solution of their discrete flows, and consequently, they provide the value of SOC^*_{cont} as the equilibrium for soil organic carbon; differently, the steady-state solution of the discrete RothC model (9) depends on the chosen stepsize Δt. In Table 1, we report the equilibrium solutions $\mathbf{c}^*_{RothC}(\Delta t)$ for decreasing values of the stepsize Δt. When we proceed with $\Delta t = 1$, we obtain the original monthly time-stepping procedure, while $\Delta t = 1/30$ represents a daily temporal updating. Notice that, as predicted by the theoretical results, $\|\mathbf{c}^*_{cont}\|_2 = 62.6516 \leq \|\mathbf{c}^*_{RothC}(1)\|_2 = 62.695$, i.e., the original discrete RothC model overestimates in norm the theoretical equilibrium value.

The reduction of the temporal stepsize of 1/30 causes approximately the same reduction of the absolute errors as $\dfrac{\|\mathbf{c}^*_{cont} - \mathbf{c}^*_{RothC}(1)\|_2}{\|\mathbf{c}^*_{cont} - \mathbf{c}^*_{RothC}(1/30)\|_2} = 31.534$, with this confirming the first-order accuracy of the approximation of the steady-state solution. In Figure 3a, we report the relative errors $\dfrac{\|\mathbf{c}^*_{cont} - \mathbf{c}^*_{RothC}(\Delta t)\|_2}{\|\mathbf{c}^*_{cont}\|_2}$ in correspondence of halved stepsizes starting from $\Delta t = 1$ together with their bounds evaluated in (10). In the same figure, we report the error on the equilibrium value of SOC normalized respect to $\mathbf{c}^*_{cont}$, i.e.,

$$\dfrac{SOC^*_{RothC}(\Delta t) - SOC^*_{cont}}{\|\mathbf{c}^*_{cont}\|_2} = \mathbf{w}^T(\Delta t) \dfrac{\mathbf{c}^*_{cont}}{\|\mathbf{c}^*_{cont}\|_2}$$

where $\mathbf{w}(\Delta t)$ is defined in (11), with respect to the same reduction of the stepsize Δt.

Table 1. Dependence of the steady-state solution of the discrete RothC model on the temporal stepsize Δt. The values in each compartment converge, with first-order accuracy, to the entries of $\mathbf{c}^*_{cont} = [0.4254, 11.1867, 1.4887, 61.6253]^T$.

Δt	1	$\frac{1}{5}$	$\frac{1}{10}$	$\frac{1}{15}$	$\frac{1}{20}$	$\frac{1}{25}$	$\frac{1}{30}$
	0.5326	0.4456	0.4354	0.4321	0.4304	0.4294	0.4287
$\mathbf{c}^*_{RothC}$	11.2653	11.2024	11.1946	11.1919	11.1906	11.1899	11.1893
	1.5118	1.4933	1.4910	1.4902	1.4898	1.4896	1.4894
	61.6541	61.6311	61.6282	61.6272	61.6267	61.6265	61.6263

4.2. Periodic Solutions

In these experiments, we compared the periodic solutions provided by the three discrete models, obtained for the functions $\rho(t)$ and $\mathbf{b}(t)$ plotted in Figure 2. Different from the evaluation of steady-state solutions, the ERE and the novel NS procedure, as well as the discrete RothC model provide long-term periodic solutions affected by their own numerical errors, also depending on the chosen stepsize Δt. As shown in Figure 3b, the RothC discrete model provides the worst performance when compared with the other procedures in terms of errors with respect to the reference solution. In Figure 4, we plot the related SOC indicator (1) in a one-period time span obtained with the three discrete procedures applied with reduced stepsizes $\Delta t = 1, 1/2, 1/4, 1/8$. In the same figure, the reference solution obtained in the long run with the ode45 MATLAB function, with tolerance set at machine precision, is plotted. Notice that all the methods need to be applied with a $\Delta t << 1$ in order to have an acceptable level of accuracy when compared to the reference solution. Again, the original discrete RothC model was confirmed as a less accurate integrator of the continuous model (1) with respect to both the ERE and NS models.

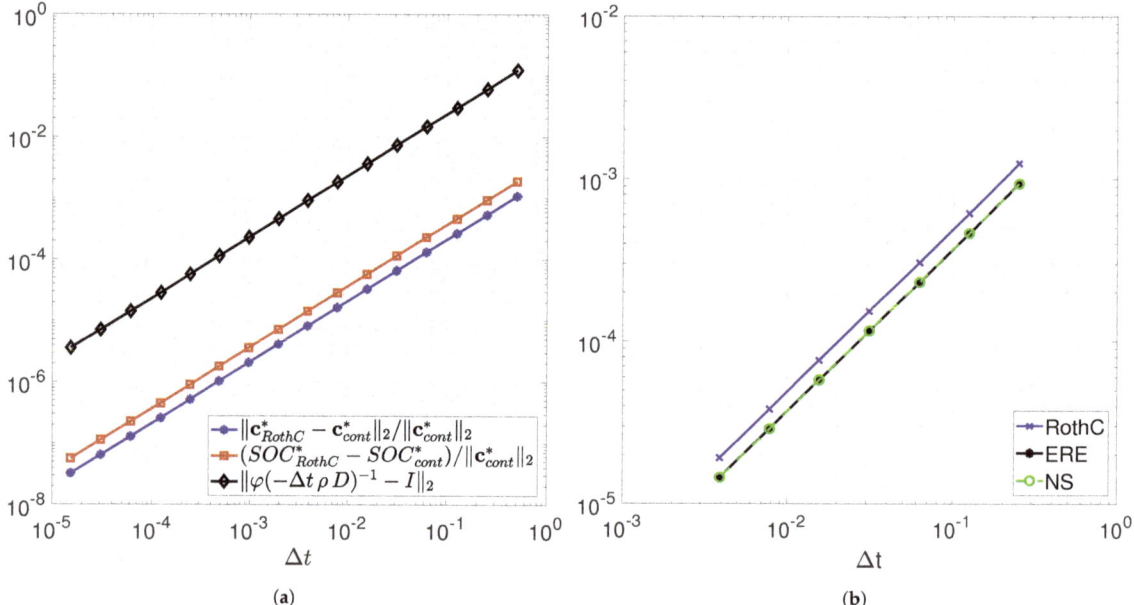

Figure 3. (**a**) Relative errors of $\mathbf{c}^*_{RothC}$ (blue) and SOC^*_{RothC} (orange) with halved stepsizes, starting from $\Delta t = 1$. The bounds on the relative error $\|\varphi(\Delta t \rho D)^{-1} - I\|_2$ in (10) are also plotted in black. The estimated order of convergence is 1.0037 for the relative error on $\mathbf{c}^*_{RothC}$ and 1.0022 for the relative error on SOC. (**b**) The log-log plot of the norm two errors of the periodic solutions of discrete RothC, Exponential Rosenbrock–Euler (ERE) and Non-Standard (NS) schemes, with respect to the reference solution obtained in the long run with the ode45 Matlab code, at $\Delta t = 1/2^i$, $i = 2, \ldots, 8$. The estimated orders of convergence are 1.0042, 1.0017 and 1.0018 for the discrete RothC, ERE and NS schemes, respectively.

In our second experiment, we started at $t_0 = 1$ with $\mathbf{c}_0 = \mathbf{c}^*_{cont}$ in (18), and we proceeded until the final time $T_f = t_0 + 15\,T$ was reached. We recall that the period was set at $T = 12$, and we used the monthly, weekly and daily update by setting $\Delta t = 1$, $\Delta t = 1/4$ and $\Delta t = 1/30$, respectively. In Figure 5, we report the values of SOC obtained for the three different procedures. Two main observations have to be made: first, the initial value $\mathbf{c}^*_{cont}$ corresponding to the equilibrium solution with mean values $\hat{\rho}$ and $\hat{\mathbf{b}}$ was higher than the attained value of SOC reference value at T_f when $\rho(t)$ and $\mathbf{b}(t)$ were not averaged. This indicates that the temporal oscillations of $\rho(t)$ and $\mathbf{b}(t)$ around their mean values cannot be neglected when evaluating, through the SOC indicator, the achievement of land degradation neutrality in the fifteen years from 2015–2030. Second, whatever discrete model is chosen, a qualitative long-term accordance between an approximate value of the SOC indicator and its theoretical solution needs, at least, a weekly update procedure.

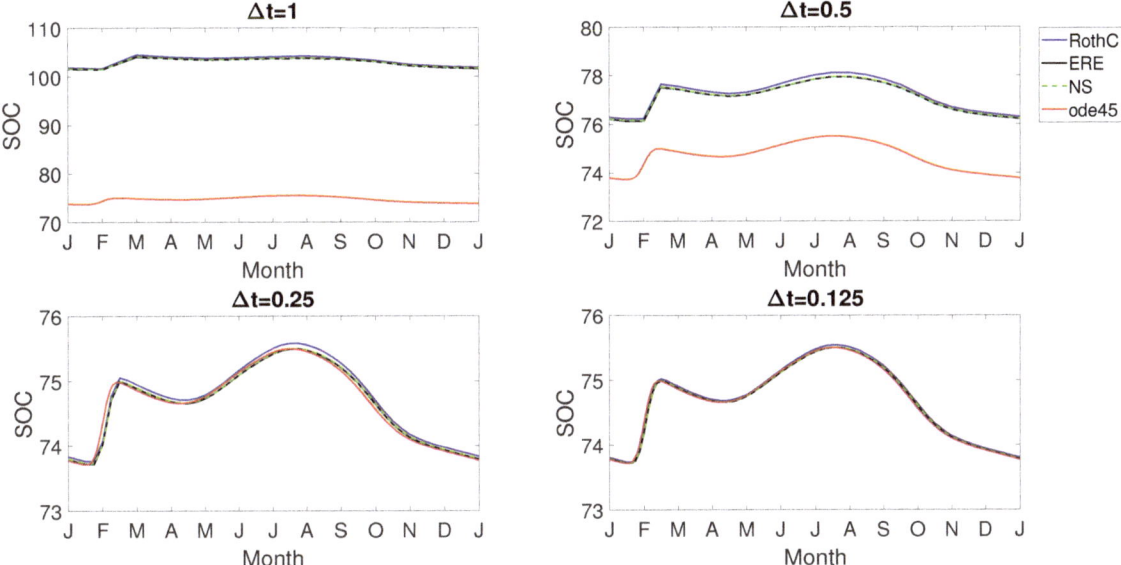

Figure 4. Periodic dynamics of the SOC indicator in a one-period span: convergence of approximated solutions to the reference solution (plotted as a continuous red line) obtained with MATLAB ode45 code, by an increasingly reduction of the stepsize Δt.

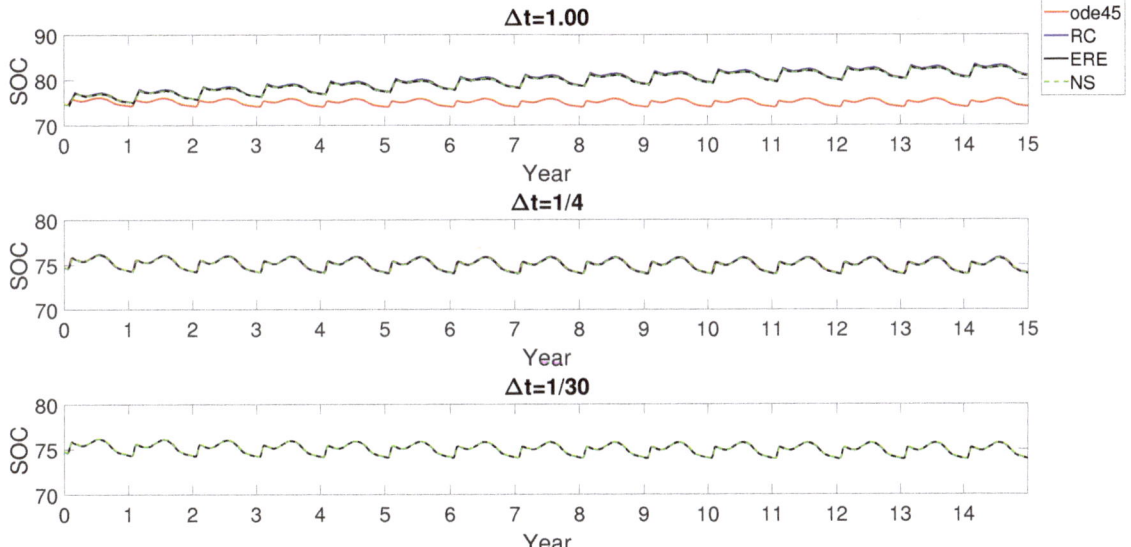

Figure 5. Long-term simulation of the SOC indicator over 15 years with the discrete RothC, ERE and NS methods. Stepsizes are set as $\Delta t = 1$ (1 month), $\Delta t = 1/4$ (1 week) and $\Delta t = 1/30$ (1 day). The reference solution obtained by ode45 with tolerance at the machine precision is also plotted.

5. The Hoosfield Spring Barley Experiment

We illustrate the model and test the three methods using data from the Hoosfield spring barley experiment, one of the classical long-term experiments carried out at the Rothamster Experimental Station [22]. The same dataset was used as an example of the use of the RothC model in [6]. The Hoosfield experiment was conducted from 1852 till 2000. Spring barley has been grown continuously in the whole period, except in the years 1912, 1933, 1943 and 1967, when the experiment was fallowed to control weeds. The initial SOC content in the soil at 1852 was measured as $33.8632\,t\,C\,ha^{-1}$, split out in the soil compartments in the following way: $2.7\,t\,C\,ha^{-1}$ in IOM, $0.1533\,t\,C\,ha^{-1}$ in DPM, $4.4852\,t\,C\,ha^{-1}$ in RMP, $0.6671\,t\,C\,ha^{-1}$ in BIO and $25.8576\,t\,C\,ha^{-1}$ in HUM.

The ρ function was estimated on a monthly basis from the weather dataset in [6], including average monthly air temperature, monthly open pan evaporation and monthly rainfall. It was also affected by the percentage of clay in the soil (23.4% in Rothamsted soil) and by the monthly soil cover factor (covered or fallow). Assuming that the soil was covered only from April to July for each crop year, the ρ function assumed the values in Table 2. The other parameters of the model were set as $\alpha = 0.10$, $\beta = 0.12$, $\gamma = 0.59$ and $\eta = 0.49$. Moreover, the monthly decomposition rate vector $\mathbf{k} = [0.8333, 0.0250, 0.0550, 0.0017]$ was considered.

Table 2. Monthly values for ρ for the Hoosfield spring barley experiment, estimated from weather data in [6], clay content of 23.4% and soil cover factors in crop and fallow years.

	Jan.	Feb.	Mar.	Apr.	May	Jun.	July	Aug.	Sep.	Oct.	Nov.	Dec.
$\rho(t_n)$ Crop years	0.3561	0.3723	0.5068	0.4471	0.7473	0.7779	0.2491	0.4151	0.6570	1.1277	0.6092	0.4594
$\rho(t_n)$ Fallow years	0.3561	0.3723	0.5068	0.7451	1.2454	1.2965	0.4151	0.4151	0.6570	1.1277	0.6092	0.4594

Three different scenarios were simulated with the three numerical methods analysed in Section 3, and the results were compared with direct observations of SOC quantity in the soil in 1882, 1913, 1946, 1975, 1982 and 1987:

Scenario 1. Unmanured treatment: For this scenario, the annual input of plant residuals was calculated to be $1.60\,t\,C\,ha^{-1}\,y^{-1}$, distributed as in Figure 6, in every year except in those that were fallow. For these years, a null plant residuals input was considered. In this scenario, FYM input was zero.

Scenario 2. Farmyard manure: The second treatment included inputs from both plant residuals and FYM, as scheduled in Figure 7. The annual input of plant residuals was calculated to be $2.8\,t\,C\,ha^{-1}\,y^{-1}$, greater than the one in Scenario 1, due to the farmyard treatment.

Scenario 3. Mixed treatment, farmyard manure till 1871: In this scenario, farmyard manure was assured every February till 1871 and nothing thereafter. This caused an annual plant residual input equal to $2.8\,t\,C\,ha^{-1}\,y^{-1}$ from 1852 till 1876 and equal to $1.60\,t\,C\,ha^{-1}\,y^{-1}$ in the following years, except for the fallow ones. Input data are shown in Figure 8.

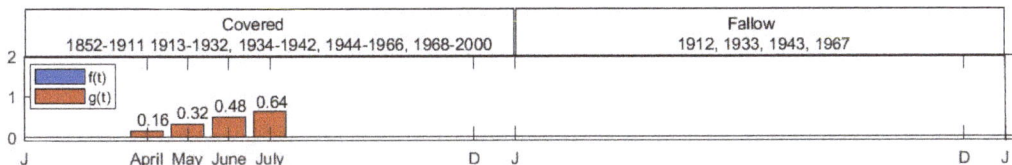

Figure 6. Hoosfield barley experiment. Scenario 1: unmanured treatment.

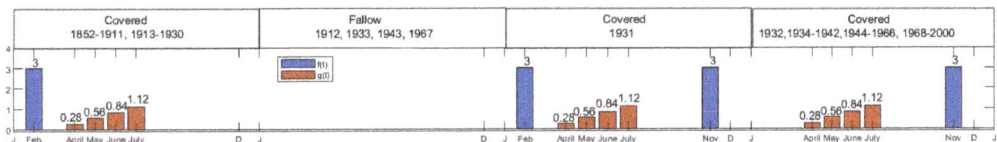

Figure 7. Hoosfield barley experiment. Scenario 2: farmyard manure annually (two times in 1931).

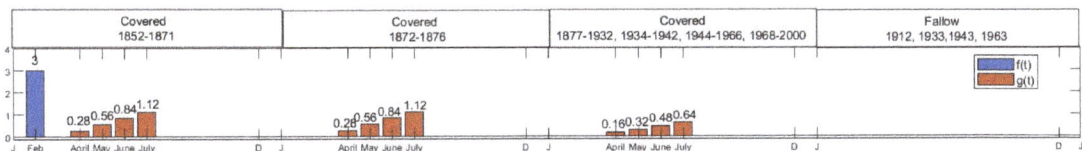

Figure 8. Hoosfield barley experiment. Scenario 3: farmyard manure 1852–1871 and nothing thereafter.

Figures 9–11 show the modelled data for total soil organic C in the three treatments, together with the measured data. The modelled results for Scenario 3 when the treatment considered FYM for only the first twenty years were considerably lower than the measurements; agreement was closer with the other two treatments (Scenarios 1 and 2). To test the models' ability to predict the achievement of land degradation neutrality, note that all the simulations gave results in agreement with measured data, i.e., the loss of neutrality in 2000 for Scenario 1 and the achievement of neutrality for Scenario 2 with respect to the initial value in 1852. Measurements indicated the achievement of land degradation neutrality also for Scenario 3; in this case, however, the three models failed in predicting for $SOC(2000)$ a value lower than the initial one, i.e., $SOC(1852) = 33.80\, t\, C\, ha^{-1}$.

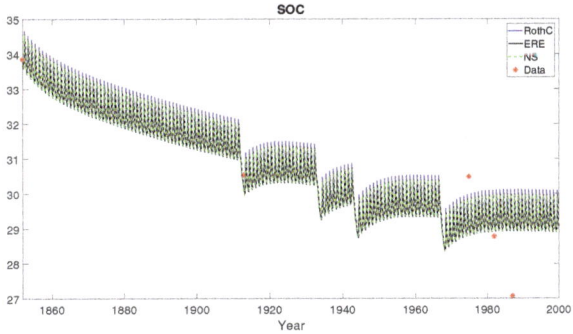

Figure 9. Hoosfield barley experiment. Scenario 1: unmanured treatment. Organic C in soil data (red bullets) modelled by the RothC (8), ERE (13) and NS (17) methods in the temporal horizon [1852, 2000] with $\Delta t = 1$. According to Table 3, ERE approximation shows the best performance with respects to both the statistical indicators RMSE and modelling Efficiency (EF).

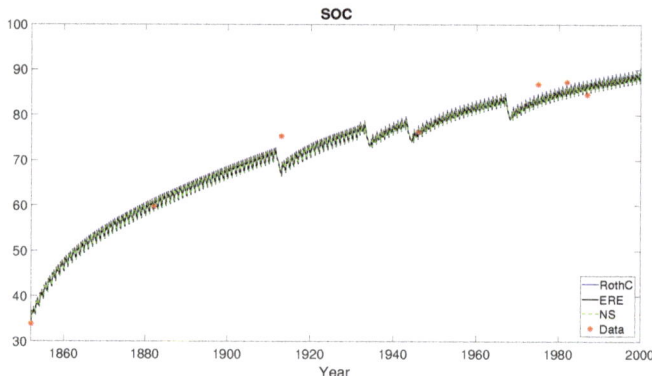

Figure 10. Hoosfield barley experiment. Scenario 2: farmyard manure annually. Organic C in soil data (red bullets) modelled by the RothC (8), ERE (13) and NS (17) methods in the temporal horizon [1852, 2000] with $\Delta t = 1$. According to Table 3, RothC approximation shows the best performance with respect to both the statistical indicators RMSE and EF.

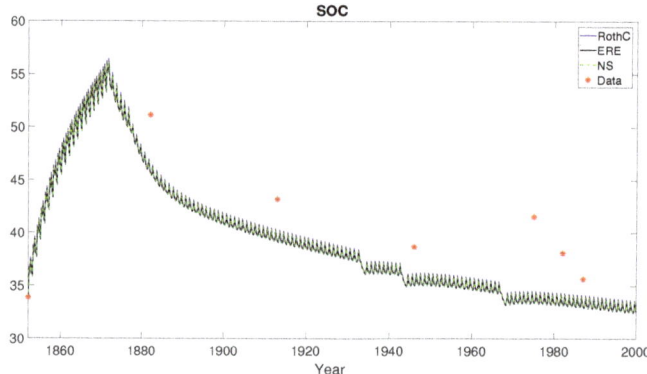

Figure 11. Hoosfield barley experiment. Scenario 3: farmyard manure 1852–1871 and nothing thereafter. Organic C in soil data (red bullets) modelled by the RothC (8), ERE (13) and NS (17) methods in the temporal horizon [1852, 2000] with $\Delta t = 1$. According to Table 3, RothC approximation shows the best performance with respect to both the statistical indicators RMSE and EF.

To assess the performance of the discrete models and compare measured and simulated valued, as in [16], we used two well-known statistical indices: RMSE (Root Mean Squared Error) and EF (modelling Efficiency):

$$RMSE = \sqrt{\frac{1}{N}\sum_{i=1}^{N}(O_i - S_i)^2} \qquad EF = 1 - \frac{\sum_{i=1}^{N}(O_i - S_i)^2}{\sum_{i=1}^{N}(O_i - \overline{O})^2}$$

where O_i and S_i are observed and simulated SOC at the ith value, $\overline{O}$ is the mean of the observed data and N is the number of observations. The closer the RMSE is to zero, the better the simulated solution describes the data. EF can range from $-\infty$ to one, with the best performance at EF = 1. Negative values of EF indicate that the observed mean is a better predictor than the model.

The results of the evaluation of the above indicators related to the three discrete models applied to approximate Hoosfield barley experiments data for the three different scenarios are reported in Table 3. The obtained values confirm that Scenario 3 was the worst case, i.e., the case when simulated data were more different from the measurements. The three discrete models provided similar values for the simulation of the three different treatments, and we could not select a method that clearly outperformed the others. However, from Table 3, we could set the ERE model as the best model for Scenario 1, while the RothC model provided better results for both Scenarios 2 and 3. Comparing the indicators for all scenarios with respect to the approximation of the experimental data, the minimum value of $RMSE = 1.1596$ was assumed by the ERE model in Scenario 1, while the maximum value of $EF = 0.9609$ was assumed by the RothC method in Scenario 2.

Table 3. RMSE and EF statistical indices for evaluating and comparing the performances of the RothC, ERE and novel non-standard NS models.

	Scenario 1			Scenario 2			Scenario 3		
	RothC	ERE	NS	RothC	ERE	NS	RothC	ERE	NS
RMSE	1.1665	1.1596	1.1599	3.5406	3.6029	3.5961	4.5772	4.6310	4.6254
EF	0.7318	0.7350	0.7348	0.9609	0.9595	0.9597	0.2564	0.2388	0.2406

6. Conclusions

Soil organic carbon is one of the key indicators of land degradation status. In this paper, we analysed the RothC model, a simple tool developed for predicting the dynamic evolution of the content of soil organic carbon under the effect of weather conditions and land use data. Both the continuous version, based on a linear, non-autonomous differential system, and the original discrete monthly time-stepping procedure were considered. The aim of our study was to compare the qualitative analysis of both continuous and discrete dynamics in approximating steady-state and long-term periodic solutions. We focused on these aspects since they provide the information concerning the possible achievement of land degradation neutrality. We found that the steady-state solutions of the original discrete model were first-order accurate approximations of the steady-state solutions of the continuous differential model. The accuracy of the approximation represents the main weakness of the RothC discrete model when considered as a first-order accurate approximation of its continuous version. We point out that well-established numerical procedures such as, for example, the Exponential Rosenbrock–Euler (ERE) method, have, by construction, the same equilibria of the approximating differential system. Conversely, the discrete RothC model overestimates the steady-state equilibrium of the continuous flow, so that, in order to obtain the correct estimate with first-order accuracy, we have to reduce the stepsize of the updating time procedure. This fact may have a negative effect in real applications. It is necessary to run the RothC model to equilibrium first, in order to align the initial content of soil organic carbon with the measured one [6]. Nevertheless, the good matching of real data and numerical approximations, together with the available implementation in the RothC 26.3 open-source interface [6], justifies the wide use in the literature of the original discrete RothC model and explains the lack of use of more accurate numerical integrators. For this reason, in this paper, our additional objective was to propose a novel non-standard first-order procedure, the so-called NS discrete model, able to approximate the decomposition dynamics in the same way as the original discrete RothC model. This procedure features the same steady-state solution of the continuous model and exhibits a computational cost lower than the one required by the ERE time-stepping procedure. We also provided a simple code that implements both the original discrete RothC model and the monthly time-stepping NS and ERE models in a MATLAB environment. That allows the reproduction of our results and facilitates future comparisons among all the approaches. The discrete models were firstly tested on a hypothetical example as first-order accurate numerical integrators, and secondly, they were tested on the classical Hoosfield barley experiment to evaluate their ability in reproducing observed data. If we consider the discrete RothC model as a difference scheme, which approximates the continuous RothC differential system, numerical tests showed its coarseness with respect to both the ERE and NS procedures. When applied to the Hoosfield barley experiment, the three discrete models provided very similar results. However, the evaluation of the statistical error indicators still revealed the discrete RothC model as the best scheme for approximating real data in both Scenarios 2 and 3, while the ERE model outperformed the discrete RothC and the novel procedure in Scenario 1. In predicting the long-term behaviours, which are useful to establish the achievement of land degradation neutrality, the three discrete models failed in giving results in accordance with theoretical values in our hypothetical test, as well as with real data in the case of Scenario 3 of the Hoosfield spring barley experiment. This motivates our future research to consider more complex SOC dynamics. In particular, we will analyse the MOMOS model [23,24], which puts the microbial community at the centre of all transformation processes in the cycle of organic matter, from assimilation by degradation to loss by mineralization. Further analysis will also consider nonlinear SOC dynamics [25,26] and spatially explicit models described by partial differential equations [27,28] in order to deal with real domains, as protected areas, covered by non-homogeneous land use patterns. Finally, using field measurements and remote sensing information for modelling land degradation will significantly improve the obtained results [29]. The final aim is to select and provide a robust modelling tool for evaluating the trends of SOC stocks in the Alta Murgia National Park, where the achievement of land degradation neutrality is hampered by a combination of anthropic pressures and climate change [30–32].

Author Contributions: The authors F.D., C.M. and A.M. contributed equally to this work. All authors have read and agreed to the published version of the manuscript.

Funding: This research was carried out within the Innonetwork Project COHECO, No. 8Q2LH28, POR Puglia FESR-FSE 2014–2020 and within the eLTER PLUS project. The eLTER PLUS project received funding from the European Union's Horizon 2020 research and innovation programme under Grant Agreement No. 871128.

Institutional Review Board Statement: Not applicable.

Informed Consent Statement: Not applicable.

Acknowledgments: We thank the anonymous referees for their helpful comments. We thank also Nicholas Tavalion for having carefully read the paper and for his help in improving its revision.

Conflicts of Interest: The authors declare no conflict of interest. The funders had no role in the design of the study; in the collection, analyses, or interpretation of data; in the writing of the manuscript; nor in the decision to publish the results.

References

1. Orr, B.; Cowie, A.; Castillo Sanchez, V.; Chasek, P.; Crossman, N.; Erlewein, A.; Louwagie, G.; Maron, M.; Metternicht, G.; Minelli, S.; et al. Scientific conceptual framework for land degradation neutrality. In *A Report of the Science-Policy Interface*; United Nations Convention to Combat Desertification (UNCCD): Bonn, Germany, 2017.
2. Aynekulu, E.; Lohbeck, M.; Nijbroek, R.P.; Ordoñez, J.C.; Turner, K.G.; Vågen, T.G.; Winowiecki, L.A. *Review of Methodologies for Land Degradation Neutrality Baselines: Sub-National Case Studies from Costa Rica and Namibia*; International Center for Tropical Agriculture (CIAT): Kampala, Uganda, 2017; 59p.
3. FAO. Measuring and modelling soil carbon stocks and stock changes in livestock production systems: Guidelines for assessment (Version 1). In *Livestock Environmental Assessment and Performance (LEAP) Partnership*; FAO: Rome, Italy, 2019; p. 170.
4. Paustian, K.; Collier, S.; Baldock, J.; Burgess, R.; Creque, J.; DeLonge, M.; Dungait, J.; Ellert, B.; Frank, S.; Goddard, T.; et al. Quantifying carbon for agricultural soil management: From the current status toward a global soil information system. *Carbon Manag.* **2019**, *10*, 567–587. [CrossRef]
5. Ponce-Hernandez, R.; Koohafkan, P.; Antoine, J. *Assessing Carbon Stocks and Modelling Win-Win Scenarios of Carbon Sequestration through Land-Use Changes*; Food & Agriculture Org.: Rome, Italy, 2004; Volume 1.

6. Coleman, K.; Jenkinson, D.S. *ROTHC-26.3: A Model for the Turnover of Carbon in Soil: Model Description and Users Guide: K. Coleman and DS Jenkinson*; IACR: Lyon, France, 1995.
7. Parton, W. The CENTURY model. In *Evaluation of Soil Organic Matter Models*; Springer: Berlin/Heidelberg, Germany, 1996; pp. 283–291.
8. Sallih, Z.; Pansu, M. Modelling of soil carbon forms after organic amendment under controlled conditions. *Soil Biol. Biochem.* **1993**, *25*, 1755–1762. [CrossRef]
9. Martin, M.P.; Cordier, S.; Balesdent, J.; Arrouays, D. Periodic solutions for soil carbon dynamics equilibriums with time-varying forcing variables. *Ecol. Model.* **2007**, *204*, 523–530. [CrossRef]
10. Parshotam, A. The Rothamsted soil-carbon turnover model—Discrete to continuous form. *Ecol. Model.* **1996**, *86*, 283–289. [CrossRef]
11. Coleman, K.; Jenkinson, D.; Crocker, G.; Grace, P.; Klir, J.; Körschens, M.; Poulton, P.; Richter, D. Simulating trends in soil organic carbon in long-term experiments using RothC-26.3. *Geoderma* **1997**, *81*, 29–44. [CrossRef]
12. Morais, T.G.; Teixeira, R.F.; Domingos, T. Detailed global modelling of soil organic carbon in cropland, grassland and forest soils. *PLoS ONE* **2019**, *14*, e0222604. [CrossRef]
13. Formaggia, L.; Scotti, A. Positivity and conservation properties of some integration schemes for mass action kinetics. *SIAM J. Numer. Anal.* **2011**, *49*, 1267–1288. [CrossRef]
14. Martiradonna, A.; Colonna, G.; Diele, F. GeCo: Geometric Conservative non-standard schemes for biochemical systems. *Appl. Numer. Math.* **2020**, *155*, 38–57. [CrossRef]
15. Diele, F.; Marangi, C. Geometric Numerical Integration in Ecological Modelling. *Mathematics* **2020**, *8*, 25. [CrossRef]
16. Francaviglia, R.; Baffi, C.; Nassisi, A.; Cassinari, C.; Farina, R. Use of the "ROTHC" model to simulate soil organic carbon dynamics on a silty-loam inceptisol in Northern Italy under different fertilization practices. *EQA-Int. J. Environ. Qual.* **2013**, *11*, 17–28. [CrossRef]
17. Farina, R.; Coleman, K.; Whitmore, A.P. Modification of the RothC model for simulations of soil organic C dynamics in dryland regions. *Geoderma* **2013**, *200*, 18–30. [CrossRef]
18. Chen, Y.J.; Ascher, U.M.; Pai, D.K. Exponential Rosenbrock–Euler integrators for elastodynamic simulation. *IEEE Trans. Vis. Comput. Graph.* **2017**, *24*, 2702–2713. [CrossRef]
19. Hochbruck, M.; Ostermann, A.; Schweitzer, J. Exponential Rosenbrock-type methods. *SIAM J. Numer. Anal.* **2009**, *47*, 786–803. [CrossRef]
20. Mickens, R.E. *Applications of Nonstandard Finite Difference Schemes*; World Scientific: Singapore, 2000.
21. Martiradonna, A. NSRothC-NonStandard RothC Models in Matlab. 2021. Available online: https://github.com/CnrIacBaGit/NSRothC (accessed on 8 February 2021).
22. Rothamsted Experimental Station, Great Britain. *Rothamsted: Guide to the Classical Field Experiments*; AFRC Institute of Arable Crops Research: Hertsfordshire, UK, 1991.
23. Pansu, M.; Sarmiento, L.; Rujano, M.; Ablan, M.; Acevedo, D.; Bottner, P. Modeling organic transformations by microorganisms of soils in six contrasting ecosystems: Validation of the MOMOS model. *Glob. Biogeochem. Cycles* **2010**, *24*, GB1008. [CrossRef]
24. Pansu, M.; Machado, D.; Bottner, P.; Sarmiento, L. Modelling microbial exchanges between forms of soil nitrogen in contrasting ecosystems. *Biogeosci. Discuss.* **2014**, *11*, 915–927. [CrossRef]
25. Hammoudi, A.; Iosifescu, O.; Bernoux, M. Mathematical analysis of a nonlinear model of soil carbon dynamics. *Differ. Equ. Dyn. Syst.* **2015**, *23*, 453–466. [CrossRef]
26. Wang, Y.; Chen, B.; Wieder, W.R.; Leite, M.; Medlyn, B.E.; Rasmussen, M.; Smith, M.J.; Agusto, F.B.; Hoffman, F.; Luo, Y. Oscillatory behaviour of two nonlinear microbial models of soil carbon decomposition. *Biogeosciences* **2014**, *11*, 1817–1831. [CrossRef]
27. Hammoudi, A.; Iosifescu, O. Mathematical Analysis of a Chemotaxis-Type Model of Soil Carbon Dynamic. *Chin. Ann. Math. Ser. B* **2018**, *39*, 253–280. [CrossRef]
28. Elzein, A.; Balesdent, J. Mechanistic simulation of vertical distribution of carbon concentrations and residence times in soils. *Soil Sci. Soc. Am. J.* **1995**, *59*, 1328–1335. [CrossRef]
29. Caspari, T.; van Lynden, G.; Bai, Z. Land Degradation Neutrality: An evaluation of methods. In *Report Commissioned by German Federal Environment Agency (UBA)*; Umweltbundesamt: Dessau-Roßlau, Germany, 2015.
30. Marangi, C.; Casella, F.; Diele, F.; Lacitignola, D.; Martiradonna, A.; Provenzale, A.; Ragni, S. Mathematical tools for controlling invasive species in Protected Areas. In *Mathematical Approach to Climate Change and its Impacts*; Springer: Berlin/Heidelberg, Germany, 2020; pp. 211–237.
31. Lacitignola, D.; Diele, F.; Marangi, C. Dynamical scenarios from a two-patch predator–prey system with human control–Implications for the conservation of the wolf in the Alta Murgia National Park. *Ecol. Model.* **2015**, *316*, 28–40. [CrossRef]
32. Baker, C.M.; Diele, F.; Lacitignola, D.; Marangi, C.; Martiradonna, A. Optimal control of invasive species through a dynamical systems approach. *Nonlinear Anal. Real World Appl.* **2019**, *49*, 45–70. [CrossRef]

The Role of Spectral Complexity in Connectivity Estimation

Elisabetta Vallarino [1], Alberto Sorrentino [1,2,*], Michele Piana [1,2] and Sara Sommariva [1]

[1] Dipartimento di Matematica, Università di Genova, via Dodecaneso 35, 16146 Genova, Italy; vallarino@dima.unige.it (E.V.); piana@dima.unige.it (M.P.); sommariva@dima.unige.it (S.S.)
[2] CNR–SPIN, Corso Ferdinando Maria Perrone 24, 16152 Genova, Italy
* Correspondence: sorrentino@dima.unige.it

Abstract: The study of functional connectivity from magnetoecenphalographic (MEG) data consists of quantifying the statistical dependencies among time series describing the activity of different neural sources from the magnetic field recorded outside the scalp. This problem can be addressed by utilizing connectivity measures whose computation in the frequency domain often relies on the evaluation of the cross-power spectrum of the neural time series estimated by solving the MEG inverse problem. Recent studies have focused on the optimal determination of the cross-power spectrum in the framework of regularization theory for ill-posed inverse problems, providing indications that, rather surprisingly, the regularization process that leads to the optimal estimate of the neural activity does not lead to the optimal estimate of the corresponding functional connectivity. Along these lines, the present paper utilizes synthetic time series simulating the neural activity recorded by an MEG device to show that the regularization of the cross-power spectrum is significantly correlated with the signal-to-noise ratio of the measurements and that, as a consequence, this regularization correspondingly depends on the spectral complexity of the neural activity.

Keywords: regularization theory; multivariate stochastic processes; cross-power spectrum; magnetoencephalography; MEG; functional connectivity; spectral complexity

1. Introduction

Magnetoencephalography (MEG) provides high temporal resolution measurements of the magnetic field associated to neural currents. The MEG device relies on superconducting sensors, named SQUIDs, organized in a helmet array close and around the scalp. MEG experimental time series can be used essentially to address two neuroscientific problems, whose solution requires both an accurate mathematical modelization based on Maxwell's equations, and the numerical reduction of such formal models [1].

The first problem is concerned with the dynamical ill-posed inverse problem of estimating parameters associated with the neural sources inducing the magnetic field signal [2–8]. The second problem is concerned with the quantification of the interactions among neural sources located in different cortical areas and intertwined by means of either anatomical or functional connectivity [9–14].

In particular, the connectivity problem can be addressed by either computing proper connectivity metrics directly from the experimental time series provided by the MEG sensors or searching for connections in the source space, i.e., among the neural time series estimated as solutions of the inversion process. This second approach has the advantages of reducing the impact of volume conduction and providing results that can be more easily interpreted in the framework of neuroscientific models [15–17]. Several approaches for identifying connectivity paths rely on physiological models assuming that the functional communication between different brain areas is regulated by the synchronization of their activity at specific temporal frequencies [18,19]. This implies that, for these models, the frequency domain represents the natural computational framework where to perform the connectivity analysis. This is the reason why, in the present paper, we focus on the analysis of the cross-power spectrum, which is the mathematical quantity of reference for the computation of most frequency-domain connectivity measures [11,20,21]. From an operational viewpoint, the computation of the cross-power spectrum in the source space typically relies on a two-step procedure: first the neural activity is estimated by applying a regularized inversion method on the recorded time series and then the cross-power spectrum is computed from the Fourier transform of the estimated neural time series [22].

This paper investigates how to optimally select the regularization parameter in the inversion procedure in order to obtain the best possible estimate of the neural cross-power spectrum. In fact, we consider the Tikhonov method (better known as Minimum Norm Estimation (MNE) in the MEG world [2]) as it is one of the most commonly used inverse methods in connectivity studies [23–25]; we study the interplay between the regularization parameter providing the reconstructed neural time series minimizing the relative error in ℓ_2-norm, and the one that allows the optimal estimate of the cross-power spectrum according to the normalized Frobenius norm. The conceptual motivation of this problem is illustrated in Figure 1, which tentatively sketches the result of recent investigations in MEG-based connectivity research, i.e., that the regularization parameter leading to the optimal estimate of the neural activity may not lead to the optimal estimate of the cross-power spectrum and vice versa. In fact, in [26] the authors used numerical simulations to compare the parameter that provides the best estimate of the power spectrum with the one that provides the best estimate of coherence and showed that the latter is in general two orders of magnitude smaller than the former. More recently, Vallarino et al. [27] addressed an analogous problem via analytical computations, considering a simplified model. Specifically, under the assumption that the neural time series are realizations of white Gaussian processes, the authors proved that the parameter providing the best neural activity estimate is more than twice as large as the one providing the best estimate of the cross-power spectrum.

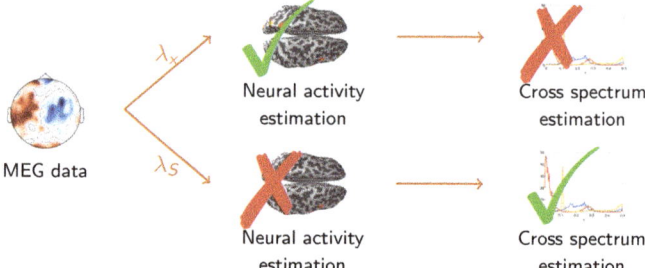

Figure 1. Schematic representation of the differences between the regularization parameter providing the best time series estimate (λ_x) and the one providing the best cross-power spectrum estimate (λ_S). The first one provides an optimal reconstruction of the neural activity, but it may not lead to an optimal estimate of the cross-power spectrum; vice versa, λ_S provides an optimal reconstruction of the cross-power spectrum at the expense of a sub-optimal estimate of the time series.

The present paper focuses on an analysis of the impact of spectral complexity of the actual neural signal on the value of the two regularization parameters. Specifically, we simulate synthetic MEG signals and discuss how the optimal parameter for the reconstruction of the cross-power spectrum depends on its signal-to-noise ratio and how this latter quantity is related to the spectral richness of the neural sources. To this aim, we considered a simulation setting in which the signal is modeled as a multivariate autoregressive process.

This paper is structured as follows. Section 2 introduces the problem in a formal way. Section 3 describes how the synthetic data are simulated and analyzed. Section 4 presents the results of the analysis. Our conclusions are offered in Section 5.

2. Definition of the Problem

2.1. Forward Model

Let $\mathbf{X}(t) = (X_1(t),\ldots,X_N(t))^\top \in R^N$ be a multivariate stationary stochastic process whose realizations $\mathbf{x}(t)$ can not be observed and let $\mathbf{Y}(t) = (Y_1(t),\ldots,Y_M(t))^\top \in R^M$ be the process whose realizations $\mathbf{y}(t)$ are used to infer information on $\mathbf{x}(t)$. Let $\mathbf{Y}(t)$ and $\mathbf{X}(t)$ be related by the following equation

$$\mathbf{Y}(t) = \mathbf{G}\mathbf{X}(t) + \mathbf{N}(t), \tag{1}$$

where $\mathbf{G} \in R^{M\times N}$ is the forward matrix and $\mathbf{N}(t) = (N_1(t),\ldots,N_M(t))^\top \in R^M$ is the measurement noise, that is here assumed to be a white Gaussian process with zero mean and covariance matrix $\alpha^2 \mathbf{I}$, i.e., $\mathbf{N}(t) \sim \mathcal{N}(0,\alpha^2 \mathbf{I})$, independent from $\mathbf{X}(t)$.

2.2. Cross-Power Spectrum

We are interested in reconstructing the cross-power spectrum of $\mathbf{X}(t)$, which describes the statistical dependencies between each pair of time series $(X_j(t), X_k(t))_{j,k\in\{1,\ldots,N\}}$. The cross-power spectrum is a one parameter family of $N \times N$ matrices $\mathbf{S}^\mathbf{x}(f)$, whose (j,k)-th element is defined as

$$S^\mathbf{x}_{j,k}(f) = \lim_{T\to\infty} \frac{1}{T} E[\hat{X}_j(f,T)\hat{X}_k(f,T)^H], \tag{2}$$

where $\hat{X}_j(f,T)$ is the Fourier transform of $X_j(t)$ over the interval $[0,T]$, defined as

$$\hat{X}_j(f,T) = \int_0^T X_j(t)e^{-2\pi ift}\,\mathrm{d}t \tag{3}$$

and X^H is the Hermitian transpose of X [28].

Given a realization $\mathbf{x}(t)$ of the process $\mathbf{X}(t)$, the cross-power spectrum $S^\mathbf{x}(f)$ can be estimated via the Welch method [29], which consists in partitioning the data in P overlapping segments multiplied by a window function, $\{w(t)\mathbf{x}^p(t)\}_{p=1}^P$, computing their discrete Fourier transform $\hat{\mathbf{x}}^p(f) = \frac{1}{L}\sum_{t=0}^{L-1} \mathbf{x}^p(t)w(t)e^{\frac{-2\pi itf}{L}}$ and averaging:

$$\mathbf{S}^\mathbf{x}(f) = \frac{L}{PW}\sum_{p=1}^P \hat{\mathbf{x}}^p(f)\hat{\mathbf{x}}^p(f)^H, \quad f=0,\ldots,L-1, \tag{4}$$

where L is the length of each segment and $W = \frac{1}{L}\sum_{t=0}^{L-1} w(t)^2$.

It is often the case that the data reach a high dimension, and visual inspection of the cross-power spectrum is not doable. In such cases a metric that describes the spectral properties of the signals would be useful. Here we use the spectral complexity coefficient, defined as follows.

Definition 1. *Given a realization $\mathbf{x}(t)$ of the process $\mathbf{X}(t)$, and the corresponding cross-power spectrum $\mathbf{S}^\mathbf{x}(f)$, we define the spectral complexity coefficient as the average of the elements of the upper triangular part of the matrix obtained by computing the squared ℓ_2-norm over the frequencies of $\mathbf{S}^\mathbf{x}_{j,k}(f)$, $j,k=1,\ldots,N$, that is*

$$c = \frac{2}{N(N+1)}\sum_{j=1}^N \sum_{k=j}^N \sum_f \left|S^\mathbf{x}_{j,k}(f)\right|^2. \tag{5}$$

The spectral complexity coefficient assumes small values if the elements of the cross-power spectrum are flat, that is when time series do not present any periodic trend and no dependencies among the pairs of time series are present. On the contrary, it assumes large values if the elements of the cross-power spectrum are peaked, that is when time series present periodic trends and complex relations among them. Finally, we observe that in Definition 1 only the elements on the upper triangular part of $\mathbf{S}^{\mathbf{x}}(f)$ are considered because $\mathbf{S}^{\mathbf{x}}(f)$ is Hermitian.

2.3. Two-Step Approach for Cross-Power Spectrum Estimation

Let us now consider a realization of Equation (1). Further than an estimate of the hidden data $\mathbf{x}(t)$, an estimate of the cross-power spectrum can be obtained from $\mathbf{y}(t)$. Such estimate can be achieved through a two-step process [22]:

i. First, a regularized estimate $\mathbf{x}_\lambda(t)$ of $\mathbf{x}(t)$ is obtained by solving the inverse problem associated to Equation (1). Here we consider the Tikhonov regularized solution [30] of the problem which is defined as

$$\mathbf{x}_\lambda(t) = \arg\min_{\mathbf{x}(t)} \left\{ \|\mathbf{G}\mathbf{x}(t) - \mathbf{y}(t)\|_2^2 + \lambda \|\mathbf{x}(t)\|_2^2 \right\}; \qquad (6)$$

where λ is a proper regularization parameter and $\|\cdot\|_2$ is the ℓ_2-norm.

ii. Then, the corresponding estimate of the cross-power spectrum $\mathbf{S}^{\mathbf{x}_\lambda}(f)$ is computed from the reconstructed time series using the Welch method, as described in the previous section.

Remark 1. *In many applied fields, Tikhonov regularization with an ℓ_2 penalty term has been outdated by more modern techniques that use sparsity-inducing penalization terms such as ℓ_1 or ℓ_p with $0 < p < 1$. Indeed, also in the M/EEG literature there has been considerable effort in developing ℓ_1 solutions [31,32], and mixed norm solutions [33]; both these approaches have proved to provide superior performances in terms of localization of neural activity. However, these newer methods are seldom used in connectivity studies, for good reasons: ℓ_1 solutions computed independently at each time point produce extremely jittering reconstructions, resulting in highly sparse time courses that are not suitable for computing connectivity metrics. Mixed norms, that have been developed precisely to overcome this jittering problem, are computationally very expensive, and this actually prevents their use with the large datasets typically involved in connectivity studies.*

When applying the described two-step process, the regularization parameter λ in Equation (6) has to be set for the computation of $\mathbf{x}_\lambda(t)$. Thus, the problem naturally arises of the choice of such parameter, which can be set in order to optimally reconstruct either $\mathbf{x}_\lambda(t)$ or $\mathbf{S}^{\mathbf{x}_\lambda}(f)$. We define optimality through the minimization of the normalized norm of the discrepancy between the true and the reconstructed time series and cross-power spectra as follows.

Definition 2. *Given the regularized solution (6) and the cross-power spectrum (4), we define the optimal regularization parameter for the reconstruction of $\mathbf{x}(t)$ as*

$$\lambda_{\mathbf{x}}^* = \arg\min_\lambda \varepsilon_{\mathbf{x}}(\lambda) \quad \text{with} \quad \varepsilon_{\mathbf{x}}(\lambda) = \frac{\sum_t \|\mathbf{x}_\lambda(t) - \mathbf{x}(t)\|_2^2}{\sum_t \|\mathbf{x}_\lambda(t)\|_2^2 + \sum_t \|\mathbf{x}(t)\|_2^2} \; ; \qquad (7)$$

and the optimal parameter for the reconstruction of $\mathbf{S}^{\mathbf{x}}(f)$ as

$$\lambda_{\mathbf{S}}^* = \arg\min_\lambda \varepsilon_{\mathbf{S}}(\lambda) \quad \text{with} \quad \varepsilon_{\mathbf{S}}(\lambda) = \frac{\sum_f \|\mathbf{S}^{\mathbf{x}_\lambda}(f) - \mathbf{S}^{\mathbf{x}}(f)\|_F^2}{\sum_f \|\mathbf{S}^{\mathbf{x}_\lambda}(f)\|_F^2 + \sum_f \|\mathbf{S}^{\mathbf{x}}(f)\|_F^2} \; ; \qquad (8)$$

where $\|\cdot\|_F$ is the Frobenius norm; $\varepsilon_{\mathbf{x}}(\lambda)$ and $\varepsilon_{\mathbf{S}}(\lambda)$ will be called reconstruction errors.

The reconstruction errors range from 0 to 1 and penalize both a too small and a too large value of λ. In fact, they assume their maximum value when either λ is very high and thus $\mathbf{x}_\lambda(t)$ is negligible with respect to $\mathbf{x}(t)$, or when λ is too small and thus, vice versa, $\mathbf{x}(t)$ is negligible with respect to $\mathbf{x}_\lambda(t)$. This definition may appear overly complex compared to, e.g., a mere ℓ_2-norm of the difference; however, in the presence of sparse data where only a few time series are non-zero, the simple ℓ_2-norm would prefer a very high regularization parameter in order to minimize the error on the null time series, at the expense of the error on the non-zero ones; our definition aims to cope with this limitation of the ℓ_2-norm. A similar definition has been introduced in [34].

In experimental contexts, where $\mathbf{x}(t)$ is not known, the choice of the optimal regularization parameter is crucial. This matter is widely discussed in literature [35–38], and many criteria have been proposed. Such criteria apply to Equation (1) and can be used to set the regularization parameter $\lambda_\mathbf{x}$. A possibility is to set the regularization parameter as a function of the signal-to-noise ratio (SNR), which describes the level of the desired signal with respect to that of the measurement noise; for Equation (1) the SNR is defined as follows.

Definition 3. *Consider the linear model (1). We define the signal-to-noise ratio of $\mathbf{X}(t)$ related to such model as*

$$\mathrm{SNR}^{\mathbf{X}} = 10\log_{10}\left(\frac{\sum_t \|\mathbf{G}\mathbf{X}(t)\|_2^2}{\sum_t \|\mathbf{N}(t)\|_2^2}\right). \tag{9}$$

To the best of our knowledge, the choice of the optimal regularization parameter for the reconstruction of the cross-power spectrum has never been related to the signal-to-noise ratio. This relation is presented in Section 4; however we first need to relate the cross-power spectrum of the unknown $\mathbf{S}^{\mathbf{X}}(f)$ with that of the data $\mathbf{S}^{\mathbf{Y}}(f)$.

By computing the cross-power spectrum of both sides of Equation (1) and from the linearity of the Fourier transform it follows that

$$\mathbf{S}^{\mathbf{Y}}(f) = \mathbf{G}\mathbf{S}^{\mathbf{X}}(f)\mathbf{G}^\top + \mathbf{S}^{\mathbf{N}}(f), \tag{10}$$

where the mixed terms $\mathbf{S}^{\mathbf{XN}}(f)$ and $\mathbf{S}^{\mathbf{NX}}(f)$ are negligible thanks to the independence between $\mathbf{X}(t)$ and $\mathbf{N}(t)$. Just like for Equation (1), we can define the signal-to-noise ratio for Equation (10) as follows.

Definition 4. *Consider the linear model (10). We define the signal-to-noise ratio of $\mathbf{S}^{\mathbf{X}}(f)$ related to such model as*

$$\mathrm{SNR}^{\mathbf{S}} = 10\log_{10}\left(\frac{\sum_f \|\mathbf{G}\mathbf{S}^{\mathbf{X}}(f)\mathbf{G}^\top\|_F^2}{\sum_f \|\mathbf{S}^{\mathbf{N}}(f)\|_F^2}\right). \tag{11}$$

This definition is in line with the definition of $\mathrm{SNR}^{\mathbf{X}}$ for the signal, the main difference being in the use of the Frobenius norm rather than the ℓ_2-norm, motivated by the fact that we are working with matrices rather than vectors.

2.4. Multivariate Autoregressive Models

To model the statistical relationships between the different components of the stochastic process $\mathbf{X}(t)$, in this work the latter is assumed to follow a stable multivariate autoregressive model of order P [39].

Definition 5. *A zero-mean stochastic process $\mathbf{X}(t) \in \mathbb{R}^N$ is said to follow a multivariate autoregressive (MVAR) model of order P if*

$$\mathbf{X}(t) = \sum_{k=1}^{P} \mathbf{A}(k)\mathbf{X}(t-k) + \boldsymbol{\varepsilon}(t) \qquad \forall t, \tag{12}$$

where $\mathbf{A}(k) \in \mathbb{R}^{N \times N}$ are fixed coefficient matrices, and $\boldsymbol{\varepsilon}(t) \in \mathbb{R}^N$ is a white Gaussian noise process.

Moreover, the MVAR model described by Equation (12) is said to be stable if

$$\det(\mathbf{I} - \sum_{k=1}^{P} \mathbf{A}(k) z^k) \neq 0 \quad \forall\, z \in \mathbb{C} \text{ s.t. } |z| \leq 1, \tag{13}$$

where $\mathbf{I}$ is the identity matrix of size N.

Remark 2. *From Equation (12) it can be easily seen that the process $\mathbf{X}(t)$ is uniquely determined by the process $\varepsilon(t)$ and by the first P time points, $\mathbf{X}(0), \ldots, \mathbf{X}(P-1)$. Indeed, consider for example an MVAR model of order 1 (a similar proof holds for the general case $P > 1$ and can be found in [39]); then, for each time point t*

$$\begin{aligned}
\mathbf{X}(t) &= \mathbf{A}(1)\mathbf{X}(t-1) + \varepsilon(t) \\
&= \mathbf{A}(1)^2 \mathbf{X}(t-2) + \mathbf{A}(1)\varepsilon(t-1) + \varepsilon(t) \\
&= \mathbf{A}(1)^t \mathbf{X}(0) + \sum_{k=0}^{t-1} \mathbf{A}(1)^k \varepsilon(t-k)\,.
\end{aligned}$$

Such a model satisfies the stability condition defined in Equation (13) if all the eigenvalues of the coefficient matrix $\mathbf{A}(1)$ have a modulus of less then one, the condition that guarantees the sequence of exponential matrices $\left\{ \mathbf{A}(1)^k \right\}_k$ is absolutely summable.

According to Equation (12), if the process $\mathbf{X}(t)$ follows an MVAR model, then at each time point the value of $\mathbf{X}(t)$ can be derived as a weighted sum of the values of the process at the previous P time points, $\mathbf{X}(t-1), \ldots, \mathbf{X}(t-P)$, plus a random perturbation $\varepsilon(t)$. In particular, the (i,j)-th elements of the coefficient matrices, $a_{ij}(1), \ldots, a_{ij}(P)$, describe how the value of the i-th component of the process depends on the past of the j-th component. Different connectivity patterns, with various levels of complexity, can thus be obtained by tuning the off-diagonal values of the coefficient matrices. Due to their flexibility and simplicity, MVAR models have been used by various authors in the framework of MEG functional connectivity estimation as a benchmark for testing and comparing different connectivity metrics [21,23,34,40–43]. Other models have been proposed to simulate different connectivity patterns, such as coherent sinusoidal time series [26], neural mass models [44,45], or Kuramoto models [46,47]. However a comprehensive comparison of all possible generative models is beyond the scope of this work.

3. Generation and Analysis Pipeline of the MEG Simulated Data

In this section we describe the numerical simulation that led to the main results of our study. First we introduce the continuous MEG forward problem and its discretized version, then we describe how we generated the data and, finally, we describe the inverse model and how we numerically computed the optimal regularization parameters.

3.1. MEG Forward Model

The MEG forward problem aims at computing the magnetic field produced outside the head by an electric current that flows inside the brain. The quasi static approximation of Maxwell's equations provides the local relationship between the recorded magnetic field and the neural currents [1,48,49]. The two equations that are of interest here read as

$$\nabla \times \mathbf{E}(\mathbf{r}, t) = 0 \tag{14}$$

$$\nabla \times \mathbf{B}(\mathbf{r}, t) = \mu_0 \mathbf{J}(\mathbf{r}, t); \tag{15}$$

where $\mathbf{E}(\mathbf{r},t)$ and $\mathbf{B}(\mathbf{r},t)$ are the electric and magnetic fields at location $\mathbf{r}$ and time t, μ_0 is the magnetic permeability in vacuum and $\mathbf{J}(\mathbf{r},t)$ is the total electric current that flows inside the brain. The latter is the sum of two contributions

$$\mathbf{J}(\mathbf{r},t) = \mathbf{J}^p(\mathbf{r},t) + \mathbf{J}^v(\mathbf{r},t), \tag{16}$$

$\mathbf{J}^p(\mathbf{r},t)$ being the primary current directly related to the brain activity, while $\mathbf{J}^v(\mathbf{r},t) = -\sigma(\mathbf{r})\nabla V(\mathbf{r},t)$ is the induced volume current due the non-null conductivity $\sigma(\mathbf{r})$ of the brain, $V(\mathbf{r},t)$ being the electric scalar potential.

The manipulation of Maxwell's Equations leads to the Biot–Savart Equation

$$\mathbf{B}(\mathbf{r},t) = \mathbf{B}_0(\mathbf{r},t) - \frac{\mu_0}{4\pi}\int_\Omega \sigma(\mathbf{r}')\nabla' V(\mathbf{r}',t) \times \frac{\mathbf{r}-\mathbf{r}'}{|r-r'|^3}dv', \tag{17}$$

where Ω is the volume occupied by the brain, the first term $\mathbf{B}_0(\mathbf{r},t) = \frac{\mu_0}{4\pi}\int_\Omega \mathbf{J}(\mathbf{r}',t)\frac{\mathbf{r}-\mathbf{r}'}{|r-r'|^3}dv'$ is the magnetic field induced by the primary current, whereas the second term is related to the volume current.

Solving the forward problem requires the computation of these two contributions knowing the primary current. Although for the first one straightforward numerical integration is feasible, for the second one it is common to model the head as the union of nested homogeneous volumes $\{\Omega_j\}_{j=1,\dots,J}$ and to replace volume integration with surface integration. In this way, the Biot–Savart equation becomes

$$\mathbf{B}(\mathbf{r},t) = \mathbf{B}_0(\mathbf{r},t) + \frac{\mu_0}{4\pi}\sum_{i,j}(\sigma_i-\sigma_j)\int_{\partial\Omega_{i,j}} V(\mathbf{r}',t)\frac{\mathbf{r}-\mathbf{r}'}{|r-r'|^3} \times \mathbf{n}_{i,j}(\mathbf{r}')ds', \tag{18}$$

where $\partial\Omega_{i,j}$ is the contact surface between regions Ω_i and Ω_j and $\mathbf{n}_{i,j}(\mathbf{r}')$ is the unit vector normal to the surface $\partial\Omega_{i,j}$ at $\mathbf{r}'$ from region i to region j.

The forward problem can now be solved by computing the second term at the right hand side of Equation (18) after having computed $V(\mathbf{r},t)$ by solving the equation

$$\nabla\cdot\mathbf{J}^p(\mathbf{r},t) - \nabla\cdot(\sigma(\mathbf{r})\nabla V(\mathbf{r},t)) = 0, \tag{19}$$

which follows from Equation (15) by applying the divergence.

For further details on the MEG forward problem, we refer the reader to [1].

3.2. The Leadfield Matrix

Experimental contexts require the discretization of the forward problem. This involves a discretization of both the volume occupied by the brain and the volume outside the head.

When using a distributed model for the primary current $\mathbf{J}^p$, the brain volume is uniformly divided in N small parcels. If N is sufficiently big and thus each parcel has a sufficient small area, the activity in each brain parcel is approximated by a point-like source, henceforth denoted as dipole. From a mathematical point of view, each dipole is a vector whose strength and direction represent the intensity and orientation of the primary current in the corresponding brain area [1].

As for the volume outside the brain, it is natural to discretize it in correspondence of the MEG sensors. Let us denote the measured magnetic field as $\mathbf{y}(t) = (y_1(t),\dots,y_M(t))$. Now, observing that the magnetic field $\mathbf{B}$ depends linearly on the primary current $\mathbf{J}^p$, the magnetic field in correspondence of the sensors of the instrument is

$$\mathbf{y}(t) = \sum_{k=1}^N G(\mathbf{r}_k)\mathbf{q}_k(t) + \mathbf{n}(t), \tag{20}$$

where $\mathbf{r}_k$, $k=1,\dots,N$, is the location of the k-th brain parcel, $G(\mathbf{r}_k) \in R^{M\times 3}$ is the corresponding leadfield matrix, and $\{\mathbf{q}_k(t)\}_{k=1,\dots,N}$ are the electric current intensities along the three orthogonal direction of the N dipoles within the brain at time t and $\mathbf{n}(t)$ is the measurement noise. The l-th column of $G(\mathbf{r}_k)$ contains the measurement at a sensor level when a unit current dipole is placed at location $\mathbf{r}_k$ and oriented along the l-th orthogonal direction.

In this work, we assume dipoles to be located only on the brain cortical mantle and their orientation to be normal to the local cortical surface [50]. In this case, the electric current intensities are scalar quantities (we refer to them as $\{q_k\}_{k=1,\ldots,N}$) and the leadfield matrices are column vectors (we refer to them as $\{G_k\}_{k=1,\ldots,N}$).

Let us define
$$\mathbf{x}(t) := (q_1(t), \ldots, q_N(t)) \tag{21}$$
and
$$\mathbf{G} := [G_1, \ldots, G_N] \in R^{M \times N}; \tag{22}$$
reassembling Equations (21) and (22) in to Equation (20), we get
$$\mathbf{y}(t) = \mathbf{G}\mathbf{x}(t) + \mathbf{n}(t), \tag{23}$$
which can be interpreted as a realization of Equation (1). From now on we refer to $\mathbf{G}$ as to the leadfield matrix.

For the simulation presented in this work, we used the leadfield matrix available in the sample dataset of MNE Python [51]. We selected magnetometers and set a fixed orientation. For computational reasons, the available source space, containing 1884 sources, was uniformly down-sampled to obtain 274 sources. Thus, our model has $M = 102$ sensors and $N = 274$ dipole sources.

3.3. Data Generation

We simulated $N_{mod} = 10$ pairs of active sources, $(z_1(t), z_2(t))^\top$, with unidirectional coupling from the first to the second; their time series follow a multivariate autoregressive (MVAR) model of order $P = 5$ [39,43]

$$\begin{pmatrix} z_1(t) \\ z_2(t) \end{pmatrix} = \sum_{k=1}^{P} \begin{pmatrix} a_{1,1}(k) & 0 \\ a_{2,1}(k) & a_{2,2}(k) \end{pmatrix} \begin{pmatrix} z_1(t-k) \\ z_2(t-k) \end{pmatrix} + \begin{pmatrix} \varepsilon_1(t) \\ \varepsilon_2(t) \end{pmatrix}, \quad t = P, \ldots, T. \tag{24}$$

The non-zero elements $a_{i,j}(k)$ of the coefficient matrices were drawn from a normal distribution of zero mean and standard deviation γ, and T = 10,000. We retained only coefficient matrices providing (i) a stable MVAR model [39] and (ii) pairs of signals $(z_1(t), z_2(t))^\top$ such that the ℓ_2-norm of the strongest one was less than three times the ℓ_2-norm of the weakest one. In order to obtain time series with different spectral complexity coefficients we set γ to N_{mod} different values randomly drawn in the interval $[0.1, 1]$. The values of the spectral complexity coefficient of the N_{mod} simulated time series are reported in Table 1. Finally, the resulting time series $(z_1(t), z_2(t))^\top$ were normalized by the mean of their standard deviations over time, so that pairs of time series drawn from different models had similar magnitude. Figure 2 shows a sample of the the cross-power spectra among the simulated pairs of time series. The figure shows that for increasing values of the spectral complexity coefficient the cross-power spectrum of the corresponding time series becomes more peaked. Each pair of simulated time series was then assigned to $N_{loc} = 20$ pairs of point like sources randomly chosen in the source space, so that the ratio of the norms of the corresponding columns of the leadfield matrix was close to one, i.e., they had similar intensity at a sensor level, and their distance was grater than 7 cm. The remaining $N - 2$ sources were set to have null activity.

Source space activity was then projected to sensor level by multiplying the simulated source activity by the leadfield matrix and white Gaussian noise was added to obtain $N_{snr} = 6$ levels of SNR$^\mathbf{x}$ evenly spaced in the interval $[-20 \text{ dB}, 5 \text{ dB}]$.

Summarizing, we generated $N_{mod} \cdot N_{loc} \cdot N_{snr} = 1200$ different sensor level configurations. The green box in Figure 3 shows a visual representation of the simulation pipeline.

Table 1. The table reports the values of the spectral complexity coefficients, c_j, associated to each simulated multivariate autoregressive (MVAR) model, m_j, $j = 1, \ldots, N_{mod}$.

Model	1	2	3	4	5	6	7	8	9	10
Spectral complexity coefficient	1.41	1.96	2.14	3.10	3.44	4.17	4.64	5.67	6.69	8.67

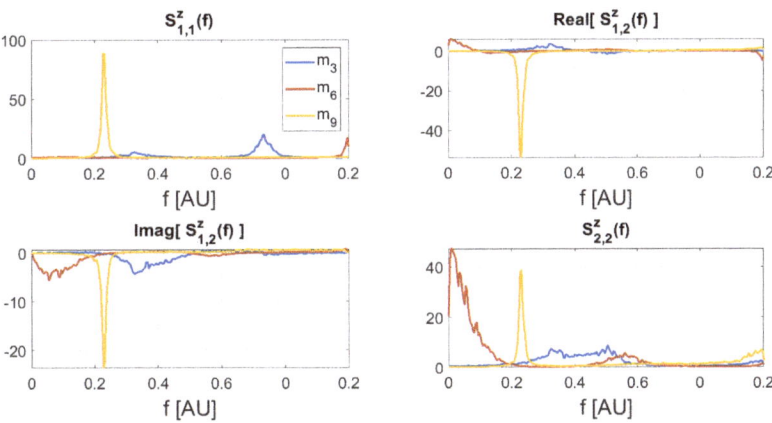

Figure 2. Real and imaginary part of the cross–power spectra of three simulated time series. Higher values of spectral complexity correspond to more peaked spectra.

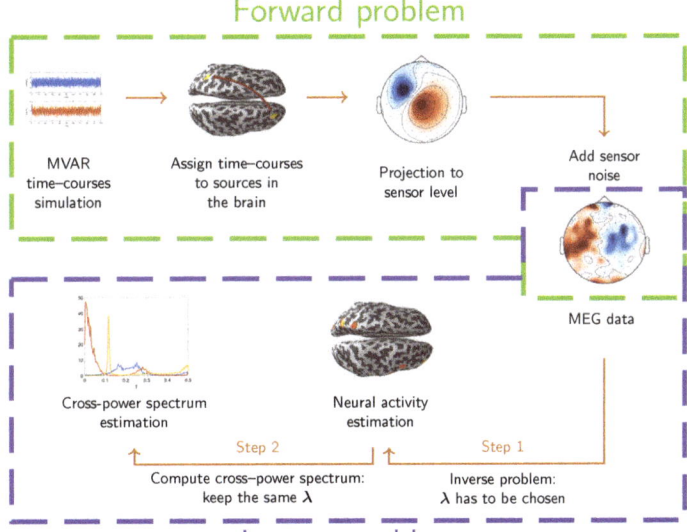

Figure 3. Pipeline of the simulation of the data (green box) and of the estimation of the cross–power spectrum (blue box).

3.4. Inverse Model

Source space time series were reconstructed using the Tikhonov method, also known as the minimum norm estimate (MNE) [2] within the MEG community. For each combination of source time series, source locations, and $\text{SNR}^{\mathbf{X}}$ level, we computed the optimal regularization parameters $\lambda_{\mathbf{x}}^*$ and $\lambda_{\mathbf{S}}^*$ by minimizing the reconstruction errors $\varepsilon_{\mathbf{x}}(\lambda)$ and $\varepsilon_{\mathbf{S}}(\lambda)$, defined in Definition 2. The minimization procedure was achieved by using the Matlab built in function fminsearch that implements an iterative procedure based on the simplex method developed by Lagarias and colleagues [52]. In more detail, $\lambda_{\mathbf{x}}^*$ and $\lambda_{\mathbf{S}}^*$ have been obtained by applying such procedure to $\varepsilon_{\mathbf{x}}(\lambda)$ and $\varepsilon_{\mathbf{S}}(\lambda)$, respectively; in both cases the starting point of the simplex method was set equal to $10^{\left(-\frac{\text{SNR}^{\mathbf{X}}}{10}\right)}$, which corresponds to the optimal value of $\lambda_{\mathbf{x}}$ in the case of white Gaussian signals [27]. The blue box in Figure 3 describes the inverse procedure to obtain an estimate of the cross-power spectrum and stresses the role of the regularization parameter in the two-step process.

4. Results

In this section we illustrate the results of our analysis. We begin with the description of the analytical dependence between $\text{SNR}^{\mathbf{X}}$ and $\text{SNR}^{\mathbf{S}}$, then we highlight how the optimal parameter for the reconstruction of the cross-power spectrum depends on $\text{SNR}^{\mathbf{S}}$ and how this implies that the spectral complexity of the signal is behind such dependence. Finally we show how the reconstruction error $\varepsilon_{\mathbf{S}}(\lambda)$ behaves for different values of the regularization parameters. As a byproduct, this analysis also confirms the results of Vallarino et al. [27] in the case of a more complex setting.

4.1. Analytical Relation between $\text{SNR}^{\mathbf{X}}$ and $\text{SNR}^{\mathbf{S}}$

From Equations (9) and (11) and reminding that $\mathbf{N}(t) \sim \mathcal{N}(0, \alpha^2 \mathbf{I})$ it follows that

$$\text{SNR}^{\mathbf{X}} = 10 \log_{10}\left(\frac{\sum_t \|\mathbf{GX}(t)\|_2^2}{MT\alpha^2}\right); \tag{25}$$

and

$$\text{SNR}^{\mathbf{S}} = 10 \log_{10}\left(\frac{\sum_f \|\mathbf{GS}^{\mathbf{X}}(f)\mathbf{G}^\top\|_F^2}{MN_f\alpha^2}\right), \tag{26}$$

where T is the number of time points and N_f is the number of frequencies used to compute the cross-power spectrum. Observe that to derive Equation (26) we used the fact that the cross-power spectrum of a white noise Gaussian process of zero mean and covariance matrix $\alpha^2 \mathbf{I}$ is $\mathbf{S}^{\mathbf{N}}(f) = \alpha^2 \mathbf{I}$.

By isolating α^2 from Equation (25) and substituting in Equation (26) we obtain

$$\text{SNR}^{\mathbf{S}} = 10 \log_{10}\left(\frac{T^2 M \sum_f \|\mathbf{GS}^{\mathbf{X}}(f)\mathbf{G}^\top\|_F^2}{N_f \sum_t \|\mathbf{GX}(t)\|_2^4}\right) + 2\text{SNR}^{\mathbf{X}}. \tag{27}$$

Equation (27) relates the signal-to-noise ratio of $\mathbf{X}(t)$ with that of $\mathbf{S}^{\mathbf{X}}(f)$. It shows that, for same levels of $\text{SNR}^{\mathbf{X}}$, $\text{SNR}^{\mathbf{S}}$ changes with the spectral complexity coefficient of the signals. In fact, the higher the spectral complexity coefficient, the higher the quantity $\|\mathbf{GS}^{\mathbf{X}}(f)\mathbf{G}^\top\|_F^2$. Intuitively, this happens because when the signal has a higher spectral complexity coefficient its cross-power spectrum is more peaked and thus it is stronger over the cross-power spectrum of the noise with respect to a signal with a lower spectral complexity coefficient.

4.2. Dependence of λ_S^* on SNR^S

As described in Section 3 we simulated several sensor level configurations, based on different combinations of spectral complexity coefficients, source locations, and SNR^X levels. For each configuration, we collected the two optimal parameters λ_x^* and λ_S^* and we investigated their dependence on the signal-to-noise-ratio. In accordance with classical results from inverse theory [36], we found that λ_x^* depends on the signal-to-noise ratio. What is novel here is the relation between λ_S^* and both SNR^X and SNR^S. Indeed, for increasing SNR^X, less regularization is needed, but such dependence varies with the MVAR models. On the other side, the dependence of λ_S^* on SNR^S is neater and does not depend on the models. Figure 4 shows this result; on the left the regularization parameters for the cross-power spectrum reconstruction versus SNR^X are shown, while on the right the same parameters are shown with respect to SNR^S. For the ease of presentation the figure shows the parameters related to one source location; while on the left lines corresponding to different MVAR models have different heights, on the right they overlap.

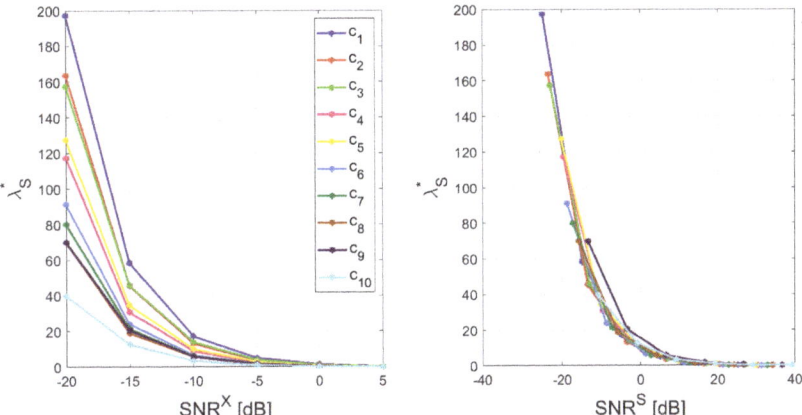

Figure 4. Optimal regularization parameters for the reconstruction of the cross–power spectrum (λ_S^*) as a function of SNR^X (**left**) and SNR^S (**right**). Different colors correspond to different MVAR models. On the left, the lines have different heights, while on the right they overlap, meaning that the dependence of λ_S^* on SNR^S is neater with respect to SNR^X.

4.3. $\lambda_S^* < \lambda_x^*$ and Dependency from the Spectral Complexity

We also investigated the relation between the two optimal regularization parameters. Figure 5 shows the ratio $\frac{\lambda_S^*}{\lambda_x^*}$ versus SNR^X for the simulated MVAR models. The ratio between the two parameters is always smaller than $\frac{1}{2}$, meaning that $\lambda_S^* < \frac{1}{2}\lambda_x^*$, as it was analytically proved in a simplified case in [27]. Further to this, the figure shows that for increasing spectral complexity coefficients this ratio gets smaller. This latter result is directly related to Equation (27). In fact, for same levels of SNR^X, signals with higher spectral complexity have higher SNR^S and, thus, need less regularization.

4.4. The Reconstruction Errors

To show the benefit of using a value of the regularization parameter different from λ_x^* when estimating the cross-power spectrum, in Figure 6 we plotted the reconstruction errors $\varepsilon_S(\lambda)$ as a function of the regularization parameter (normalized by λ_x^*) obtained when considering two illustrative realizations of the simulated sensor data. Specifically, we fixed the locations and time courses of the pair of interacting sources and we considered the corresponding simulated MEG data for two levels of SNR^X, namely $SNR^X = -20$ dB and $SNR^X = 5$ dB. Similar results where obtained when considered the other source configurations.

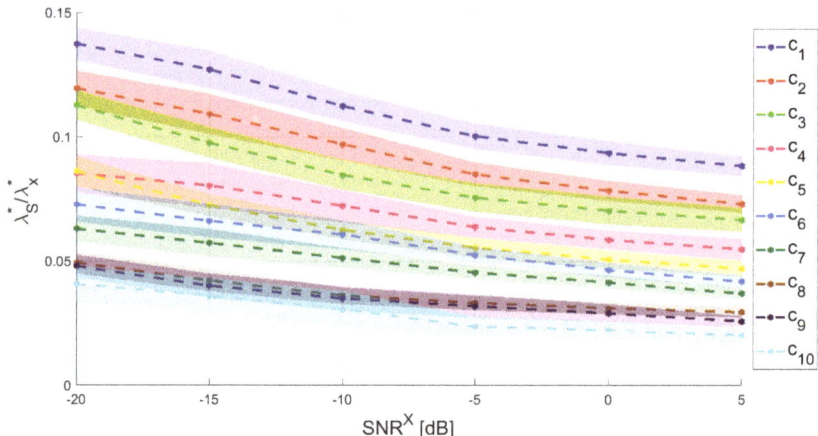

Figure 5. Ratio between the optimal parameters ($\frac{\lambda_S^*}{\lambda_x^*}$) as a function of SNR^X. Different colors correspond to MVAR model with different spectral complexities. Dashed lines are the mean of the ratio over the different sources location; solid colors correspond to the standard deviation of the mean.

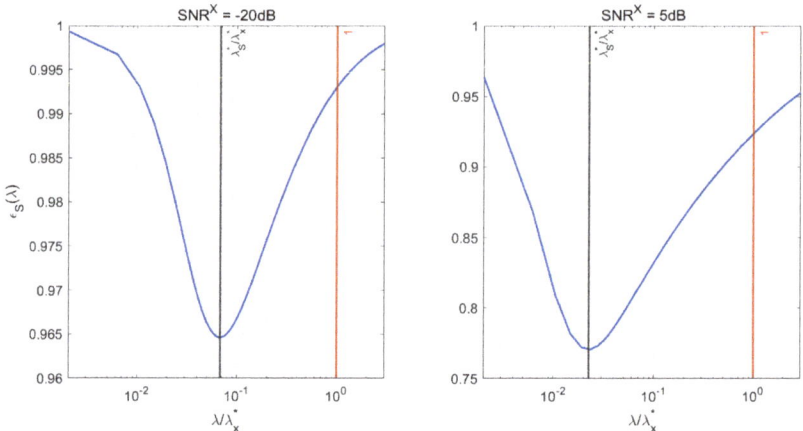

Figure 6. Reconstruction error $\varepsilon_S(\lambda)$ for two simulated data mimicking MEG signals with SNR^X = −20 dB (lowest considered signal–to–noise ratio (SNR), **left** panel) and SNR^X = 5 dB (highest considered SNR, **right** panel). In each panel, black and red vertical lines highlight the values of $\varepsilon_S(\lambda)$ in correspondence of λ_S^* and λ_x^*, respectively.

As shown by Figure 6, for both the values of SNR^X the value of the reconstruction error significantly decreases when λ_S^* is used instead of λ_x^*. Specifically in this simulation, $\varepsilon_S(\lambda)$ drops from 0.99 to 0.96 when SNR^X = −20 dB, and from 0.92 to 0.77 when SNR^X = 5 dB.

Notably, one may observe that the relative reconstruction errors shown in Figure 6 are rather large, being above 90% in the low-SNR case and remaining above 75% even in the high-SNR scenario. We point out that this fact is mainly due to the combined effect of two factors: first, Tikhonov regularization tends to produce reconstructions that are small but non-zero almost everywhere, as it reduces but does not cancel entirely backprojection of noise; second, in our simulations the true activity is zero everywhere but in two points. These two facts inevitably lead to large relative errors that, however, pleasantly decrease for increasing values of SNR^X.

5. Discussion and Conclusions

In the present work, we investigated the role of the spectral complexity of a time series, $\mathbf{x}(t)$, in the design of an optimal inverse technique for estimating its cross-power spectrum, $\mathbf{S}^{\mathbf{x}}(f)$, from indirect measurements of the time series itself. Motivated by an analysis pipeline widely used for estimating brain functional connectivity from MEG data, we reconstructed the cross-power spectrum in two steps: first, we estimated the unknown time series by using the Tikhonov method, then, we computed the cross-power spectrum of the reconstructed time series. In the present work, we used numerical simulations to study how the spectral complexity of $\mathbf{x}(t)$ impacts the value of the regularization parameter that provides the best reconstruction of the cross-power spectrum.

As a first analytical result, we related $\text{SNR}^{\mathbf{X}}$ to $\text{SNR}^{\mathbf{S}}$, i.e., the signal-to-noise ratio of the time series and the signal-to-noise ratio of the corresponding cross-power spectra. The obtained formula suggests that, for a fixed level of $\text{SNR}^{\mathbf{X}}$, $\text{SNR}^{\mathbf{S}}$ depends on the spectral complexity of $\mathbf{x}(t)$: the higher the spectral complexity coefficient the higher $\text{SNR}^{\mathbf{S}}$. Intuitively this happens because a higher value of the spectral complexity coefficient corresponds to a more peaked cross-power spectrum that will emerge over the cross-power spectrum of the noise.

To test the effect of this result on the choice of the Tikohonov regularization parameter in a practical scenario, we simulated a large set of MEG data and applied the described two-step approach for estimating the cross-power spectrum of the underlying neural sources. In details, we simulated 1200 synthetic MEG data with varying $\text{SNR}^{\mathbf{X}}$ generated by pairs of coupled point-like sources at varying locations and with different spectral complexities. For each simulated data, we computed the two parameters providing the best estimates of the time series ($\lambda_{\mathbf{x}}^*$) and of the cross-power spectrum ($\lambda_{\mathbf{S}}^*$), defined as the ones minimizing the relative ℓ_2 norm of the difference between the true and the reconstructed time series/cross-power spectrum according to Definition 2. As shown by Figure 4, the results of our simulations highlighted a high correlation between the values of $\lambda_{\mathbf{S}}^*$ and of $\text{SNR}^{\mathbf{S}}$.

Eventually, we focused on the relationship between the two parameters $\lambda_{\mathbf{x}}^*$ and $\lambda_{\mathbf{S}}^*$, whose ratio is shown in Figure 5. The figure points out that this ratio depends on the spectral complexity of the simulated time series. This fact may be understood in lights of the previous results, as $\lambda_{\mathbf{S}}^*$ depends on $\text{SNR}^{\mathbf{S}}$ that in turns depends on the spectral complexity coefficient. Additionally, we found that, for all the simulated data, $\frac{\lambda_{\mathbf{S}}^*}{\lambda_{\mathbf{x}}^*} < \frac{1}{2}$, in line with the results shown in [27] for a simplified model where the neural time series were assumed to be white Gaussian processes. Moreover, when the spectral complexity coefficient increases ($c > 5$ in our simulations) the ratio between the two parameters approaches 0.01. This agrees with the results shown in [26] where, by simulating sinusoidal signals, the authors suggested to use for connectivity estimation a parameter of two orders of magnitude lower. In fact, our numerical results indicate that the use of $\lambda_{\mathbf{S}}^*$ results in a substantially lower reconstruction error on the cross-power spectrum, particularly when the data have a high SNR.

The present work focuses on the cross-power spectrum as a connectivity metric. Even though the cross-power spectrum is the starting point for the computation of many connectivity metrics it would be interesting to directly investigate the behavior of the Thikonov regularization parameters when using such metrics. Future works will be devoted to this. It is also worth noticing that the definition of optimality when defining the regularization parameters is not univocal, since many metrics can be used. A common example is the area under the curve (AUC), which is the metric that was used in [26]. The use of different metrics would firstly strengthen our results and would also allow a more straightforward comparison with the results of [26]. Finally, the dependence of $\lambda_{\mathbf{S}}^*$ on $\text{SNR}^{\mathbf{S}}$ suggests that an analysis of such dependence could be considered for the definition of a rule for choosing $\lambda_{\mathbf{S}}^*$ in practical scenarios.

Author Contributions: Conceptualization: M.P., S.S., and A.S.; methodology: S.S. and E.V.; software: E.V.; validation and writing: all authors. All authors have read and agreed to the published version of the manuscript.

Funding: E.V., A.S., S.S. and M.P. have been partially supported by Gruppo Nazionale per il Calcolo Scientifico.

Institutional Review Board Statement: Not applicable.

Informed Consent Statement: Not applicable.

Data Availability Statement: Not applicable.

Conflicts of Interest: The authors declare no conflict of interest.

References

1. Hämäläinen, M.; Hari, R.; Ilmoniemi, R.J.; Knuutila, J.; Lounasmaa, O.V. Magnetoencephalography—theory, instrumentation, and applications to noninvasive studies of the working human brain. *Rev. Mod. Phys.* **1993**, *65*, 413. [CrossRef]
2. Hämäläinen, M.S.; Ilmoniemi, R.J. Interpreting magnetic fields of the brain: Minimum norm estimates. *Med. Biol. Eng. Comput.* **1994**, *32*, 35–42. [CrossRef]
3. Calvetti, D.; Pascarella, A.; Pitolli, F.; Somersalo, E.; Vantaggi, B. A hierarchical Krylov–Bayes iterative inverse solver for MEG with physiological preconditioning. *Inverse Probl.* **2015**, *31*, 125005. [CrossRef]
4. Costa, F.; Batatia, H.; Oberlin, T.; D'Giano, C.; Tourneret, J.Y. Bayesian EEG source localization using a structured sparsity prior. *NeuroImage* **2017**, *144*, 142–152. [CrossRef] [PubMed]
5. Sorrentino, A.; Piana, M. Inverse Modeling for MEG/EEG data. In *Mathematical and Theoretical Neuroscience*; Springer: Cham, Switzerland, 2017; pp. 239–253.
6. Bekhti, Y.; Lucka, F.; Salmon, J.; Gramfort, A. A hierarchical Bayesian perspective on majorization-minimization for non-convex sparse regression: Application to M/EEG source imaging. *Inverse Probl.* **2018**, *34*, 085010. [CrossRef]
7. Luria, G.; Duran, D.; Visani, E.; Sommariva, S.; Rotondi, F.; Sebastiano, D.R.; Panzica, F.; Piana, M.; Sorrentino, A. Bayesian multi-dipole modelling in the frequency domain. *J. Neurosci. Methods* **2019**, *312*, 27–36. [CrossRef] [PubMed]
8. Ilmoniemi, R.J.; Sarvas, J. *Brain Signals: Physics and Mathematics of MEG and EEG*; The MIT Press: Cambridge, MA, USA, 2019
9. Geweke, J. Measurement of linear dependence and feedback between multiple time series. *J. Am. Stat. Assoc.* **1982**, *77*, 304–313. [CrossRef]
10. Baccalá, L.A.; Sameshima, K. Partial directed coherence: A new concept in neural structure determination. *Biol. Cybern.* **2001**, *84*, 463–474. [CrossRef]
11. Nolte, G.; Bai, O.; Wheaton, L.; Mari, Z.; Vorbach, S.; Hallett, M. Identifying true brain interaction from EEG data using the imaginary part of coherency. *Clin. Neurophysiol.* **2004**, *115*, 2292–2307. [CrossRef]
12. Pereda, E.; Quiroga, R.Q.; Bhattacharya, J. Nonlinear multivariate analysis of neurophysiological signals. *Prog. Neurobiol.* **2005**, *77*, 1–37. [CrossRef] [PubMed]
13. Sakkalis, V. Review of advanced techniques for the estimation of brain connectivity measured with EEG/MEG. *Comput. Biol. Med.* **2011**, *41*, 1110–1117. [CrossRef] [PubMed]
14. Chella, F.; Marzetti, L.; Pizzella, V.; Zappasodi, F.; Nolte, G. Third order spectral analysis robust to mixing artifacts for mapping cross-frequency interactions in EEG/MEG. *NeuroImage* **2014**, *91*, 146–161. [CrossRef] [PubMed]
15. Schoffelen, J.M.; Gross, J. Source connectivity analysis with MEG and EEG. *Hum. Brain Mapp.* **2009**, *30*, 1857–1865. [CrossRef] [PubMed]
16. Barzegaran, E.; Knyazeva, M.G. Functional connectivity analysis in EEG source space: The choice of method. *PLoS ONE* **2017**, *12*, e0181105. [CrossRef] [PubMed]
17. Van de Steen, F.; Faes, L.; Karahan, E.; Songsiri, J.; Valdes-Sosa, P.A.; Marinazzo, D. Critical comments on EEG sensor space dynamical connectivity analysis. *Brain Topogr.* **2019**, *32*, 643–654. [CrossRef]
18. Fries, P. A mechanism for cognitive dynamics: Neuronal communication through neuronal coherence. *Trends Cogn. Sci.* **2005**, *9*, 474–480. [CrossRef] [PubMed]
19. Fries, P. Rhythms for cognition: Communication through coherence. *Neuron* **2015**, *88*, 220–235. [CrossRef] [PubMed]
20. Nunez, P.L.; Silberstein, R.B.; Shi, Z.; Carpenter, M.R.; Srinivasan, R.; Tucker, D.M.; Doran, S.M.; Cadusch, P.J.; Wijesinghe, R.S. EEG coherency II: Experimental comparisons of multiple measures. *Clin. Neurophysiol.* **1999**, *110*, 469–486. [CrossRef]
21. Nolte, G.; Ziehe, A.; Nikulin, V.V.; Schlögl, A.; Krämer, N.; Brismar, T.; Müller, K.R. Robustly estimating the flow direction of information in complex physical systems. *Phys. Rev. Lett.* **2008**, *100*, 234101. [CrossRef]
22. Schoffelen, J.M.; Gross, J. Studying dynamic neural interactions with MEG. In *Magnetoencephalography: From Signals to Dynamic Cortical Networks*; Springer: Cham, Switzerland, 2019; pp. 1–23.
23. Anzolin, A.; Presti, P.; Van De Steen, F.; Astolfi, L.; Haufe, S.; Marinazzo, D. Quantifying the effect of demixing approaches on directed connectivity estimated between reconstructed EEG sources. *Brain Topogr.* **2019**, *32*, 655–674. [CrossRef]

24. Mahjoory, K.; Nikulin, V.V.; Botrel, L.; Linkenkaer-Hansen, K.; Fato, M.M.; Haufe, S. Consistency of EEG source localization and connectivity estimates. *NeuroImage* **2017**, *152*, 590–601. [CrossRef]
25. Hincapié, A.S.; Kujala, J.; Mattout, J.; Pascarella, A.; Daligault, S.; Delpuech, C.; Mery, D.; Cosmelli, D.; Jerbi, K. The impact of MEG source reconstruction method on source-space connectivity estimation: A comparison between minimum-norm solution and beamforming. *NeuroImage* **2017**, *156*, 29–42. [CrossRef]
26. Hincapié, A.S.; Kujala, J.; Mattout, J.; Daligault, S.; Delpuech, C.; Mery, D.; Cosmelli, D.; Jerbi, K. MEG Connectivity and Power Detections with Minimum Norm Estimates Require Different Regularization Parameters. *Comput. Intell. Neurosci.* **2016**, *2016*. [CrossRef]
27. Vallarino, E.; Sommariva, S.; Piana, M.; Sorrentino, A. On the two-step estimation of the cross-power spectrum for dynamical linear inverse problems. *Inverse Probl.* **2020**, *36*, 045010. [CrossRef]
28. Bendat, J.S.; Piersol, A.G. *Random Data: Analysis and Measurement Procedures*; John Wiley & Sons: Hoboken, NJ, USA, 2011; Volume 729.
29. Welch, P. The use of fast Fourier transform for the estimation of power spectra: A method based on time averaging over short, modified periodograms. *IEEE Trans. Audio Electroacoust.* **1967**, *15*, 70–73. [CrossRef]
30. Tikhonov, A.N.; Goncharsky, A.; Stepanov, V.; Yagola, A.G. *Numerical Methods for the Solution of Ill-Posed Problems*; Springer Science & Business Media: Dordrecht, The Netherlands, 2013; Volume 328.
31. Matsuura, K.; Okabe, Y. Selective minimum-norm solution of the biomagnetic inverse problem. *IEEE Trans. Biomed. Eng.* **1995**, *42*, 608–615. [CrossRef] [PubMed]
32. Uutela, K.; Hämäläinen, M.; Somersalo, E. Visualization of magnetoencephalographic data using minimum current estimates. *NeuroImage* **1999**, *10*, 173–180. [CrossRef]
33. Gramfort, A.; Kowalski, M.; Hämäläinen, M. Mixed-norm estimates for the M/EEG inverse problem using accelerated gradient methods. *Phys. Med. Biol.* **2012**, *57*, 1937–1961. [CrossRef] [PubMed]
34. Chella, F.; Marzetti, L.; Stenroos, M.; Parkkonen, L.; Ilmoniemi, R.J.; Romani, G.L.; Pizzella, V. The impact of improved MEG–MRI co-registration on MEG connectivity analysis. *NeuroImage* **2019**, *197*, 354–367. [CrossRef] [PubMed]
35. Thompson, A.M.; Brown, J.C.; Kay, J.W.; Titterington, D.M. A study of methods of choosing the smoothing parameter in image restoration by regularization. *IEEE Trans. Pattern Anal. Mach. Intell.* **1991**, *13*, 326–339. [CrossRef]
36. Hanke, M.; Hansen, P.C. Regularization methods for large-scale problems. *Surv. Math. Ind* **1993**, *3*, 253–315.
37. Hansen, P.C. *Rank-Deficient and Discrete Ill-Posed Problems: Numerical Aspects of Linear Inversion*; SIAM: Philadelphia, PA, USA, 2005.
38. Vogel, C.R. *Computational Methods for Inverse Problems*; SIAM: Philadelphia, PA, USA, 2002.
39. Lütkepohl, H. *New Introduction to Multiple Time Series Analysis*; Springer-Verlag: Berlin/Heidelberg, Germany, 2005.
40. Haufe, S.; Nikulin, V.V.; Müller, K.R.; Nolte, G. A critical assessment of connectivity measures for EEG data: A simulation study. *NeuroImage* **2013**, *64*, 120–133. [CrossRef]
41. Haufe, S.; Ewald, A. A simulation framework for benchmarking EEG-based brain connectivity estimation methodologies. *Brain Topogr.* **2019**, *32*, 625–642. [CrossRef]
42. Liuzzi, L.; Quinn, A.J.; O'Neill, G.C.; Woolrich, M.W.; Brookes, M.J.; Hillebrand, A.; Tewarie, P. How sensitive are conventional MEG functional connectivity metrics with sliding windows to detect genuine fluctuations in dynamic functional connectivity? *Front. Neurosci.* **2019**, *13*, 797. [CrossRef] [PubMed]
43. Sommariva, S.; Sorrentino, A.; Piana, M.; Pizzella, V.; Marzetti, L. A comparative study of the robustness of frequency-domain connectivity measures to finite data length. *Brain Topogr.* **2019**, *32*, 675–695. [CrossRef]
44. Wendling, F.; Bartolomei, F.; Bellanger, J.; Chauvel, P. Epileptic fast activity can be explained by a model of impaired GABAergic dendritic inhibition. *Eur. J. Neurosci.* **2002**, *15*, 1499–1508. [CrossRef]
45. Astolfi, L.; Cincotti, F.; Mattia, D.; Marciani, M.G.; Baccala, L.A.; de Vico Fallani, F.; Salinari, S.; Ursino, M.; Zavaglia, M.; Ding, L.; et al. Comparison of different cortical connectivity estimators for high-resolution EEG recordings. *Hum. Brain Mapp.* **2007**, *28*, 143–157. [CrossRef]
46. Acebrón, J.A.; Bonilla, L.L.; Vicente, C.J.P.; Ritort, F.; Spigler, R. The Kuramoto model: A simple paradigm for synchronization phenomena. *Rev. Mod. Phys.* **2005**, *77*, 137. [CrossRef]
47. Cabral, J.; Luckhoo, H.; Woolrich, M.; Joensson, M.; Mohseni, H.; Baker, A.; Kringelbach, M.L.; Deco, G. Exploring mechanisms of spontaneous functional connectivity in MEG: How delayed network interactions lead to structured amplitude envelopes of band-pass filtered oscillations. *NeuroImage* **2014**, *90*, 423–435. [CrossRef] [PubMed]
48. Baillet, S.; Mosher, J.C.; Leahy, R.M. Electromagnetic brain mapping. *IEEE Signal Process. Mag.* **2001**, *18*, 14–30. [CrossRef]
49. Sorrentino, A. Particle filters for magnetoencephalography. *Arch. Comput. Methods Eng.* **2010**, *17*, 213–251. [CrossRef]
50. Dale, A.M.; Sereno, M.I. Improved localization of cortical activity by combining EEG and MEG with MRI cortical surface reconstruction: A linear approach. *J. Cogn. Neurosci.* **1993**, *5*, 162–176. [CrossRef] [PubMed]
51. Gramfort, A.; Luessi, M.; Larson, E.; Engemann, D.A.; Strohmeier, D.; Brodbeck, C.; Parkkonen, L.; Hämäläinen, M.S. MNE software for processing MEG and EEG data. *NeuroImage* **2014**, *86*, 446–460. [CrossRef] [PubMed]
52. Lagarias, J.C.; Reeds, J.A.; Wright, M.H.; Wright, P.E. Convergence properties of the Nelder–Mead simplex method in low dimensions. *SIAM J. Optim.* **1998**, *9*, 112–147. [CrossRef]

Article
A Cellular Potts Model for Analyzing Cell Migration across Constraining Pillar Arrays

Marco Scianna* and Luigi Preziosi

Department of Mathematical Sciences, Politecnico di Torino, Corso Duca degli Abruzzi 24, 10129 Torino, Italy; luigi.preziosi@polito.it
* Correspondence: marco.scianna@polito.it

Abstract: Cell migration in highly constrained environments is fundamental in a wide variety of physiological and pathological phenomena. In particular, it has been experimentally shown that the migratory capacity of most cell lines depends on their ability to transmigrate through narrow constrictions, which in turn relies on their deformation capacity. In this respect, the nucleus, which occupies a large fraction of the cell volume and is substantially stiffer than the surrounding cytoplasm, imposes a major obstacle. This aspect has also been investigated with the use of microfluidic devices formed by dozens of arrays of aligned polymeric pillars that limit the available space for cell movement. Such experimental systems, in particular, in the designs developed by the groups of Denais and of Davidson, were here reproduced with a tailored version of the Cellular Potts model, a grid-based stochastic approach where cell dynamics are established by a Metropolis algorithm for energy minimization. The proposed model allowed quantitatively analyzing selected cell migratory determinants (e.g., the cell and nuclear speed and deformation, and forces acting at the nuclear membrane) in the case of different experimental setups. Most of the numerical results show a remarkable agreement with the corresponding empirical data.

Keywords: Cellular Potts model; cell migration; nucleus deformation; microchannel device

MSC: 34K34; 37N25; 92C17

1. Introduction

The ability of cells to move within different environments is crucial in a diverse array of processes. For instance, during development, the coordinated movement of cells of different origin is fundamental for both shaping the growing embryo and organogenesis: migratory defects at all stages may in fact lead to severe malformations [1]. In mature organisms, immune cells are mobilized from the bloodstream to enter sites of infection, and then into the lymph nodes for effector functions [2]. Moreover, the migration of epithelial cells and fibroblasts is vital for proper wound healing and the repair of basement membranes and connective tissues. In pathological conditions, cell migration is involved in chronic inflammatory diseases, such as arteriosclerosis, and in cancer invasion and metastasization [3]. The process of cell migration is finally exploited in biomedical engineering applications for the regeneration of various tissues, such as cartilage, skin, or peripheral nerves in vivo or in vitro [4–7].

The migratory efficacy of cells is determined, to a large extent, by their capacity to squeeze through strictly confined environments. For instance, tissue membranes and vessel walls, as well as dense regions of structural extracellular matrices (ECMs), represent physical barriers characterized by significantly small openings and pores [8,9]. Under these conditions, cells can achieve substantial movement by degrading/modifying their surroundings to create sufficient space, for example, by the secretion of matrix metalloproteinases (MMPs) or by squeezing to fit through the available space [10–12]. In the latter option, the elasticity of the cell becomes an important factor. In this respect, the cytoplasm is very flexible and can undergo large deformations. On the other hand, the voluminous nucleus is much stiffer: it is therefore a hampering factor for cell movement [13–16].

The relation between cell mobility and remodeling ability is efficiently studied with two experimental systems. On the one hand, cultured cells are stimulated to move in engineered fibrous scaffolds, which mimic highly confined in vivo connective tissues [17]. On the other hand, they are seeded and allowed to locomote within microfluidic-based devices, characterized by the presence of constrictions formed by fixed and insoluble polymeric structures (e.g., extended walls or arrays of pillars) [18–21].

Such a second type of experimental system, in particular, in the design version proposed by Denais and coworkers in [19] and by Davidson and colleagues in [18], was here reproduced and simulated using an extended Cellular Potts model (CPM, [22–27]). This is a grid-based Monte Carlo technique that employs a stochastic energy minimization principle to determine system evolution. The approach proposed in this work is similar to those proposed in [28–30], which employed a compartmental representation of cells to analyze selected aspects of their movement within matrix environments. However, it is characterized by some relevant novelties and features, namely:

- The definition of a Boltzmann law for lattice updates specific for each type of proposed cell dynamics (e.g., cytosolic extension or retraction, and nucleus deformation);
- An analysis of the forces that act on the nuclear envelope at different stages of individual locomotion;
- A continuous and close feedback and feedforward between computational and experimental results, in a perspective of a data-driven model refinement.

In this respect, as an outcome, we focused on experimentally addressable characteristics of cell shape and locomotion (e.g., the velocity and transit time within a constriction, and nucleus deformation ratio). In particular, we predicted how these quantities were affected by selected manipulations either of cell properties or of channel layout. In this respect, we successfully replicated most of the experimental results proposed in the two reference papers [18,19]. For the sake of completeness, we investigated the rationale underlying the few discrepancies that emerged between the computational and in vitro outcomes.

The rest of the paper is organized as follows. In Section 2, we clarify the assumptions on which our approach was based and describe each model component. In this part of the work, we also introduce and provide an estimation of the model parameters and define how we quantified and characterized cell movement. The computational findings are then presented in Section 3, where we separately deal with simulations relative to the channel layouts proposed in [18,19], respectively. Finally, the proposed results are discussed in the last Section 4, which also contains some hints for future perspectives.

2. Mathematical Model

The Cellular Potts model (CPM) is a grid-based stochastic approach that realistically preserves the identity of cell-scale elements and describes their behavior and mutual interactions in energetic terms and constraints. An extended version of the CPM was employed here to schematically reproduce cell migratory behavior within the two types of microfluidic-based devices developed in [18,19]. As shown in Figure 1A, both experimental systems are composed of dozens of arrays of aligned polymeric pillars, which form a series of parallel channels. Such structural elements are fixed and represent a constriction for cell movement because they limit the free space and form bottlenecks: their dimensions and distributions can be varied to mimic different patterns of spatial limitations. An extended group of cells, initially disposed just outside the entrance of the channels, then individually moves up to a chemical gradient, which is established by the diffusion of a molecular factor from a sink reservoir located on the opposite side of the devices; see, again, Figure 1A.

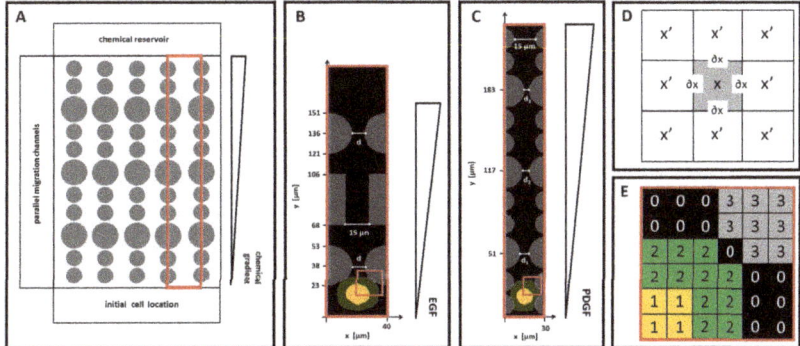

Figure 1. (**A**) Schematic representation of the microfluidic devices proposed in [18,19]. (**B**) Simulation domain Ω replicating a representative channel of the migration device employed in [19]. (**C**) Simulation domain Ω reproducing a representative migratory channel of the migration device employed in [18]. (**D**) Portion of a 2D Cellular Potts model (CPM) lattice with a generic lattice site **x**, its border $\partial \mathbf{x}$, and its first-nearest neighbors **x**′. (**E**) The virtual cell is compartmentalized into a central nuclear cluster, the object Σ_1 of type N (in yellow), and in the surrounding cytosolic region, the element Σ_2 of type C (in green). The rigid pillars are reproduced by CPM objects $\Sigma_{\sigma \geq 3}$ of type P (in gray). The rest of the domain is formed by an extended undifferentiated element Σ_0 of type M (in black).

In order to reduce the computational complexity of the problem, a two-dimensional CPM domain $\Omega \subset \mathbb{R}^2$ is used to reproduce a planar section of a single migratory channel, taken as representative of each of the two reference devices; see Figure 1B,C. In both cases, Ω is a regular lattice, formed by identical square grid sites that, with an abuse of notation, are identified by their center $\mathbf{x} = (x, y) \in \mathbb{R}^2$. Each grid site is assigned a unique index, i.e., an integer number $\sigma(\mathbf{x}) \in \mathbb{N}$, that can be interpreted as a degenerate *spin*, a name originally inherited from statistical physics [31,32]. The border of a lattice site **x** is identified as $\partial \mathbf{x}$, and one of its neighbors, as **x**′, while its overall Moore neighborhood is identified as $\Omega'_\mathbf{x}$, i.e., $\Omega'_\mathbf{x} = \{\mathbf{x}' \in \Omega : \mathbf{x}' \text{ is a neighbor of } \mathbf{x}\}$; see Figure 1D. The subdomains of contiguous sites with identical spin form discrete objects Σ_σ (e.g., $\Sigma_{\bar{\sigma}} = \{\mathbf{x} \in \Omega : \sigma(\mathbf{x}) = \bar{\sigma}\}$), which have an associated type $\tau(\Sigma_\sigma)$.

A single representative cell is then included in Ω. Following the approach proposed in [27], the simulated agent, labeled by the integer $\eta = 1$, is defined as a compartmentalized element. It is composed of two subregions, which, in turn, are classical CPM objects Σ_σ: the nucleus, a central cluster $\Sigma_{\sigma=1}$ of type $\tau = N$, and the surrounding cytosol $\Sigma_{\sigma=2}$ of type $\tau = C$; see Figure 1E. Both cell compartments share, as an additional attribute, the individual identification number $\eta = 1$. In other words, the entire cell is identified by $\eta = 1$, whereas its internal compartments are identified by the pairs $(\eta = 1, \sigma = 1)$ (the nucleus) and $(\eta = 1, \sigma = 2)$ (the cytoplasm).

The extracellular environment is then differentiated into a polymeric component, $\tau = P$, and a medium component, $\tau = M$, as done in [30,33,34]. The polymeric state is assigned to identify the rigid pillars forming the migratory channel. In particular, each of them is a disconnected CPM element Σ_σ, identified by its own identification number $\sigma \geq 3$ (since σ-values $\in \{1, 2\}$ are used for the intracellular compartments; see above). The dimensions and positions of such structures are specified later on. The medium-like state instead identifies the free surface of the microchannel, i.e., where the cell moves and the chemical substances diffuse: it is conventionally assumed to be a single object $\Sigma_{\sigma=0}$ isotropically distributed throughout the simulation domain, as shown in panel (E) of Figure 1.

Cell movement results from an iterative and stochastic minimization of a free energy, defined by a *Hamiltonian* functional H (whose components are defined below). The employed algorithm consists of a series of elementary steps of a modified Metropolis method for Monte Carlo–Boltzmann dynamics [25,35], which is able to implement the natural exploratory behavior of biological individuals. Procedurally, at each time step t, called a Monte Carlo step (MCS, the basic unit of time of the model), a lattice site x_{so} (so for *source*) belonging to a cell compartment $\Sigma_{\sigma(x_{so})}$ is selected at random and attempts to copy its spin, $\sigma(x_{so})$, into one of its unlike neighbors, $x_{ta} \in \Omega'_{x_{so}} : x_{ta} \notin \Sigma_{\sigma(x_{so})}$ (ta for *target*), also randomly selected. In particular, if $\sigma(x_{so}) = 2$ and $\sigma(x_{ta}) = 0$, the cell is protruding, i.e., extending its motile membrane structures within the extracellular space, whereas if $\sigma(x_{so}) = 0$ and $\sigma(x_{ta}) = 2$, the cell is retracting. Finally, if $\sigma(x_{so}) = 1$ and $\sigma(x_{ta}) = 2$ or if $\sigma(x_{so}) = 2$ and $\sigma(x_{ta}) = 1$, the cell is reorganizing, i.e., it is undergoing internal remodeling. Polymeric pillars are instead fixed and immutable, i.e., they are not allowed to move or be deformed by the cell. The proposed trial of spin update is finally accepted with a Boltzmann-like probability function $P(\sigma(x_{so}) \to \sigma(x_{ta}))$, whose form is slightly changed here with respect to the general version given in [27]:

$$P(\sigma(x_{so}) \to \sigma(x_{ta})) = \tanh(T(\sigma(x_{so}), \sigma(x_{ta}))) \min\left\{1, \exp\left(\frac{-\Delta H}{T(\sigma(x_{so}), \sigma(x_{ta}))}\right)\right\}. \quad (1)$$

In Equation (1), ΔH is the net difference of the system energy due to the proposed change of domain configuration, whereas the parameter $T(\sigma(x_{so}), \sigma(x_{ta})) \in R_+$ is a Boltzmann temperature that accounts for the trial cell behavior. The retraction/protrusion dynamics are in fact dictated by the intensity and the frequency of plasma membrane (PM) ruffles, which, on a molecular level, are determined by polarization/depolarization processes in the actin cytoskeleton (refer to [36–38] and references therein). Cell internal reorganization instead depends on the agitation rate for the nuclear cluster. According to the above considerations, we indeed have

$$T(\sigma(x_{so}), \sigma(x_{ta})) = \begin{cases} T_{\tau(\sigma(x_{so}))}, & \text{if } \sigma(x_{so}) = 2 \text{ and } \sigma(x_{ta}) = 0; \\ T_{\tau(\sigma(x_{ta}))}, & \text{if } \sigma(x_{so}) = 0 \text{ and } \sigma(x_{ta}) = 2; \\ T_{\tau(\sigma(x_{so}))}, & \text{if } \sigma(x_{so}) = 1 \text{ and } \sigma(x_{ta}) = 2; \\ T_{\tau(\sigma(x_{ta}))}, & \text{if } \sigma(x_{so}) = 2 \text{ and } \sigma(x_{ta}) = 1, \end{cases} \quad (2)$$

with the stochastic law regulating cell movement that can be finally specified as

$$P(\sigma(x_{so}) \to \sigma(x_{ta})) = \begin{cases} \tanh(T_C) \min\left\{1, \exp\left(\frac{-\Delta H}{T_C}\right)\right\}, \\ \quad \text{if the cell is protruding or retracting;} \\ \tanh(T_N) \min\left\{1, \exp\left(\frac{-\Delta H}{T_N}\right)\right\}, \\ \quad \text{if the cell is reorganizing.} \end{cases} \quad (3)$$

In particular, we set a sufficiently high $T_C > 1$ since cells moving in confined environments are widely shown to have an active fluid-like cytoplasm. A lower $T_N < 1 < T_C$ is instead fixed since the nucleus does not have active movement dynamics (i.e., self-propulsion), but it only displaces passively, i.e., is dragged by the surrounding cytoskeleton elements; see, also, [39] for a more detailed mechanical explanation.

Remark. The acceptance probability resulting from Equations (2) and (3) differs from the general version given in [27], and used in [28–30], as a consequence of the fact that the Boltzmann temperature T depends both on the type of moving cell compartment, as in our previous papers, and on the characteristics of the target grid element. According to us, this is a significant improvement of the CPM algorithm. For instance, it allows differentiating cases of the same cell cytosolic element that tries either to extend in the free extracellular domain or to occupy the space belonging to the nucleus. In fact, in the former case, the possibility of cell morphological updates is biologically determined by the cytoskeletal agitation rate (i.e., by a determinant of the source site), whereas, in the latter case, it relies on the resistance to movement exerted by the stiff organelle (i.e., on a determinant of the target site). The transition probability functions employed in our previous papers did not allow capturing such aspects.

The *Hamiltonian* functional establishing the system energy is then given by the sum of three contributions:

$$H(t) = H_{\text{adhesion}}(t) + H_{\text{shape}}(t) + H_{\text{chemotaxis}}(t). \tag{4}$$

H_{adhesion} is the general extension of Steinberg's differential adhesion hypothesis (DAH) [25,40,41]. In particular, it is differentiated in the contributions due to either the generalized contact tension between the nucleus and the cytoplasm within the cell, or the effective adhesion between the migrating individual and an extracellular component:

$$H_{\text{adhesion}}(t) = H_{\text{adhesion}}^{\text{int}}(t) + H_{\text{adhesion}}^{\text{ext}}(t) = \sum_{\substack{(\partial x \in \partial \Sigma_1) \cap \\ (\partial x' \in \partial \Sigma_2)}} J_{\text{N,C}}^{\text{int}} + \sum_{\substack{(\partial x \in \partial \Sigma_2) \cap \\ (\partial x' \in \partial \Sigma_0)}} J_{\text{C,M}}^{\text{ext}} + \sum_{\substack{(\partial x \in \partial \Sigma_2) \cap \\ (\partial x' \in \partial \Sigma_{\sigma \geq 3})}} J_{\text{C,P}}^{\text{ext}}, \tag{5}$$

where x and x' are two neighboring sites (i.e., $x' \in \Omega'_x$) and $\Sigma_{\sigma(x)}$ and $\Sigma_{\sigma(x')}$, two neighboring elements. $\partial \Sigma_\sigma$ is instead intended as the border of Σ_σ (i.e., $\partial \Sigma_\sigma = \bigcup_{x \in \Sigma_\sigma} \partial x$). The coefficients Js $\in \mathbb{R}$ are the binding forces per unit area and are obviously symmetric w.r.t. their indices. In particular, $J_{\text{N,C}}^{\text{int}}$ implicitly models the forces exerted by intermediate actin filaments and microtubules to anchor the nucleus to the cell cytoskeleton. In the perspective of energy minimization, $J_{\text{N,C}}^{\text{int}} < 0$ is set to prevent cell fragmentation, as done in [28–30]. $J_{\text{C,M}}^{\text{ext}}$ and $J_{\text{C,P}}^{\text{ext}}$, in principle, evaluate the heterophilic contact interactions between the cell and the extracellular elements: however, both are here fixed equal to zero. This choice, successfully employed in [29,39], was made to directly analyze the influence of cell deformability on the motile behavior. The experimental literature also demonstrates that most cell lines display a sustained ameboid movement, characterized by a poorly adhesive mode, when crawling in confined environments [21,42,43].

H_{shape} models the geometrical attributes of the subcellular compartments, which are written as nondimensional relative deformations in the following quadratic form:

$$H_{\text{shape}}(t) = H_{\text{surface}}(t) + H_{\text{perimeter}}(t) = \sum_{\sigma=1,2} \left[\kappa_{\Sigma_\sigma} \left(\frac{s_{\Sigma_\sigma}(t) - s_{\Sigma_\sigma}(0)}{s_{\Sigma_\sigma}(t)} \right)^2 + \nu_{\Sigma_\sigma} \left(\frac{p_{\Sigma_\sigma}(t) - p_{\Sigma_\sigma}(0)}{p_{\Sigma_\sigma}(t)} \right)^2 \right]. \tag{6}$$

Such an energy term indeed depends on the actual surface and perimeter of the subcellular units, i.e., $s_{\Sigma_\sigma}(t)$ and $p_{\Sigma_\sigma}(t)$, respectively, as well as on the corresponding target quantities, i.e., $s_{\Sigma_\sigma}(0)$ and $p_{\Sigma_\sigma}(0)$, respectively, which are here assumed to be characteristic of the relaxed/initial individual configuration. κ_{Σ_σ} and $\nu_{\Sigma_\sigma} \in R_+$ represent, instead, mechanical moduli in units of energy: in particular, κ_{Σ_σ} refer to surface changes of the subcellular compartments, while ν_{Σ_σ} relate to their deformability/elasticity, i.e., to the ease with which they are able to remodel, changing their perimeter. Both parameters are here taken to be constant: however, they may vary in time as a consequence, for example, of intracellular chemical dynamics, as commented on in the conclusive section of the work.

The fluctuations of the cell surface are kept negligible by setting high constant values $\kappa_{\Sigma_1} = \kappa_N = \kappa_{\Sigma_2} = \kappa_C > 1$. This choice was based on the assumptions that the migrating cell has an adequate amount of nutrients to avoid volume loss and that it does not significantly grow during movement, as confirmed by experimental images in [18,19]. A low $\nu_{\Sigma_2} = \nu_C < 1$ then allows the large cytosolic deformations experimentally observed in the cases of cell movement in confined environments. Empirical evidence also shows that the nucleus is able (when needed) to undergo morphological reorganization but to a lower extent than the surrounding cytosolic compartment. Accordingly, we opted to set $1 > \nu_{\Sigma_1} = \nu_N > \nu_{\Sigma_2} = \nu_C$.

$H_{\text{chemotaxis}}$ reproduces the effect of cell preferential migration towards zones with higher concentrations of a diffusing chemoattractant, which was constantly used in the experiments to obtain a sustained cell movement [18,19]. Such an energy contribution was implemented by a local linear-type relation of the form that was firstly used in [44] to reproduce Dictyostelium discoideum aggregation and then constantly adopted in most CPM-based approaches:

$$\Delta H_{\text{chemotaxis}} = \mu(\sigma(\mathbf{x}_{\text{so}}), \sigma(\mathbf{x}_{\text{ta}}))[c_t(\mathbf{x}_{\text{ta}}, t) - c_t(\mathbf{x}_{\text{so}}, t)], \qquad (7)$$

where $\mathbf{x}_{\text{so}}$ and $\mathbf{x}_{\text{ta}}$ are, respectively, the source and the final lattice site randomly selected during a trial update in an MCS and $c_t(\mathbf{x}, t) = c(\mathbf{x}, t) + \sum_{\mathbf{x}' \in \Omega'_\mathbf{x}} c(\mathbf{x}', t)$, where $\mathbf{x} \in \{\mathbf{x}_{\text{so}}, \mathbf{x}_{\text{ta}}\}$ is a nonlocal measure of the molecular substance sensed by the moving cell site, since c denotes the chemical concentration; see Equation (9). Finally, $\mu \in \mathbb{R}^+$ represents the strength of the chemotactic response: in particular, we set

$$\mu(\sigma(\mathbf{x}_{\text{so}}), \sigma(\mathbf{x}_{\text{ta}})) = \begin{cases} \mu_{\tau(\sigma(\mathbf{x}_{\text{so}}))} = \mu_C, & \text{if } \sigma(\mathbf{x}_{\text{so}}) = 2 \text{ and } \sigma(\mathbf{x}_{\text{ta}}) = 0; \\ \mu_{\tau(\sigma(\mathbf{x}_{\text{ta}}))} = \mu_C, & \text{if } \sigma(\mathbf{x}_{\text{so}}) = 0 \text{ and } \sigma(\mathbf{x}_{\text{ta}}) = 2; \\ 0, & \text{else.} \end{cases} \qquad (8)$$

Equation (8) implies that only cytosolic dynamics are affected by molecular signals. Finally, the full expression and activity of cell chemical receptors was set by a high $\mu_C > 1$.

Evolution of the molecular variable. According to the experimental designs in [18,19], we assume that the virtual molecular substance is released (i.e., produced) at a constant rate from the top edge of the domain (denoted as $\partial\Omega^{\text{prod}}$), homogeneously, and constantly diffuses and decays, being eventually taken up by the cell. In mathematical terms, we have the following reaction–diffusion (RD) law:

$$\begin{cases} \dfrac{\partial c(\mathbf{x},t)}{\partial t} = \underbrace{D_c \Delta c(\mathbf{x},t) \delta_{\sigma(\mathbf{x}),\{0\}}}_{\text{diffusion}} - \underbrace{\lambda_c c(\mathbf{x},t)}_{\text{decay}} - \underbrace{\min\{c_{\max}, \chi_c c(\mathbf{x},t)\} \delta_{\sigma(\mathbf{x}),\{1,2\}}}_{\text{cell uptake}} \\ \quad \text{in } \mathbf{x} \in \Omega; \\ c(\partial \mathbf{x}) = c_{\text{prod}} \quad \text{at } \partial \mathbf{x} \in \partial\Omega^{\text{prod}}; \\ c(\partial \mathbf{x}) = 0 \quad \text{at } \partial \mathbf{x} \in \partial\Omega \setminus \partial\Omega^{\text{prod}}, \end{cases} \qquad (9)$$

where $\delta_{x,y} = \{1, x = y; 0, x \neq y\}$ is the Kronecker delta. Equation (9) indeed states that the chemical substance (i) diffuses only through the domain grid sites not occupied by the cell or by a rigid pillar, (ii) locally decays everywhere, and (iii) undergoes consumption only at the domain grid sites occupied by a cell compartment. In particular, cell chemical absorption follows a piecewise-linear approximation of a Michaelis–Menten law. This simplification is realistic since cells' capacity to internalize diffusing substances is limited. We finally set $\lambda_c < \chi_c$, as the natural decay of a molecule is typically negligible compared to the cell uptake. We remark that Equation (9) neglects the diffusion of the chemical *within* the cell after its uptake: such dynamics would imply the definition of specific coefficients, i.e., characterizing the diffusion of the substance within each of the two intracellular compartments. However, the inclusion of this aspect would not have had an impact on the topic of our study, which was, rather, the dependence of the migratory potential of an individual on its remodeling ability.

3. Results

Model parameters and computational details. The characteristic size (lateral edge) of the domain grid sites is hereafter denoted by $|\partial x|$ and was fixed equal to 0.5 µm. The temporal resolution of the model was, as seen, the MCS, which was constantly set to correspond to 2 s, as previously done in [28–30]. The PDE for the evolution of the chemical factors was numerically solved with a finite difference scheme on a grid with the same spatial resolution as Ω, characterized by 30 diffusion steps per MCS. This temporal scale was sufficiently small to guarantee the stability of the numerical method.

In all the simulations, the representative motile cell was initially seeded at the bottom region of the channel, displaying a nonpolarized morphology, with a perimeter and surface given by $p_{\Sigma_i}(0), s_{\Sigma_i}(0)$, for $i = 1, 2$, respectively. In particular, its initially round nucleus lay in the center of the individual. All the forthcoming computational realizations started with 100 annealed MCSs to have a realistic arrangement of the cell body within the structure. Such system configuration updates obeyed the following rule:

$$P(\sigma(\mathbf{x}_{\text{so}}) \to \sigma(\mathbf{x}_{\text{ta}}))(t) = \begin{cases} 0, & \text{if } \Delta H > 0; \\ 0.5, & \text{if } \Delta H = 0; \\ 1, & \text{if } \Delta H < 0. \end{cases} \quad (10)$$

During such annealed MCSs, chemical kinetics did not occur yet. The entire set of model parameters, labeled as $\mathcal{P}$, can be divided in two groups:

$$\mathcal{P} = \mathcal{P}_1 \cup \mathcal{P}_2 = \left\{ s_{\Sigma_1}(0), p_{\Sigma_1}(0), s_{\Sigma_2}(0), p_{\Sigma_2}(0), D_c, \lambda_c, \chi_c, c_{\max}, c_{\text{prod}}, \mu_C \right\} \cup \left\{ J_{\text{N,C}}^{\text{int}}, \kappa_N, \kappa_C, \nu_N, \nu_C, T_N, T_C \right\}. \quad (11)$$

$\mathcal{P}_1$ is composed of coefficients that directly relate to biological quantities and therefore depend on the specific experimental system. In this respect, the cell dimensions and channel measures were derived from images and movies presented in [18,19] (both in the text and in the Supplementary Material). The kinetic coefficients of the chemicals used in that work were instead evaluated using data from the literature. In particular, the maximal cell uptake was calculated as in [33,39]. The chemotactic response μ_C was finally established by comparing experimental and numerical cell velocities in open spaces (i.e., in regions of the channels far from structural constrictions). The parameters belonging to $\mathcal{P}_2$, listed in Table 1, are instead more technical and do not depend on the specific empirical device. In this respect, they were taken from previous published CPMs dealing with cell migration within two- and three-dimensional matrix environments. However, preliminary simulations showed that the behavior of the model proposed in this paper was fairly robust in large regions of the parameter space around this estimate.

Table 1. Values of the Cellular Potts model (CPM) technical parameters, i.e., those included in the set $\mathcal{P}_2$ (see Equation (11)), which were used for both experimental settings.

CPM Parameter	Value	Reference(s)
Contact force between the nucleus and the cytosol	$J_{\text{N,C}}^{\text{int}} = -20$	[28–30]
Surface rigidity of the nucleus	$\kappa_N = 10$	In the range of values used in [28–30]
Surface rigidity of the cytosol	$\kappa_C = 10$	In the range of values used in [28–30]
Stiffness of the nucleus	$\nu_N = 0.9$	[28]
Stiffness of the cytosol	$\nu_C = 0.5$	In the range of values used in [28–30]
Motility of the nucleus	$T_N = 0.5$	In the range of values used in [28–30]
Motility of the cytosol	$T_C = 10$	In the range of values used in [28–30]

Quantification of the numerical results. The *position* of the cell $\eta = 1$ at any time t was established by the position of its center of mass $\mathbf{x}_\eta^{\text{CM}}(t) = (x_\eta^{\text{CM}}(t), y_\eta^{\text{CM}}(t))$. Coherently, the *position* of its nucleus was established by the position of its center of mass $\mathbf{x}_{\Sigma_1}^{CM}(t) = (x_{\Sigma_1}^{CM}(t), y_{\Sigma_1}^{CM}(t))$.

A cell was denoted as *invasive* if at least one of its membrane sites touched the top border of the domain. It was denoted as invasive with respect to a constriction if the center of mass of its nucleus passed the midpoint of that constriction.

The *instantaneous directional velocity* of the cell at a given location $\tilde{y}$ along the representative microchannel was defined as $v_\eta(y_\eta^{CM} = \tilde{y}) = \left[\tilde{y} - y_\eta^{CM}(\tilde{t} - \Delta t)\right]/\Delta t$, where $\tilde{t}$ is such that $y_\eta^{CM}(\tilde{t}) = \tilde{y}$ and $\Delta t = 1$ MCS (2 s). Similarly, the *instantaneous directional velocity* of the nucleus was given by $v_N(y_{\Sigma_1}^{CM} = \tilde{y}) = \left[\tilde{y} - y_{\Sigma_1}^{CM}(\tilde{t} - \Delta t)\right]/\Delta t$.

Morphological changes of the nucleus were quantified by its *deformation ratio*. This was given, for any given location $\tilde{y}$ within the representative microchannel, by the ratio between its geometrical moments of inertia evaluated with respect to the horizontal and vertical axes that passed through its center of mass, i.e.,

$$r_N(y_{\Sigma_1}^{CM} = \tilde{y}) = i_{x_{\Sigma_1}^{CM}}(\tilde{t})/i_{y_{\Sigma_1}^{CM}}(\tilde{t}) = \sum_{\substack{\mathbf{x}=(x,y)\in\Sigma_1 \\ \text{at time } \tilde{t}}} (y-\tilde{y})^2 / \sum_{\substack{\mathbf{x}=(x,y)\in\Sigma_1 \\ \text{at time } \tilde{t}}} (x - x_{\Sigma_1}^{CM}(\tilde{t}))^2, \qquad (12)$$

where $\tilde{t}$ has the same meaning as before. In particular, $r_N \approx 1$ corresponded to an almost round shape of the nucleus, whereas $r_N \ll 1$ (or $\gg 1$) corresponded to its horizontal (or vertical) elongation.

The *transit time* of the cell within a constriction was evaluated as the period of time from its first to its last contact with one of the two pillars that formed that constriction.

We finally calculated the *force* acting on any nuclear border site by adapting the algorithmic procedure described and employed in [45]. In particular, we started with the consideration that local forces can be related to the negative gradient of the Hamiltonian H, i.e., $\mathbf{F}(\mathbf{x}) = (F_x(\mathbf{x}), F_y(\mathbf{x})) = -\nabla H = -(\partial H/\partial x, \partial H/\partial y)$, $\mathbf{x} \in \Omega$ being a generic lattice site. We then employed a centered approximation to the first partial derivative, also observing that the nuclear cluster could extend or retract in any of the two principal directions by a small step $|\partial \mathbf{x}|$ (the characteristic size of the domain grid elements). As a result, we obtained that, for each $\mathbf{x}$ such that $\sigma(\mathbf{x}) = 1$,

$$-F_x(\mathbf{x}) \approx \left(H_N(\Sigma_{\sigma(\mathbf{x})} + \triangle_x \Sigma_{\sigma(\mathbf{x})}) - H_N(\Sigma_{\sigma(\mathbf{x})} - \triangle_x \Sigma_{\sigma(\mathbf{x})})\right)/(2|\partial \mathbf{x}|), \qquad (13)$$

where H_N includes only the energetic contributions relative to the organelle and $\triangle_x \Sigma_{\sigma(\mathbf{x})}$ denotes the one-site-large possible variation of its extension along the horizontal axis. The y-component of the force analogously read as

$$-F_y(\mathbf{x}) \approx \left(H_N(\Sigma_{\sigma(\mathbf{x})} + \triangle_y \Sigma_{\sigma(\mathbf{x})}) - H_N(\Sigma_{\sigma(\mathbf{x})} - \triangle_y \Sigma_{\sigma(\mathbf{x})})\right)/(2|\partial \mathbf{x}|). \qquad (14)$$

3.1. Cell Motion between Structural Elements with Different Geometries

In this section, we specifically focus on cell migration within a channel representative of the device developed in [19]. In this respect, the simulation lattice Ω was formed by 80×400 (respectively, in x and y) sites, which corresponded to an experimental domain of 40 μm $\times$ 200 μm. As shown in Figure 1B, it contained only three pairs of pillars. The distance between the rectangular elements was kept fixed and equal to 15 μm, whereas the distance between the two pairs of round pillars, hereafter denoted by d, was varied in the different simulation settings. All the performed simulations lasted 1.8×10^4 MCSs (≈ 600 min). The estimate of the model parameters belonging to $\mathcal{P}_1$ here refers to breast cancer cells and to epidermal growth factor (EGF), in accordance with the experimental materials used in [19] (cf. Table 2). In particular, the cell had an initial nuclear diameter $d_N = 14$ μm and an overall size equal to 24 μm.

Table 2. Values of the model parameters whose estimates relate to the experimental system proposed in [19]. They belong to the set $\mathcal{P}_1$ defined in Equation (11).

Cell dimensions	Value	Reference(s)
Initial surface of the nucleus	$s_{\Sigma_1}(0) = 155\ \mu m^2$	[19]
Initial perimeter of the nucleus	$p_{\Sigma_1}(0) = 44\ \mu m$	[19]
Initial surface of the cytosol	$s_{\Sigma_2}(0) = 452\ \mu m^2$	[19]
Initial perimeter of the cytosol	$p_{\Sigma_2}(0) = 116\ \mu m$	[19]
Device dimensions	**Value**	**Reference(s)**
Horizontal width	40 μm	[19]
Vertical length	200 μm	[19]
Width between each pair of rectangular structures	15 μm	[19]
Width between each pair of round structures	d (variable)	[19]
Coefficients of EGF kinetics	**Value**	**Reference(s)**
EGF diffusion rate	$D_c = 16 \cdot 10^{-7}\ cm^2 s^{-1}$	[46]
EGF decay rate	$\lambda_c = 1.8 \cdot 10^{-4}\ s^{-1}$	[47]
EGF production rate	$c_{prod} = 0.78\ h^{-1}$	[47]
EGF internalization rate	$\chi_c = 4.3 \cdot 10^{-4}\ s^{-1}$	[48]
Maximal EGF internalization	$c_{max} = 1.2 \cdot 10^{-3}\ \mu M\ s^{-1}$	Estimated as in [33,39]
Cell chemotactic strength	$\mu_C = 8 \cdot 10^3\ \mu M^{-1}$	Fitting with empirical measures in [19]

As shown in Figure 2, when the constriction between the round pillars was large enough (i.e., $d \geq 12$ μm, so that $d/d_N \geq 0.85$), the cell was constantly able to crawl along the entire structure, guided by the chemical signals. Reductions of d then resulted in decrements in the cell invasive capacity, which was completely lost when d fell below 4 μm (i.e., for $d/d_N < 0.28$).

The top panels in Figure 3 show the cell dynamics in the case of complete channel invasion. First, a long and thin cytoplasmic pseudopodium emerged at the front of the cell, towards the chemical source, and infiltrated between the first pair of round pillars. Such a membrane protrusion then dragged the rest of the cell to enter the pore. In particular, the nucleus adopted a cigar-like shape to overcome the spatial constriction, which was possible since it had a certain degree of elasticity.

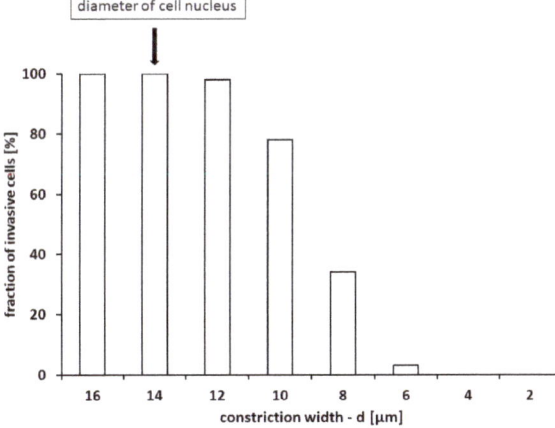

Figure 2. Percentage of invasive cells (i.e., of cells able to touch the upper border of the domain) in the case of reproduction of the device developed in [19]. Values were calculated over 100 numerical realizations.

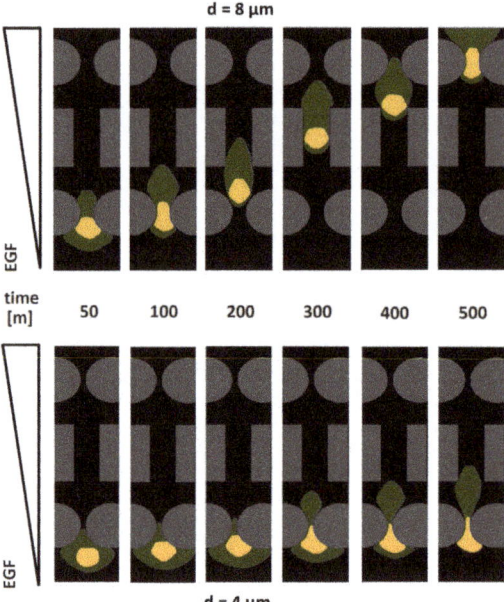

Figure 3. Simulation image sequences of the cell invading the virtual migratory channel in the representative cases $d = 8$ μm (top panels) and $d = 4$ μm (bottom panels), d being the distance between each pair of round pillars. We remark that, for $d = 8$ μm, the cell was able to fully invade the microfluidic structure in only ~30% of cases, while for $d = 4$ μm, this never happened (cf. Figure 2). We remark that the initial diameter of the nucleus was $d_N = 14$ μm.

When the cell had passed the first pair of round elements, it relaxed, and its nucleus stabilized in a quasi-spheroidal shape. A constant sustained migration was then maintained by the individual along the rectangular structures: the space between them did not in fact require further morphological deformation. Both cell compartments had to finally squeeze again when the individual approached the second pair of round pillars.

From a modeling perspective, the migratory behavior of the virtual cell was the result of a sequence of action/reaction mechanisms. First, the exogenous chemical stimulus caused the border sites of the cell cytosol to locally protrude in the direction of increasing EGF gradients, with a speed of protrusion that was approximately proportional to the modulus of the local chemotactic strength μ_C. Dragged by the leading front, the overall cytosolic region then moved forward (eventually deforming) and pulled onto the nucleus with the same force, transmitted by the contact energy $H_{\text{adhesion}}^{\text{int}}$. However, as a consequence of the higher rigidity and lower motility (i.e., $\nu_N > \nu_C$ and $T_N < T_C$, respectively), the nuclear cluster took more time to deform and displace than the surrounding compartment and, therefore, constantly lay at the trailing part of the individual body.

Such a mechanistic explanation is consistent with experimental and modeling observations presented in [49]. Therein, the authors in fact comment that a cell usually translocates almost the entire cytosol before effective nuclear transmigration, mainly in the case of small-enough pores. They also claim that the hourglass shape adopted by the nucleus in the case of passage within small constrictions is due to the pulling forces exerted by the frontal actomyosin networks. The pushing from the rear part of the cytoskeleton would instead result in an inverted bolt shape, which would not allow successful individual passage within the pore.

As captured in Figure 3 (bottom panels), in the case of small-enough interpillar distances d, the front end of the cell cytoplasm extended, as usual, between the first pair of round structures. However, the deformability of the nucleus was no longer sufficient for it to pass through such a confined space. The cell therefore remained stuck, being unable to invade.

These results are indicative of the fact that the presence of the voluminous nucleus represents a steric hindrance for the entire cell and that the degree of nuclear deformability determines its capacity to move within confined spaces. Our numerical outcomes are in remarkable agreement with the experimental evidence provided in the reference work [19]. Denais and coworkers in fact demonstrated that cells of different lineages can pass within subnuclear constrictions by only temporarily rupturing the integrity of the nuclear envelope (NE), so that the organelle becomes as fluid as the cytoplasm. Interestingly, the nuclear membrane can be restored during migration: this is the reason why subsequent NE ruptures are observed within the same cell.

Cell speed and nucleus deformation have complementary behavior, as shown by their time evolution plotted in Figure 4 (for three representative values of d). In the case of complete channel invasion (for $d = 8$ and 12 µm, i.e., for $d/d_N \approx 0.57$ and 0.85), the cell reached and maintained its maximal velocity when crawling between the rectangular elements, whose spacing required minimal nuclear deformation (i.e., $r_N \approx 1$). The cell speed was instead reduced in the proximity of the pairs of round pillars. In particular, the closer they were to each other, the more time the cell took to pass the constriction (and then to relax), as a consequence of the necessarily larger nuclear deformations. Finally, in the case of minimal space between the round structures (for $d = 4$ µm, i.e., for $d/d_N \approx 0.28$), the nuclear deformation quickly went to a maximum threshold, whereas the cell speed dropped to almost zero (the cell remained stuck).

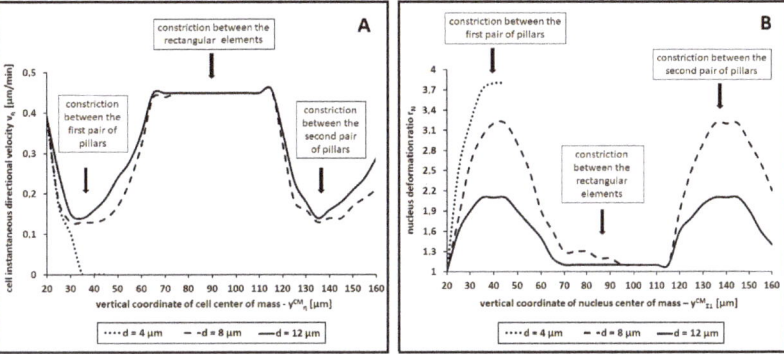

Figure 4. Quantification of cell migratory behavior in the experimental design employed in [19]. Time evolution of cell instantaneous directional velocity v_η (**panel (A)**) and of nucleus deformation ratio r_N (**panel (B)**) for three representative values of the distance d between the pairs of round pillars. In both graphs, each value is the mean over 10 simulations. We have not plotted error bars, to avoid unnecessary graphical overcomplication. However, the standard deviations were very small (of the order of 10^{-2}). We also remark that, in the simulation settings employed in this part, the initial nuclear diameter was $d_N = 14$ µm.

We then turned to analyze the force field at the nuclear boundary at different stages of cell migration. As shown in Figure 5 (left panel), when the nucleus was squeezing through a constriction, its side edges were characterized by significant inward stresses. Outward forces were instead active at the trailing and leading borders, due to the fact that it had to preserve its surface without perimeter shrinking. As soon as the nucleus had overcome the midpoint of the constriction, the inward stresses momentarily pointed almost towards the top edge of the domain, thereby acting as an instantaneous push for cell movement (see the middle panel in Figure 5). Finally, when the cell crawled within the rectangular elements, its nucleus was in a rounded relaxed configuration. In particular, its leading edge was subjected to cytosolic adhesive-based dragging forces, whereas its lateral and trailing borders were subjected only to the forces necessary to keep the surface constant while maximizing the contact with the surrounding cell compartment; see Figure 5 (right panel).

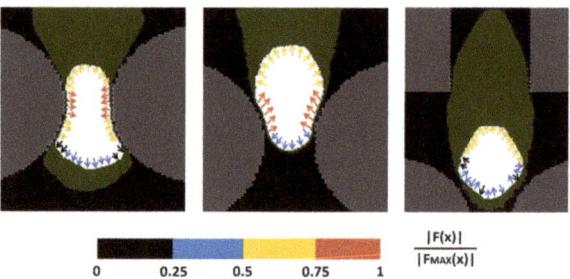

Figure 5. Representative force field at the nuclear boundary at different cell migratory stages. (**Left panel**) cell squeezing within a pair of round pillars. (**Middle panel**) cell overcoming the midpoint of the constriction. (**Right panel**) cell moving within the rectangular elements. For graphical purposes, we have only plotted selected force vectors, which are magnified, with intensity normalized with respect to the maximal value. Force components are defined in Equations (13) and (14).

Our numerical outcomes are in remarkable agreement with the analysis of the spatial distribution of nuclear envelope ruptures provided in [49] in the case of breast adenocarcinoma cells. Cao and colleagues, in fact, showed that, when a malignant individual is passing within a small pore, damage mainly occurs at the front and at the back edge of the nuclear envelope (NE), due to the significant tension. In particular, higher chances of ruptures characterize the leading border, since it is even more stretched than the trailing part. These authors also observed that NE buckling is instead located at the side regions of the organelle, i.e., those subjected to compression. Our results are also in line with those obtained in [45], where static CPM cells were shown to have inward forces at boundary convex sites and outward forces at concave border grid elements.

3.2. Cell Movement between Round Pillars with Different Spacing

Focusing on the microfluidic device used in [18], the CPM lattice Ω had 60×468 (respectively, in x and y) grid elements that replicated a 30 µm × 234 µm representative channel. The polymeric pillars located in the structure were round, with a diameter of either 15 or 30 µm. The space between pairs of smaller elements was kept fixed and equal to 15 µm, whereas the distances between the three couples of larger pillars were identified by d_i (with $i = 1, 2, 3$) and varied to reproduce different channel designs; see Figure 1C. All the forthcoming simulations lasted nearly 24 h (4.3×10^4 MCSs), in accordance with the temporal scale of the corresponding experiments in [18].

The biologically related parameters, i.e., those grouped in $\mathcal{P}_1$ in Equation (11), here refer to human fibroblasts and to platelet-derived growth factor (PDGF), in accordance with the materials mainly used in [18] (refer, also, to the Supplementary Material). They are summarized in Table 3. In particular, the cell nucleus had an initial diameter $d_N = 16$ µm while the extension of the overall individual amounted to 28 µm. We finally remark that the CPM technical parameters, i.e., those included in the set $\mathcal{P}_2$ introduced in Equation (11), were kept unaltered with respect to the values fixed in the previous section and listed in Table 1. In particular, we maintained $\nu_N = 0.9$.

Table 3. Values of the model parameters whose estimates were specifically designed to reproduce the experimental setup used in [18]. They belonged to the set $\mathcal{P}_1$ introduced in Equation (11).

Cell Dimensions	Value	Reference(s)
Initial surface of the nucleus	$s_{\Sigma_1}(0) = 200\ \mu m^2$	[18]
Initial perimeter of the nucleus	$p_{\Sigma_1}(0) = 50\ \mu m$	[18]
Initial surface of the cytosol	$s_{\Sigma_2}(0) = 416\ \mu m^2$	[18]
Initial perimeter of the cytosol	$p_{\Sigma_2}(0) = 138\ \mu m$	[18]
Channel Dimensions	**Value**	**Reference(s)**
Horizontal width	$30\ \mu m$	[18]
Vertical length	$234\ \mu m$	[18]
Width between each pair of smaller pillars	$15\ \mu m$	[18]
Widths between the three pairs of larger pillars	d_i, with $i = 1, 2, 3$ (variable)	[18]
Coefficients of PDGF Kinetics	**Value**	**Reference(s)**
PDGF diffusion rate	$D_c = 1.13 \cdot 10^{-10}\ m^2 s^{-1}$	[18] (Supplementary Material)
PDGF decay rate	$\lambda_c = 4 \cdot 10^{-6}\ s^{-1}$	Fitting with empirical measures in [18]
PDGF production rate	$c_{prod} = 0.2\ h^{-1}$	Fitting with empirical measures in [18]
PDGF internalization rate	$\chi_c = 1 \cdot 10^{-5}\ s^{-1}$	Fitting with empirical measures in [18]
Maximal PDGF internalization	$c_{max} = 5 \cdot 10^{-4}\ \mu M\ s^{-1}$	Estimated as in [33,39]
Cell chemotactic strength	$\mu_C = 4 \cdot 10^2\ \mu M^{-1}$	Fitting with empirical measures in [18]

We first assessed the effectiveness of chemotactic-driven cell migration in the case of a channel design characterized by a sequence of constrictions with decreasing widths ($d_1 = 5\ \mu m$, $d_2 = 3\ \mu m$, and $d_3 = 2\ \mu m$, which resulted in $d_1/d_N \approx 0.31$, $d_2/d_N \approx 0.18$, and $d_3/d_N \approx 0.12$). Such a domain layout was used hereafter unless explicitly said. As it is possible to see in Figure 6A, the virtual cell was constantly unable to invade the entire structure: it in fact overcame the first (largest) constriction only in a few cases. Such numerical results are not surprising if compared with those summarized in Figure 2. In fact, glioblastoma cells and fibroblasts (used as representative cell lines for the simulations of this and of the previous section, respectively) have almost the same dimensions (compared to the spacing between the pairs of rigid pillars), and their characteristic model parameters were kept unchanged. However, this set of numerical outcomes disagrees with the corresponding empirical evidence. As shown in the Supplementary Figure S4b in [18], a significant number of experimental cells (i.e., nearly 40%) are able to penetrate the entire channel, in the case of stable chemical gradients. The underlying reason relies on the fact that the cell lines used in [18] have a more deformable nucleus than those used in [19] and simulated in the previous section, as a consequence of their deficiency of lamins A and C. These molecules are, in fact, the primary components of the nuclear lamina, the dense protein meshwork underlying the nuclear membrane that has been largely shown to determine the stiffness of the organelle [19,50–52]. To have a closer replication of the in vitro evidence in [18], we indeed reduced the rigidity of the nucleus of our virtual cell by decreasing the corresponding parameter ν_N, which, however, had to remain larger than $\nu_C = 0.5$. As shown in Figure 6B, a remarkable data fitting was obtained when $\nu_N \leq 0.7$.

A representative time sequence of a cell able to invade the entire channel, owing to the increased nuclear elasticity, is then proposed in Figure 7: it clearly shows the enhanced nuclear squeezing necessary to promote full invasion. A definitive confirmation in this respect is provided by Figure 8A, which quantifies the nucleus remodeling for different values of ν_N. From the same graph, we observe a residual deformation of the organelle after its passage through a constriction, as also captured in Figure 4B.

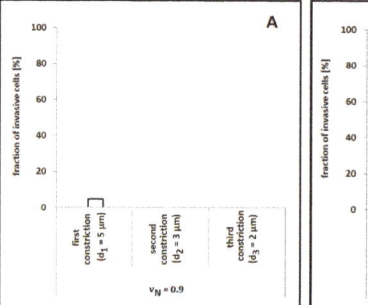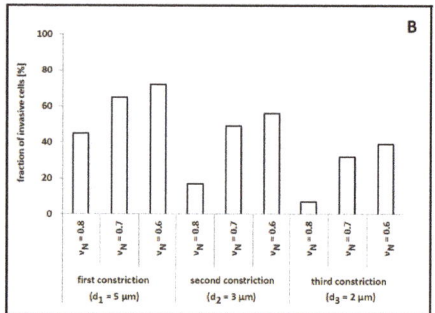

Figure 6. Quantification of chemotactic-driven cell migratory behavior in the case of the experimental setup used in [18]. (**A**) Percentage of cells that were able to overcome each constriction for the default value of the nucleus stiffness, i.e., for $\nu_N = 0.9$. (**B**) Percentage of cells that were able to overcome each constriction upon variations in the nuclear stiffness. Values were calculated over 100 numerical realizations. We remark that, in the simulation settings proposed in this paragraph, the initial diameter of the nucleus is $d_N = 16$ μm.

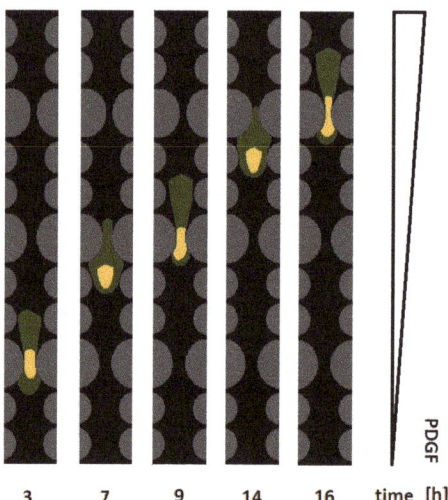

Figure 7. Simulation image sequences of cell invasion within a representative migratory channel characterized by $d_1 = 5$, $d_2 = 3$, and $d_3 = 2$ μm. The virtual individual had an enhanced nuclear elasticity (i.e., $\nu_N = 0.7$) that allowed full invasion in approximately 40% of cases (see Figure 6B). The initial diameter of the cell nucleus was $d_N = 16$ μm.

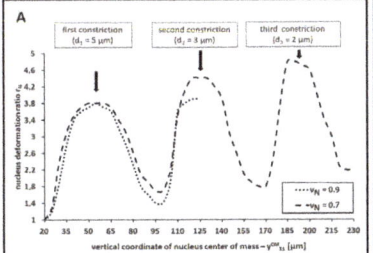 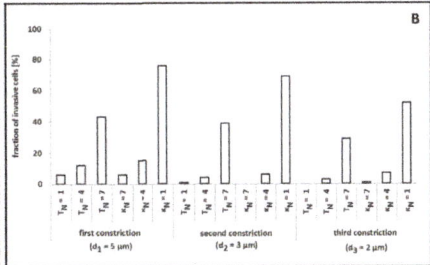

Figure 8. Quantification of cell migratory behavior in the case of the experimental setup used in [18]. (**A**) Time evolution of the deformation ratio of the nucleus, r_N, defined in Equation (12), for two different values of its elasticity ν_N. Each value is given as the mean over 10 simulations. We have not plotted error bars, to avoid unnecessary graphical overcomplication. However, standard deviations were very small (of the order of 10^{-2}). (**B**) Percentage of cells that were able to overcome each constriction upon independent variations either in the nuclear motility T_N or in the nuclear compressibility κ_N (in the case of $\nu_N = 0.9$). As usual, values were calculated over 100 numerical realizations. We again remark that the initial diameter of the cell nucleus was $d_N = 16$ µm.

With a predictive perspective, we then asked if cell migratory behavior could be promoted by variations of other biophysical determinants of the nucleus. In this respect, we no longer reduced the rigidity of the organelle (we kept a high $\nu_N = 0.9$) but independently enhanced either its motility or its compressibility (which meant, in this context, the possibility of reducing its surface) by altering T_N or κ_N, respectively. Some reasonable parameter constraints (i.e., $\kappa_N > \nu_N > \nu_C$ and $T_C > T_N$) were, however, maintained. As shown in Figure 8B, significant increments in cell invasive potential were observed only for substantial variations of the two coefficients (i.e., of at least one order of magnitude from their default values employed up to that point and listed in Table 1). However, such parametric changes led to unrealistic cell dynamics: too high values of T_N, in fact, resulted in implausibly high cell and nuclear velocities, whereas too high values of κ_N allowed an unreasonable shrinking of the organelle (that, in a realistic scenario, would cause the pathological death of the individual).

These results are experimentally confirmed in [18], where the invasive cells were not observed to undergo significant volumetric changes when passing through small constrictions. Substantial variations in nuclear shape not accompanied by similar changes in nuclear volume were also captured in [49] in the case of breast adenocarcinoma cells. Analogously, in [13], glioma cell lines were shown to transmigrate through narrow locations in a brain model in vivo, thereby increasing their metastatic potential, by only a significant squeezing of their nucleus due to a recruitment of nonmuscle myosin II (NMMII). Moreover, very recently, Irimia and Toner, in [53], demonstrated that the directional persistence of cancer cells in microsized structures is completely dependent on the steric hindrance represented by the presence of a rigid and voluminous nucleus.

We further quantified the migration profile of a cell with an enhanced nuclear elasticity (i.e., with $\nu_N = 0.7$). As shown in panel (A) of Figure 9, an asymmetry emerged between the velocity of the overall cell, v_η, and the velocity of its organelle, v_N. On one hand, as expected, the cell had a maximal speed when crawling between the pairs of smaller pillars. Velocity reductions were instead observed when it approached and passed between the three pairs of larger pillars: in particular, the decrements depended on the width of the constrictions. On the other hand, the speed of the nucleus (i) was constant in the case of locomotion within larger spaces, (ii) completely stalled as a constriction impeded its forward movement, (iii) reached an instantaneous peak once the center of the intracellular compartment had passed the midpoint of the pore, and (iv) finally decreased back to the regime value. The underlying rationale, supported by the numerical results summarized in Figure 5 (middle panel), is the following: once the organelle had passed the midpoint of a constriction, the lateral inward compressive forces temporarily aligned in the direction of cell movement. An instantaneous push then emerged, which allowed the nucleus to rapidly slip out from the constriction.

We then assessed whether successful passages through subsequent equal constrictions facilitated cell migration. We indeed compared the transit time within each pore in the case of a channel characterized by $d_1 = d_2 = d_3 = 3$ μm (i.e., by $d_1/d_N = d_2/d_N = d_3/d_N \approx 0.18$). Additionally, in this case, we fixed $\nu_N = 0.7$. As it is possible to see from Figure 9B (left graph), the virtual cell showed a trend towards faster dynamics in the case of transmigration between the second and the third pair of large pillars. A possible explanation relies on the residual deformation that characterized the nucleus after its passage within a constriction, as previously captured in Figures 5B and 9A.

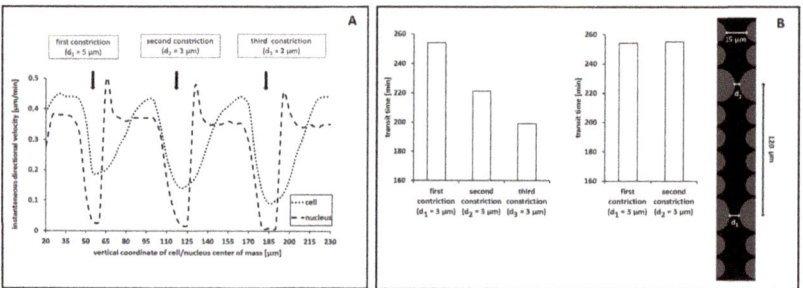

Figure 9. Quantification of cell migratory behavior in the case of reproduction of the device developed in [18]. (**A**) Time evolution of the instantaneous directional velocity of the cell, v_η, and of its nucleus, v_N, in the case of enhanced elasticity of the intracellular organelle, given by $\nu_N = 0.7$. In the graph, each value is the mean over 10 simulations. We have not plotted error bars, to avoid unnecessary graphical overcomplication. However, standard deviations were very small (of the order of 10^{-2}). (**B**) Transit time, i.e., time needed by the cell to overcome a constriction, in the case of a channel characterized by three (**left**) or two (**right**) equal pores (each 3 μm-wide). In the latter case, the two constrictions were largely-spaced, as shown in the inset reproducing the employed domain. In both plots, values were calculated over 100 numerical realizations. We recall that the initial diameter of the cell nucleus was $d_N = 16$ μm.

To further support this hypothesis, we ran a series of simulations based on a channel characterized by two 3 μm-wide constrictions that were separated by nearly 120 μm (i.e., by a sufficient spacing for the nucleus to relax and recover its original shape; see the inset in Figure 9B). The rest of the parameter settings were kept unaltered, with $\nu_N = 0.7$. As shown in Figure 9B (right graph), the transit time was the same for the passage within both pores. Such computational outcomes support our prediction but are in partial contrast with the corresponding experimental evidence. In [18], the authors in fact claim that the facilitated cell movement observed in the case of subsequent constrictions is not due to temporary residual nuclear deformations, since a reduction in the transit time was also captured in the case of spaced-enough pores. They indeed suggest that migrating cells may undergo long-lasting biochemical adaptations such as, for instance, further degradation of lamin proteins or reorganization of the cytoskeletal elements to which the nucleus is anchored. In this respect, Cao and coworkers, in [49], found that (i) the nuclei of lamin A/C-deficient cells (as those used in [18]) behave as plastic materials undergoing large irreversible deformation in the case of passages within small pores and that (ii) the nuclei of malignant cells with expressed and active A/C lamins are instead characterized by the coexistence of elastic dynamics in their envelopes and of plastic dynamics in their interiors. Such two competing effects often result in an ellipsoidal configuration of the nucleus after the exit from a pore, which is then followed by its relaxation towards a more round shape (as captured by our model).

4. Discussion

The analysis of the mechanisms underlying cell migration within confined environments has recently become a major topic in experimental research, due to cell migration's recognized importance in physiopathological phenomena and its exploitation for tissue engineering. In this respect, an increasing number of in vitro models have been developed: they mainly consist of the use either of matrix scaffolds, which mimic in vivo fibrous connective tissues, or of micropatterned devices characterized by predefined channel-like structures.

The resulting evidence has, first, provided insight into selected adhesive and proteolytic mechanisms that motile individuals activate to achieve efficient locomotion in narrow spaces. Furthermore, it has been largely shown that pivotal regulators of cell movement, and, therefore, potential targets for pharmacological interventions in human diseases, are represented by the elastic properties of the cell and of its internal organelles. More specifically, experimental outcomes have revealed the implications of nuclear deformation capacity in the migratory behavior of the overall individual (see, for example, [13–16,20,21,53]).

However, despite the development of such a variety of empirical approaches, little has been done, to our knowledge, from a theoretical point of view, except for a pair of chemomechanical approaches [49,54]. We recently tackled this shortcoming by a series of ad hoc versions of the Cellular Potts model (CPM), which analyzed selected aspects of single cell locomotion within matrix environments [28–30]. As a common feature of the proposed approaches, the moving individuals have been represented as physical objects compartmentalized into the nucleus and cytoplasm, whereas the extracellular domain has been, in turn, differentiated into a medium and a polymeric component. In particular, the introduction of distinct subcellular units has been a fundamental aspect for achieving a detailed description of cell motile behavior within structures of microsized dimensions.

Such models were, here, improved by (i) the use of a tailored Boltzmann transition probability, able to account for the specific type of cell configuration update (i.e., the retraction/extension of the cytosol or reorganization of the nuclear cluster), and (ii) the definition of a procedure to evaluate the force field that acted on the nuclear boundary during the different phases of cell migration.

The resulting CPM was then employed to reproduce the experimental systems used in [18,19], which consisted of microfluidic-based devices composed of dozens of arrays of polymeric pillars with different geometries and dimensions.

Taken together, our results first confirm that mobile cells are able to overcome the effects of size exclusion in the case of small-enough pores by only a substantially high deformability of their nucleus, in accordance with a wide range of empirical studies, e.g., [10,13–16,18,19] and references therein. The proposed numerical outcomes further reveal that, during the passage within a constriction, (i) inward stresses are active along the compressed side edges of the organelle and (ii) outward forces act at its leading and trailing borders. Interestingly, our simulations also showed that, as soon as the nucleus had overcome the midpoint of a pore, inward forces temporarily aligned with the direction of cell movement, representing, therefore, an instantaneous push for nuclear locomotion. This, according to us, is the rationale underlying the peak in the nuclear velocity that was experimentally captured in [18].

We also observed that passages within successive constrictions were facilitated by residual nuclear deformation. This result, as commented in the text, is in contrast with the corresponding empirical evidence in [18]: such a discrepancy may be due to the fact CPM objects have a full elastic behavior, whereas biological cells can instead undergo long-lasting biochemical adaptations that impact the deformation capacity of their internal organelles, an aspect not included in the CPM used here.

Summing up, our computational results are mostly characterized by a remarkable agreement with their experimental counterparts, representing further complementary determinations as well. In the case of discrepancy between numerical and empirical outcomes, we either adapted our approach in order to tackle the issue or found out a plausible underlying rationale.

Of course, a more realistic reproduction of the proposed experimental settings would be obtained by three-dimensional simulations. This computational refinement would allow having a closer *quantitative* comparison between in vitro and in silico results, mainly in terms of observables such as the cell and nuclear speed and time taken by the cell to squeeze between channel constrictions. In this respect, it is reasonable to hypothesize that the displacement and the deformation of a 3D cell body would be slower than the corresponding 2D dynamics. However, the *qualitative* fitting of our results w.r.t. their empirical counterparts would not change. In fact, the cells seeded within the microfluidic devices used in [18,19] did not experience spatial limitations in the direction orthogonal to the plane of movement, as the rigid pillars (and, therefore, the entire channels) seemed to be "tall" enough to allow a comfortable arrangement of the cell body in this respect. This aspect is confirmed by the fact that, in the experimental images and videos, the cells were constantly adherent to the substrate and subjected only to lateral deformations. For the sake of completeness, we finally remark that a 3D extension of the model would be straightforward: it would only amount to a revision of the parameter estimates, whereas the main aspects of the proposed approach (i.e., the tailored Boltzmann probability, the terms in the Hamiltonian and the law regulating the chemical kinetics) would not require any modification.

Despite the limitations typical of the theoretical modeling, it would be biologically relevant to apply our approach to scenarios not strictly related to experimental assays. First, our study may contribute to a more detailed understanding of how cancer cells invade surrounding confined tissues, permeating through the stroma and eventually entering the vasculature. In fact, these processes mainly involve the ability of single metastatic malignant cells to squeeze and crawl within confined environments with a limited space available due to the presence of dense matrices and cell linings.

It would also be interesting to analyze if the relation between nuclear deformability and cell invasive potential varies, in terms of relevance, in the case of collective migration, which is typically involved in most in vivo phenomena. In these cases, a differentiation may in fact occur among individuals within the same ensemble, such as the emergence of tip and stalk cells during angiogenic processes, which may imply differentiated motile phenotypes [55,56].

The proposed approach could be finally applied to the design of synthetic implant materials, i.e., acellular scaffolds with optimal values of pore size that may accelerate cell in-growth, critical for regenerative treatments [4,5,57].

However, to increase the realism of the future model applications, some mechanisms/processes, here disregarded but that play a major role in establishing cell migratory ability, should be included, such as (i) matrix digestion and deposition by moving individuals, which alter the surrounding space by opening paths, generating traction, increasing adhesion, and contact guidance, and (ii) possible pressure-driven displacements of tissue walls, which result in an adjustment of the geometry of the surrounding environment that may facilitate individual locomotion [56].

A significant model improvement would finally amount to the inclusion of intracellular chemical pathways, triggered by external stimuli of distinct natures. For instance, chemotactic substances (e.g., EGF and PDGF) typically activate PM receptors and therefore initiate downstream cascades that involve the biosynthesis, in the sub-plasma-membrane regions, of molecular mediators such as PI3K and MAPK [58]. Such molecules in turn induce the production of small GTPases [59], which are able to regulate several cell responses, including adhesion, migration, and, eventually, proliferation (which is not relevant for our study). From a modeling perspective, one could first focus on a subgroup of these endogenous chemicals and describe their interconnected kinetics by a system of PDEs, solved within the cytosolic compartment of the virtual cell. Constitutive laws should then be set to establish the dependence of CPM parameters, such as the Boltzmann temperatures T and the mechanical moduli κ and ν (i.e., those that describe cell properties), on the amount and the distribution of the endogenous chemicals included in the picture. We employed such a strategy to analyze the role of intracellular calcium signals, stimulated by an exogenous chemoattractant, in the process of vascular formation; see [39,60].

Intracellular pathways can be also activated by mechanical stimuli. In this respect, it would be relevant to relate lamin dynamics, which, as seen, regulate nuclear deformability, to the intensity of the stresses to which the nuclear envelope is subjected (that can be measured in terms of the moduli of the forces F_x and F_y or of the deformation ratio r_N). Coherently, the parameter ν_N should be defined as a function of the actual amount of lamins present within the cell.

Molecular inside-out signaling also occurs between moving cells and ECM elements, which are able to change the activity of intracellular molecular motors such as the already-cited GTP proteins, thereby mediating cytoskeletal contractility (Rac and Rho) [11].

It is useful to remark that the inclusion of one or more of the above-described intracellular dynamics is facilitated by the compartmentalization approach at the basis of our cell representation: it in fact allows a proper localization of the endogenous pathways of interest. However, we have to underline that such model refinements would be computationally expensive (especially in the case of the inclusion of complex-enough chemical cascades).

Author Contributions: Both authors equally contributed to the work. Both authors have read and agreed to the published version of the manuscript.

Funding: This research was partially funded by the Italian Ministry of Education, University and Research (MIUR) through the "Dipartimenti di Eccellenza" Programme (2018-2022)—Department of Mathematical Sciences "G. L. Lagrange", Politecnico di Torino (CUP: E11G18000350001). Both authors are members of GNFM (Gruppo Nazionale per la Fisica Matematica) of INdAM (Istituto Nazionale di Alta Matematica), Italy.

Institutional Review Board Statement: Not applicable.

Informed Consent Statement: Not applicable.

Data Availability Statement: All data are contained within the article.

Conflicts of Interest: The authors declare no conflict of interest.

References

1. Kurosaka, S.; Kashina, A. Cell biology of embryonic migration. *Birth Defects Res. C Embryo Today* **2008**, *84*, 102–122. [CrossRef] [PubMed]
2. Friedl, P.; Weigelin, B. Interstitial leukocyte migration and immune function. *Nat. Immunol.* **2008**, *9*, 960–969. [CrossRef] [PubMed]
3. Sahai, E. Illuminating the metastatic process. *Nat. Rev. Cancer* **2007**, *7*, 737–749. [CrossRef] [PubMed]
4. Capito, R.M.; Spector, M. Scaffold-based articular cartilage repair. *IEEE Eng. Med. Biol. Mag.* **2003**, *22*, 42–50. [CrossRef]
5. Harley, B.A.; Spilker, M.H.; Wu, J.W.; Asano, K.; Hsu, H.P.; Spector, M.; Yannas, I.V. Optimal degradation rate for collagen chambers used for regeneration of peripheral nerves over long gaps. *Cells Tissues Organs* **2008**, *176*, 153–165. [CrossRef]
6. Spadaccio, C.; Rainer, A.; De Porcellinis, S.; Centola, M.; De Marco, F.; Chello, M.; Trombetta, M.; Genovese, J.A. AG-CSF functionalized PLLA scaffold for wound repair: An in vitro preliminary study. *Conf. Proc. IEEE Eng. Med. Biol. Soc.* **2010**, *2010*, 843–846.
7. Yannas, I.V.; Lee, E.; Orgill, D.P.; Skrabut, E.M.; Murphy, G.F. Synthesis and characterization of a model extracellular matrix that induces partial regeneration of adult mammalian skin. *Proc. Natl. Acad. Sci. USA* **1989**, *86*, 933–937. [CrossRef]
8. Hotary, K.B.; Allen, E.D.; Brooks, P.C.; Datta, N.S.; Long, M.W.; Weiss, S.J. Membrane type I matrix metalloproteinase usurps tumor growth control imposed by the three-dimensional extracellular matrix. *Cell* **2003**, *114*, 33–45. [CrossRef]
9. Wolf, K.; Wu, Y.I.; Liu, Y.; Geiger, J.; Tam, E. Multi-step pericellular proteolysis controls the transition from individual to collective cancer cell invasion. *Nat. Cell.Biol.* **2007**, *9*, 893–904. [CrossRef]
10. Friedl, P.; Brocker, E.B. The biology of cell locomotion within three-dimensional extracellular matrix. *Cell. Mol. LifeSci.* **2000**, *57*, 41–64. [CrossRef]
11. Friedl, P.; Wolf, K. Tumour-cell invasion and migration: Diversity and escape mechanisms. *Nat. Rev. Cancer* **2003**, *3*, 362–374. [CrossRef]
12. Friedl, P.; Wolf, K. Plasticity of cell migration: A multiscale tuning model. *J. Cell Biol.* **2009**, *188*, 11–19. [CrossRef]
13. Beadle, C.; Assanah, M.C.; Monzo, P.; Vallee, R.; Rosenfeld, S.; Canoll, P. The role of myosin ii in glioma invasion of the brain. *Mol. Biol. Cell* **2008**, *19*, 3357–3368 [CrossRef]
14. Friedl, P.; Wolf, K.; Lammerding, J. Nuclear mechanics during cell migration. *Curr. Opin. Cell Biol.* **2011**, *23*, 253. [CrossRef]

15. Wolf, K.; Mazo, I.; Leung, H.; Engelke, K.; von Andrian, U.H.; Deryugina, E.I.; Strongin, A.Y.; Brocker, E.B.; Friedl, P. Compensation mechanism in tumor cell migration mesenchymal amoeboid transition after blocking of pericellular proteolysis. *J. Cell Biol.* **2003**, *160*, 267–277. [CrossRef] [PubMed]
16. Wolf, K.; Te Lindert, M.; Krause, M.; Alexander, S.; Te Riet, J.; Willis, A.L.; Hoffman, R.M.; Figdor, C.G.; Weiss, S.J.; Friedl, P. Physical limits of cell migration: Control by ECM space and nuclear deformation and tuning by proteolysis and traction force. *J. Cell Biol.* **2013**, *201*, 1069–1084. [CrossRef]
17. Wolf, K.; Alexander, S.; Schacht, V.; Coussens, L.M.; von Andrian, U.H.; van Rheenen, J.; Deryugina, E.; Friedl, P. Collagen-based cell migration models in vitro and in vivo. *Semin. Cell Dev. Biol.* **2009** *20*, 931–941. [CrossRef]
18. Davidson, P.M.; Sliz, J.; Isermann, P.; Denais, C.; Lammerding, J. Design of a microfluidic device to quantify dynamic intra-nuclear deformation during cell migration through confining environments. *Integr. Biol.* **2015**, *7*, 1534–1546. [CrossRef]
19. Denais, C.M.; Gilbert, R.M.; Isermann, P.; McGregor, A.L.; te Lindert, M.; Weigelin, B.; Davidson, M.; Friedl, P.; Wolf, K.; Lammerding, J. Nuclear envelope rupture and repair during cancer cell migration. *Science* **2016**, *352*, 353–358. [CrossRef] [PubMed]
20. Irimia, D.; Charras, G.; Agrawal, N.; Mitchison, T.; Toner, M. Polar stimulation and constrained cell migration in microfluidic channels. *Lab A Chip* **2007**, *7*, 1783. [CrossRef]
21. Rolli, C.G.; Seufferlein, T.; Kemkemer, R.; Spatz, J.P. Impact of tumor cell cytoskeleton organization on invasiveness and migration: A microchannel-based approach. *PLoS ONE* **2010**, *5*, e8726. [CrossRef]
22. Balter, A.; Merks, R.M.; Poplawski, N.J.; Swat, M.; Glazier, J.A. The Glazier-Graner-Hogeweg model: Extensions, future directions, and opportunities for further study. In *Single-Cell-Based Models in Biology and Medicine, Mathematics and Biosciences in Interactions*; Anderson, A.R.A., Chaplain, M.A.J., Rejniak, K.A., Eds.; Birkhaüser: Basel, Switzerland, 2007; pp. 157–167.
23. Glazier, J.A.; Balter, A.; Poplawski, N.J. Magnetization to morphogenesis: A brief history of the Glazier-Graner-Hogeweg model. In *Single-Cell-Based Models in Biology and Medicine, Mathematics and Biosciences in Interactions*; Anderson, A.R.A., Chaplain, M.A.J., Rejniak, K.A., Eds.; Birkhaüser: Basel, Switzerland, 2007; pp. 79–106.
24. Glazier, J.A.; Graner, F. Simulation of the differential adhesion driven rearrangement of biological cells. *Phys. Rev. E Stat. Phys. Plasmas Fluids Relat. Interdiscip. Top.* **1993**, *47*, 2128–2154. [CrossRef]
25. Graner, F.; Glazier, J.A. Simulation of biological cell sorting using a two-dimensional extended Potts model. *Phys. Rev. Lett.* **1992**, *69*, 2013–2016. [CrossRef]
26. M arée, A.F.; Grieneisen, V.A.; Hogeweg, P. The Cellular Potts Model and biophysical properties of cells, tissues and morphogenesis. In *Single-Cell-Based Models in Biology and Medicine, Mathematics and Biosciences in Interactions*; Anderson, A.R.A., Chaplain, M.A.J., Rejniak, K.A., Eds.; Birkhaüser: Basel, Switzerland, 2007; pp. 107–136.
27. Scianna, M.; Preziosi, L. Multiscale developments of the Cellular Potts Model. *Multiscale Model. Simul.* **2012**, *10*, 342–382. [CrossRef]
28. Scianna, M.; Preziosi, L. Modelling the influence of nucleus elasticity on cell invasion in fiber networks and microchannels. *J. Theor. Biol.* **2013**, *317*, 394–406. [CrossRef]
29. Scianna, M.; Preziosi, L. A cellular Potts model for the MMP-dependent and-independent cancer cell migration in matrix microtracks of different dimensions. *Comp. Mech.* **2014**, *53*, 485–497. [CrossRef]
30. Scianna, M.; Preziosi, L.; Wolf, K. A Cellular Potts Model simulating cell migration on and in matrix environments. *Math. Biosci. Eng.* **2013**, *10*, 235–261. [PubMed]
31. Ising, E. Beitrag zur theorie des ferromagnetismus. *Z. Physik.* **1925**, *31*, 253. [CrossRef]
32. Potts, R.B. Some generalized order-disorder transformations. *Proc. Camb. Phil. Soc.* **1952**, *48*, 106–109. [CrossRef]
33. Bauer, A.L.; Jackson, T.L.; Jiang, Y. A cell-based model exhibiting branching and anastomosis during tumor-induced angiogenesis. *Biophys. J.* **2007**, *92*, 3105–3121. [CrossRef] [PubMed]
34. Rubenstein, B.M.; Kaufman, L.J. The role of extracellular matrix in glioma invasion: A Cellular Potts Model approach. *Biophys. J.* **2006**, *95*, 5661–5680. [CrossRef]
35. Metropolis, N.; Rosenbluth, A.W.; Rosenbluth, M.N.; Teller, A.H.; Teller, E. Equation of state calculations by fast computing machines. *J. Chem. Phys.* **1953**, *21*, 1087–1092. [CrossRef]
36. Mogilner, A.; Oster, G. Polymer motors: Pushing out the front and pulling up the back. *Curr. Biol.* **2003**, *13*, R721–R733. [CrossRef] [PubMed]
37. Pollard, T.D.; Borisy, G.G. Cellular motility driven by assembly and disassembly of actin filaments. *Cell* **2003**, *112*, 453–465. [CrossRef]
38. Ridley, A.J.; Schwartz, M.A.; Burridge, K.; Firtel, R.A.; Ginsberg, M.H.; Borisy, G.; Parsons, J.T.; Horwitz, A.R. Cell migration: Integrating signals from front to back. *Science* **2003**, *302*, 1704–1709. [CrossRef] [PubMed]
39. Scianna, M. A multiscale hybrid model for pro-angiogenic calcium signals in a vascular endothelial cell. *Bull. Math. Biol.* **2011**, *76*, 1253–1291. [CrossRef] [PubMed]
40. Steinberg, M.S. Reconstruction of tissues by dissociated cells. Some morphogenetic tissue movements and the sorting out of embryonic cells may have a common explanation. *Science* **1963**, *141*, 401–408. [CrossRef] [PubMed]
41. Steinberg, M.S. Does differential adhesion govern self-assembly processes in histogenesis? Equilibrium configurations and the emergence of a hierarchy among populations of embryonic cells. *J. Exp. Zool.* **1970**, *171*, 395–433. [CrossRef]

42. Guck, J.; Lautenschläger, F.; Paschke, S.; Beil, M. Critical review: Cellular mechanobiology and amoeboid migration. *Integr. Biol.* **2010**, *2*, 575–583. [CrossRef]
43. Lämmermann, T.; Bader, B.L.; Monkley, S.J.; Worbs, T.; Wedlich-Soldner, R. Rapid leukocyte migration by integrin-independent flowing and squeezing. *Nature* **2008**, *453*, 51–55. [CrossRef]
44. Savill, N.J.; Hogeweg, P. Modelling morphogenesis: From single cells to crawling slugs. *J. Theor. Biol.* **1997**, *184*, 118–124. [CrossRef]
45. Rens, E.R.; Edelstein-Keshet, L. From energy to cellular forces in the Cellular Potts Model: An algorithmic approach. *PLoS Comput. Biol.* **2019**, *15*, e1007459. [CrossRef]
46. Thorne, R.G.; Hrabetova, S.; Nicholson, C. Diffusion of Epidermal Growth Factor in Rat Brain Extracellular Space Measured by Integrative Optical Imaging. *J. Neurophysiol.* **2004**, *92*, 3471–3481. [CrossRef]
47. Serini, G.; Ambrosi, D.; Giraudo, E.; Gamba, A.; Preziosi, L.; Bussolino, F. Modeling the early stages of vascular network assembly. *EMBO J.* **2003**, *22*, 1771–1779. [CrossRef]
48. Wang, D.; Lehman, R.E.; Donner, D.B.; Matli, M.R.; Warren, R.S.; Welton, M.L. Expression and endocytosis of VEGF and its receptors in human colonic vascular endothelial cells. *Am. J. Physiol. Gastrointest. Liver Physiol.* **2002**, *282*, 1088–1096. [CrossRef]
49. Cao, X.; Moeendarbary, E.; Isermann, P.; Davidson, P.M.; Wang, X.; Chen, M.B.; Burkart, A.K.; Lammerding, J.; Kamm, R.D.; Shenoy, V.B. A Chemomechanical Model for Nuclear Morphology and Stresses during Cell Transendothelial Migration. *Biophys. J.* **2016**, *111*, 1541–1552. [CrossRef] [PubMed]
50. Harada, T.; Swift, J.; Irianto, J. Nuclear lamin stiffness is a barrier to 3D migration, but softness can limit survival. *J. Cell Biol.* **2014**, *204*, 669–682. [CrossRef] [PubMed]
51. Lammerding, J.; Fong, L.G.; Ji, J.Y.; Reue, K.; Stewart, C.L.; Young, S.G.; Lee, R.T. Lamins A and C but Not Lamin B1 Regulate Nuclear Mechanics. *J. Biol. Chem.* **2006**, *281*, 25768–25780. [CrossRef] [PubMed]
52. Rowat, A.C.; Jaalouk, D.E.; Zwerger, M.; Ung, W.L.; Eydelnant, I.A.; Olins, D.E.; Olins, A.L.; Herrmann, H.; Weitz, D.A.; Lammerding, J. Nuclear Envelope Composition Determines the Ability of Neutrophil-Type Cells to Passage Through Micron-Scale Constrictions. *J. Biol. Chem.* **2013**, *288*, 8610–8618. [CrossRef]
53. Irimia, D.; Toner, M. Spontaneous migration of cancer cells under conditions of mechanical confinement. *Integr. Biol.* **2009**, *1*, 506–512. [CrossRef] [PubMed]
54. Giverso, C.; Arduino, A.; Preziosi, L. How nucleus mechanics and ECM microstructure influence the invasion of single cells and multicellular aggregates. *Bull. Math. Biol.* **2018**, *80*, 1017–1045. [CrossRef]
55. Friedl, P.; Gilmour, D. Collective cell migration in morphogenesis, regeneration and cancer. *Nat. Rev. Mol. Cell Biol.* **2009**, *10*, 445–457. [CrossRef]
56. Ilina, O.; Bakker, G.-J.; Vasaturo, A.; Hoffman, R.M.; Friedl, P. Two-photon laser-generated microtracks in 3D collagen lattices: Principles of MMP-dependent and -independent collective cancer cell invasion. *Phys. Biol.* **2011**, *8*, 015010. [CrossRef] [PubMed]
57. Behring, J.; Junker, R.; Walboomers, X.F.; Chessnut, B.; Jansen, J.A. Toward guided tissue and bone regeneration: Morphology, attachment, proliferation, and migration of cells cultured on collagen barrier membranes. A systematic review. *Odontology* **2008**, *96*, 1–11. [CrossRef]
58. Szczur, K.; Xu, H.; Atkinson, S.; Zheng, Y.; Filippi, M.-D. 2006. Rho GTPase CDC42 regulates directionality and random movement via distinct MAPK pathways in neutrophils. *Blood* **2006**, *108*, 4205–4213. [CrossRef]
59. Zhang, Y.; Vande Woude, G.F. HGF/SF-Met signaling in the control of branching morphogenesis and invasion. *J. Cell. Biochem.* **2003**, *88*, 408–417. [CrossRef]
60. Scianna, M.; Munaron, L.; Preziosi, L. A multiscale hybrid approach for vasculogenesis and related potential blocking therapies. *Prog. Biophys. Mol. Biol.* **2011**, *106*, 450–462. [CrossRef] [PubMed]

Article

Analysis of the Transient Behaviour in the Numerical Solution of Volterra Integral Equations

Eleonora Messina [1,2,*,†] **and Antonia Vecchio** [2,3,†]

1. Department of Mathematics and Applications, University of Naples "Federico II", Via Cintia, I-80126 Napoli, Italy
2. Gruppo Nazionale per il Calcolo Scientifico, Istituto Nazionale di Alta Matematica, 00185 Roma, Italy; antonia.vecchio@cnr.it
3. C.N.R. National Research Council of Italy, Institute for Computational Application "Mauro Picone", Via P. Castellino, 111-80131 Napoli, Italy
* Correspondence: eleonora.messina@unina.it
† These authors contributed equally to this work.

Abstract: In this paper, the asymptotic behaviour of the numerical solution to the Volterra integral equations is studied. In particular, a technique based on an appropriate splitting of the kernel is introduced, which allows one to obtain vanishing asymptotic (transient) behaviour in the numerical solution, consistently with the properties of the analytical solution, without having to operate restrictions on the integration steplength.

Keywords: Volterra integral equations; asymptotic-preserving; numerical stability

Citation: Messina, E.; Vecchio, A. Analysis of the Transient Behaviour in the Numerical Solution of Volterra Integral Equations. *Axioms* **2021**, *10*, 23. https://doi.org/10.3390/axioms10010023

Academic Editor: Luigi Brugnano

Received: 20 January 2021
Accepted: 18 February 2021
Published: 23 February 2021

Publisher's Note: MDPI stays neutral with regard to jurisdictional claims in published maps and institutional affiliations.

Copyright: © 2021 by the authors. Licensee MDPI, Basel, Switzerland. This article is an open access article distributed under the terms and conditions of the Creative Commons Attribution (CC BY) license (https://creativecommons.org/licenses/by/4.0/).

1. Introduction

Volterra integral equations (VIEs) of the type

$$x(t) = g(t) + \int_0^t K(t,s)x(s)ds, \ t \in [0,+\infty), \qquad (1)$$

and their discrete version,

$$x(t) = g(t) + \sum_{s=0}^t K(t,s)x(s), \ t = 1,2,\ldots, \ x(0) \text{ given}, \qquad (2)$$

are significative mathematical models for representing real-life problems involving feedback and control [1,2]. The analysis of their dynamics allows one to describe the phenomena they represent. In [3], the two equations were analysed in the unifying notation of time scales, and some results were obtained under linear perturbation of the kernel. Here, we revise this approach to obtain results on classes of linear discrete equations whose kernel can be split into a well-behaving part (the unperturbed kernel) plus a term that acts as a perturbation. The implications for numerical methods are, in general, not straightforward and pass through some restrictions on the step length. Nevertheless, here, we overcome this problem and obtain some results on the stability of numerical methods for VIEs.

For Equations (1) and (2), we assume that $x(t) = (x_1(t),\ldots,x_d(t))^T \in \mathbb{R}^d$, $g(t) = (g_1(t),\ldots,g_d(t))^T \in \mathbb{R}^d$, and $K(t,s) = (K_{ij}(t,s))_{i,j=1,\ldots,d}$ is a $d \times d$ matrix.

The paper is organised as follows. In Section 2, we introduce the split kernel for Equation (2) and, using a new formulation of Theorem 2 in [3], we provide sufficient conditions for the above-mentioned splitting that imply that the solution vanishes. In Section 3, we propose a reformulation of the (ρ,σ) methods for (1) as discrete Volterra equations and exploit the theory developed in the previous section in order to investigate its numerical stability properties. In Section 4, some applications are described and analysed through the tools developed in Sections 2 and 3, for which we obtain new and more general results on the asymptotic behaviour for the numerical solutions of both linear and nonlinear equations. In Section 5, some numerical examples are reported.

2. Asymptotics for Discrete Equations

Consider the discrete Volterra Equation (2) with $K(t,s) = P(t,s) + Q(t,s)$, where P and Q are $d \times d$ matrices. Let $r_Q(t,s)$ be the resolvent kernel associated with $Q(t,s)$, which is defined as the solution of the equation:

$$r_Q(t,s) = Q(t,s) + \sum_{l=s+1}^{t} r_Q(t,l)Q(l,s). \tag{3}$$

The following theorem, which we proved in [3], represents the starting point of our investigation.

Theorem 1. *Assume that for Equation (2) with $K(t,s) = P(t,s) + Q(t,s)$, $s = 0, \ldots, t$, it holds that:*

(i) $\sum_{s=0}^{n} \|r_Q(t,s)\| \leq R, t \geq 0$,
(ii) $\lim_{t \to \infty} \|r_Q(t,s)\| = 0$, for any fixed $s \geq 0$,
(iii) $\lim_{t \to \infty} \sum_{s=0}^{t} \|P(t,s)\| = 0$.

Then, if $\lim_{t \to \infty} g(t) = 0$,

$$\lim_{t \to \infty} x(t) = 0.$$

This theorem is particularly interesting when the matrix P in the splitting of kernel K is such that $P(t,s) = 0$ for $s = M+1, \ldots, t-N-1$, with M and N positive constants and $t \geq N + M + 1$. Then, Equation (2) can be rewritten as

$$x(t) = g(t) + \sum_{s=0}^{M} P(t,s)x(s) + \sum_{s=0}^{t} Q(t,s)x(s) + \sum_{s=t-N}^{t} P(t,s)x(s), \tag{4}$$

and the following result holds. Here and in the following, the limit of matrices is intended element-wise.

Theorem 2. *Consider Equation (4), and assume that:*

(1) $\lim_{t \to \infty} P(t,s) = 0$, for $s = 0, \ldots, M$, and $\lim_{t \to \infty} P(t,t) = 0$.
(2) $\sum_{s=0}^{t} \|Q(t,s)\| \leq \alpha < 1$, $\lim_{t \to \infty} Q(t,s) = 0$, for any fixed $s \geq 0$.

If there exists a constant $G > 0$ such that $\|g(t)\| \leq G$, then

$$\|x(t)\| \leq X, \text{ with } X > 0,$$

and if $\lim_{t \to \infty} g(t) = 0$, then

$$\lim_{t \to \infty} x(t) = 0.$$

Proof. Assumption (1) implies (iii) of Theorem 1 and, applying a well-known result (see, for example, [2] (Section 6) and [4]), for assumption (2), we have that (i) and (ii) hold. □

The case $\lim_{t \to \infty} P(t,s) = P_\infty(s) \neq 0$ and $\lim_{t \to \infty} g(t) = g_\infty \neq 0$ can be treated by the same technique if it is known that $\sum_{s=0}^{t} Q(t,s) \to E_\infty \neq I$, for $t \to \infty$ (I is the $d \times d$ identity matrix). In this case, we have the following corollary.

Corollary 1. *Assume that, for Equation (4), there exists $M > 0$ such that*

(1) $\lim_{t \to \infty} P(t,s) = P_\infty(s)$, for $j = 0, \ldots, M$, and $\lim_{t \to \infty} P(t,t) = 0$;
(2) $\sum_{s=0}^{t} \|Q(t,s)\| \leq \alpha < 1$, $\lim_{t \to \infty} Q(t,s) = 0$, for $s = 0, \ldots, t$;
(3) $\lim_{t \to \infty} \sum_{s=0}^{t} Q(t,s) = E_\infty \neq I$.

If $\lim_{t \to \infty} g(t) = g_\infty$, then

$$\lim_{t \to \infty} x(t) = (I - E_\infty)^{-1} \left(g_\infty + \sum_{s=0}^{M} P_\infty(s) x(s) \right).$$

Proof. Set $x_\infty = (I - E_\infty^{-1}) \left(g_\infty + \sum_{s=0}^{M} P_\infty(s) x(s) \right)$. A manipulation of (4) gives

$$\tilde{x}(t) = \tilde{g}(t) + \sum_{s=0}^{t} \tilde{Q}(t,s) \tilde{x}(s) + \sum_{s=t-N}^{t} \tilde{P}(t,s) \tilde{x}(s),$$

with $\tilde{x}(t) = x(t) - x_\infty$, $\tilde{P}(t,s) = 0$, for $s = 0, \ldots, M$, $\tilde{P}(t,s) = P(t,s)$, for $s = t - N, \ldots, t$, $\tilde{Q}(t,s) = Q(t,s)$, and

$$\tilde{g}(t) = g(t) - g_\infty + \left(\sum_{s=0}^{t} Q(t,s) - E_\infty \right) x_\infty + \sum_{s=0}^{M} (P(t,s) - P_\infty(s)) x_\infty + \sum_{s=t-N}^{t} P(t,s) x_\infty.$$

$\tilde{P}, \tilde{Q}$, and $\tilde{g}$ play the roles of P, Q, and g in (4). Recalling Theorem 1, all of the assumptions are satisfied, which implies that $\lim_{t \to +\infty} \tilde{x}(t) = 0$, and then $\lim_{t \to +\infty} x(t) = x_\infty$. □

3. Background Material on (ρ, σ) Methods

The analysis carried out in the previous section can be effectively applied to (ρ, σ) methods for the systems of VIEs:

$$y(t) = f(t) + \int_0^t k(t,s) y(s) ds, \quad t \in [0, +\infty). \tag{5}$$

Here, we consider the numerical solution to (5) obtained by the (ρ, σ) methods with Gregory convolution weights (see, for example, [5–7]):

$$y_n = f(t_n) + h \sum_{j=0}^{n_0-1} w_{nj} k(t_n, t_j) y_j + h \sum_{j=n_0}^{n} \omega_{n-j} k(t_n, t_j) y_j, \tag{6}$$

$n = n_0, n_0 + 1, \ldots$, where $y_n \simeq y(t_n)$, with $t_n = nh$ for $n = 0, 1, \ldots$, $h > 0$ is the step size and w_{nj}, ω_j are the weights. We assume that the weights are non-negative and that $y_0 = f(0), y_1 \ldots, y_{n_0-1}, n_0 \geq 1$, are given starting values.

The weights w_{nj}, $n = 0, 1, \ldots, j = 0, 1, \ldots, n_0 - 1$ are called the starting weights and satisfy (see [5]):

$$\sup_{n \geq 0} w_{nj} \leq W < +\infty, \quad j = 0, \ldots, n_0 - 1, \text{ and } \lim_{n \to \infty} w_{nj} = \bar{w}_j. \tag{7}$$

Moreover, we want to underline some properties of the Gregory convolution weights, ω_n (see, for example [5,7]), which will be useful in the subsequent sections:

$$\sup_n \omega_n = \Omega < +\infty, \tag{8}$$

$$\omega_i = 1, \text{ for } i \geq n_0. \tag{9}$$

From now on, we assume that h satisfies

$$\det(I - h\omega_0 k(t_n, t_n)) \neq 0, \tag{10}$$

where I is the identity matrix of size d.

Choose $n^* > n_0$ and let

$$P(n,j) = \begin{cases} 0, & j = 0,\ldots, n_0 - 1, \\ hk(t_n, t_j), & j = n_0, \ldots, n^* - 1, \\ h\omega_{n-j} k(t_n, t_j), & j = n - n_0 + 1, \ldots, n, \end{cases}$$

and

$$Q(n,j) = \begin{cases} 0, & j = 0, \ldots, n^* - 1, \ j = n - n_0 + 1, \ldots, n, \\ hk(t_n, t_j), & j = n^*, \ldots, n - n_0. \end{cases}$$

The (ρ, σ) method (6) can be written, for $n = n^* + n_0$, $n^* + n_0 + 1, \ldots$, as follows:

$$y_n = g(n) + \sum_{j=n_0}^{n^*-1} P(n,j) y_j + \sum_{j=n^*}^{n-n_0} Q(n,j) y_j + \sum_{j=n-n_0+1}^{n} P(n,j) y_j, \tag{11}$$

with $g(n) = f(t_n) + h \sum_{j=0}^{n_0-1} w_{nj} k(t_n, t_j) y_j$. This alternative formulation of the method in terms of matrices P and Q allows us to analyse its asymptotic properties using the theory developed in the previous paragraph for Equation (4). So, (11) corresponds to the discrete Equation (4) with M and N equal to $n^* - 1$ and $n_0 - 1$, respectively, and $Q(n,j) = 0$, for $j = 0, \ldots, M$.

4. Dynamic Behaviour of Numerical Approximations and Applications

In [8], we carried out an analysis of Volterra equations on time scales that allowed us to obtain results on the asymptotic behaviour of the analytical solution of (5) and on its discrete counterpart in $h\mathbb{Z}$, under the assumptions:

$$\sup_{t \geq \bar{t}} \int_{\bar{t}}^{t} \|k(t,s)\| ds \leq \alpha < 1, \tag{12}$$

and

$$\sup_{n \geq \bar{n}} h \sum_{j=\bar{n}}^{n} \|k(t_n, t_j)\| \leq \alpha < 1, \tag{13}$$

respectively, where $h > 0$, $t_n = nh$, $n = 0, 1, \ldots$, and $\bar{t} = \bar{n} h$. If $\|k(t,s)\|$ is non-increasing with respect to s, the bound (13) is certainly implied by (12) for those values of the parameter h such that

$$\sup_{n \geq \bar{n}} \left(h \|k(t_n, \bar{t})\| + \int_{\bar{t}}^{t_n} \|k(t_n, s)\| ds \right) \leq \alpha < 1.$$

This relation, which allows one to establish a connection between the behaviour of the analytical solution of (5) and of its discrete counterpart in $h\mathbb{Z}$, does not straightforwardly apply to numerical methods due to the presence of the weights w_{nj} and ω_j of the (ρ, σ) methods. This is because the weights can cause the loss of monotonicity, and they may also be greater than 1; then, (13) is not satisfied. In [8], it was proved that if (12), $\sup_{t \in [s, +\infty)} \|k(t,s)\| < +\infty$, and $\lim_{t \to \infty} k(t,s) = 0$, $\forall s \geq 0$, are satisfied, the analytical solution $y(t)$ of Equation (5) vanishes at infinity as $\lim_{t \to \infty} f(t) = 0$. Moreover, in [9], it was shown that, if $\sup_{t>0} \int_0^t \|\partial k(t,s)/\partial s\| ds < +\infty$, then there exists a positive constant A such that

$$h \sum_{j=\bar{n}}^{n} \omega_{n-j} \|k(t_n, t_j)\| \leq \int_{\bar{t}}^{t_n} \|k(t_n, s)\| ds + hA, \quad \forall n \geq \bar{n}. \tag{14}$$

The bound (14) assures that, when (12) is satisfied, the numerical solution y_n tends to zero for $n \to \infty$ if the step size h is small enough, consistently with the behaviour of $y(t)$.

Theorem 2 in Section 2 allows us to remove the restriction on h given by (14). In order to show this result, which states, in fact, the unconditional stability of the (ρ, σ) methods, we need the following preparatory lemma.

Lemma 1. *Assume that:*

(i) $\lim_{t \to +\infty} k(t,s) = 0$, for any fixed $0 \leq s \leq t$,

(ii) *there exists* $\bar{s} \geq 0$ *such that* $\partial \|k(t,s)\|/\partial s \leq 0$, *for* $s > \bar{s}$,

(iii) *there exists* $\bar{t} \geq 0$ *such that* $\int_{\bar{t}}^{t} \|k(t,s)\| ds \leq \alpha < 1$.

Then, for any $h > 0$, $\exists t^* > \max\{\bar{s}, \bar{t}\}$ *such that* $h \sum_{j=n^*}^{n} \|k(t_n, t_j)\| \leq \beta < 1$, *where* n^* *is such that* $n^* h > t^*$.

Proof. Let $h > 0$ be a fixed value of the step size. Assumption (ii) implies that $\|k(t,s)\| \leq \|k(t,\bar{t})\|$ for any $\bar{t} \leq s \leq t$. Moreover, $\lim_{t \to +\infty} \|k(t,\bar{t})\| = 0$ because of (i); thus, for any $\epsilon > 0$, we choose $t^* > \max\{\bar{t}, \bar{s}\}$ such that

$$\|k(t,\bar{t})\| < \epsilon. \tag{15}$$

Now, we choose ϵ such that $h\epsilon + \alpha \leq \beta < 1$ and n^* such that $n^* h \geq t^*$. Since, for ii), $\|k(t_n, s)\|$ is a non-increasing function in s for each $s > t^*$, we have $h \sum_{j=n^*}^{n} \|k(t_n, t_j)\| \leq h\|k(t_n, t^*)\| + \int_{t^*}^{t_n} \|k(t_n, s)\| ds \leq h\epsilon + \alpha \leq \beta < 1$. □

Theorem 3. *Assume that all the hypotheses of Lemma 1 hold for the kernel k of Equation (5); then, for the numerical approximation to its solution $y(t)$ obtained by the (ρ, σ) method (6), if $\lim_{t \to +\infty} f(t) = 0$, one has*

$$\lim_{n \to +\infty} y_n = 0.$$

Proof. For a fixed $h > 0$, Lemma 1 provides a value $n^* > n_0$ for which $h \sum_{j=n^*}^{n} \|k(t_n, t_j)\| \leq \beta < 1$, with β positive constant. Referring to the reformulation (11) of the method, all the assumptions of Theorem 2 are satisfied. Thus, because of property (7) on the asymptotic behaviour of starting weights and of assumption i) of Lemma 1, $g(n) = f(t_n) + h \sum_{j=0}^{n_0-1} w_{nj} k(t_n, t_j) y_j$ tends to zero for $n \to +\infty$. So, in view of Theorem 2, y_n also vanishes. □

Other applications of Theorem 2 are concerned with the equation

$$y(t) = f(t) + \int_0^t k(t-s)(y(s) + G(s, y(s))) ds, \tag{16}$$

which has been the subject of great attention in the literature (see, for example, [10–12]). Here and in the following, we assume that Equation (16) is scalar ($d = 1$), the kernel $k = k(t-s)$ is of convolution type, $G(t, y)$ is a continuous function for $t \in [0, +\infty)$, and $y \in \mathbb{R}$. A main assumption (see, for example, [13]) that is generally made on the nonlinear term G is that it represents a small perturbation, that is, there exists a function $p(t) > 0$ such that

$$|G(t, y)| \leq p(t)|y|. \tag{17}$$

For Equation (16), the (ρ, σ) methods with Gregory convolution weights read, for $n = n_0, n_0 + 1, \ldots$,

$$y_n = f(t_n) + h \sum_{j=0}^{n_0-1} w_{nj} k(t_n - t_j)(y_j + G(t_j, y_j)) + h \sum_{j=n_0}^{n} \omega_{n-j} k(t_n - t_j)(y_j + G(t_j, y_j)). \tag{18}$$

In order to describe the asymptotic behaviour of y_n, we prove the following theorem.

Theorem 4. *Assume that, for Equation (16), the following assumptions hold:*

Hypothesis 1. $\int_0^{+\infty} |k(t)|dt \leq \alpha < 1$,

Hypothesis 2. $\exists \bar{t} > 0$ such that $\forall t > \bar{t}$, $\frac{d|k(t)|}{dt} < 0$,

Hypothesis 3. $\lim_{t \to \infty} f(t) = 0$,

Hypothesis 4. $\lim_{t \to \infty} p(t) = 0$.

Then, for the numerical solution to (16) obtained with the method (18), one has

$$\lim_{n \to \infty} y_n = 0.$$

Proof. From (17) and (18), with $p_j = p(t_j)$, $j = 0, 1, \ldots$,

$$|y_n| \leq |f(t_n)| + h \sum_{j=0}^{n_0-1} w_{nj}|k(t_n - t_j)|(1 + p_j)|y_j| + h \sum_{j=n_0}^{n} \omega_{n-j}|k(t_n - t_j)|(1 + p_j)|y_j|.$$

Now, consider the equation

$$\zeta_n = |f(t_n)| + h \sum_{j=0}^{n_0-1} w_{nj}|k(t_n - t_j)|(1 + p_j)\zeta_j + h \sum_{j=n_0}^{n} \omega_{n-j}|k(t_n - t_j)|(1 + p_j)\zeta_j.$$

Since, from (8), ω_n are bounded, we have that $\sum_{j=n_0}^{n} \omega_{n-j}|k(t_n - t_j)|p_j \leq \Omega \sum_{j=n_0}^{n} |k(t_n - t_j)|p_j$, which is the convolution product of an l_1 (k_n) and a vanishing (p_n) sequence and, therefore, tends to zero as $n \to +\infty$. Therefore, $\forall \epsilon > 0$, $\exists \nu : \forall n > \nu$, $h \sum_{j=n_0}^{n} \omega_{n-j}|k(t_n - t_j)|p_j < \epsilon$. We choose $\epsilon > 0$ such that $\alpha + \epsilon \leq \beta < 1$ and $\bar{n}$ such that $\bar{n}h \geq \bar{t}$ in assumption h2). With $n^* \geq \max\{\nu, \bar{n}\}$, the equation for ζ_n can be written in the more convenient form, (11), for which all the assumptions of Theorem 2 hold. Thus, because of property (7) on the asymptotic behaviour of starting weights and because of the vanishing behaviour of the kernel k, $g(n) = f(t_n) + h \sum_{j=0}^{n_0-1} w_{nj}k(t_n - t_j)(1 + p_j)y_j$ tends to zero for $n \to +\infty$. Therefore, $\lim_{n \to \infty} \zeta_n = 0$. This ends the proof because, using the comparison theorem in [14], $|y_n| \leq \zeta_n$. □

This theorem states that the numerical solution y_n of (16) vanishes when the forcing term f tends to zero for any step size $h > 0$. The result is, of course, more interesting if we know that the analytical solution to (16) tends to zero. This can be proved by means of a result that the authors proved in [8]. To be more specific, the assumptions of Theorem 4 here assure that all the hypotheses of Theorem 9 in [8] are satisfied, thus implying that $\lim_{t \to \infty} y(t) = 0$.

The following result, which we prove in the case of scalar equations, represents a generalisation of Theorem 3.1 in [15], where the numerical stability of the (ρ, σ) methods up to order 3 was proved under some restriction on the step length h. In this paper, by applying Theorem 2 to the (ρ, σ) (6), we remove the constraint on the step size, and extend the investigation to any method in the class of (ρ, σ).

Theorem 5. Assume that, for Equation (5), with $d = 1$, it holds that:

(i) $\exists \bar{t} > 0$ such that $\forall s > \bar{t}$, $\frac{\partial}{\partial t}|k(t, s)| \leq 0$,

(ii) $|k(t, t)| = \varphi(t) \in L^1[0, +\infty)$,

(iii) $\varphi'(t) \leq 0$, $\left(\int_0^{+\infty} |\varphi'(t)|dt \leq \Phi < \infty, \text{ and } \lim_{t \to \infty} \varphi(t) = 0\right)$,

(iv) $\lim_{t \to \infty} f(t) = 0$.

Then, for the numerical solution y_n, obtained with the (ρ, σ) (6), it holds that

$$\lim_{n \to \infty} y_n = 0. \quad \left(\lim_{n \to \infty} y_n = 0, \text{ for } \int_{\bar{n}h}^{+\infty} \varphi(t)dt + h \int_{\bar{n}h}^{+\infty} |\varphi'(t)|dt < 1. \right)$$

Proof. Due to hypotheses (i) and (ii), there exists $\bar{s} > \bar{t}$ such that

$$\lim_{t \to \infty} |k(t,s)| = 0, \text{ for all } s > \bar{s}. \tag{19}$$

Let us fix $h > 0$. From the assumptions, it is clear that $\varphi(t) \leq \varphi_{max}$, with $\varphi_{max} > 0$. So, (ii) implies that $\int_{\bar{n}h}^{+\infty} \varphi(t)dt + h\varphi(\bar{n}h) \leq \int_{\bar{n}h}^{+\infty} \varphi(t)dt + h\varphi_{max} \leq \alpha < 1$, for some $\bar{n} = \bar{n}(h)$, which we choose such that $\bar{n}h > \bar{s}$. Since, for (iii), φ is a non-increasing function, we have

$$h \sum_{j=\bar{n}}^{n-n_0} |k(t_n, t_j)| \leq h \sum_{j=\bar{n}}^{n-n_0} |k(t_j, t_j)| = h \sum_{j=\bar{n}}^{n-n_0} \varphi(t_j) \leq \int_{\bar{n}h}^{+\infty} \varphi(t)dt + h\varphi(\bar{n}h) \leq \alpha < 1. \tag{20}$$

Then, referring to formulation (11) of the numerical method with $n^* = \bar{n}$, we want to prove that

$$\sup_n h \sum_{j=\bar{n}}^{n-n_0} |Q(n,j)| = \sup_n h \sum_{j=\bar{n}}^{n-n_0} \omega_{n-j} |k_{n,j}| < 1. \tag{21}$$

Here, $n - j > n_0$; thus, $\omega_{n-j} = 1$. So, (21) is guaranteed by (20). Furthermore, in view of (19), it is $\lim_{n \to \infty} Q(n,j) = 0$, for any fixed $j \geq \bar{n}$, and, because of (ii) and (iii), also $\lim_{n \to \infty} P(n,n) = 0$. Hence, as all the assumptions of Theorem 2 are accomplished, $\lim_{n \to \infty} y_n = 0$, without imposing any restriction on the step size h.

If, however, assumption (iii)$_2$, holds instead of (iii)$_1$, the step size h has to be chosen such that $h\Phi \leq \beta_1 < 1$ and $\int_{\bar{n}h}^{+\infty} \varphi(t)dt \leq \beta_2$, with $\beta_1 + \beta_2 \leq \alpha < 1$. So, by Lemma 1 in [9],

$$h \sum_{j=\bar{n}}^{n-n_0} |Q(n,j)| \leq \int_{\bar{n}h}^{+\infty} \varphi(t)dt + h \int_{\bar{n}h}^{+\infty} |\varphi'(t)|dt < \beta_2 + \beta_1 \leq \alpha < 1.$$

□

Consider now the following convolution equation:

$$x(t) = f(t) - \int_0^t a(t-s)x(s)ds. \tag{22}$$

Its solution has the form

$$x(t) = f(t) - \int_0^t R(t-s)f(s)ds,$$

where the resolvent kernel R is the solution of the equation:

$$R(t) = a(t) - \int_0^t a(t-s)R(s)ds, \tag{23}$$

$t \geq 0$. If the kernel $a(t)$ of Equation (22) is completely monotone, that is, $(-1)^j a^{(j)}(t) > 0$, $j = 0, 1, \ldots, t \geq 0$, then (see, e.g., [16]) the resolvent $R(t)$ is also completely monotone. Furthermore, the analytical solution $x(t)$ and its numerical approximation x_n obtained by a (ρ, σ) method both tend to zero as $t \to \infty$ and $n \to \infty$, respectively, when the forcing term $f(t)$ tends to zero (see [6]). We point out that if $\lim_{t \to \infty} a(t) = 0$, then $\lim_{t \to \infty} R(t) = 0$ as well, as $R(t)$ is the solution of a Volterra Equation (23) where the kernel is completely monotone and the forcing tends to zero. The significance of completely monotone kernels in Volterra equations is underlined in [13] (p. 27).

A nonlinear perturbation to (22) yields

$$y(t) = g(t) - \int_0^t a(t-s)(y(s) + G(s, y(s)))ds. \tag{24}$$

This equation can be written in terms of the unperturbed solution as (see [13]):

$$y(t) = x(t) - \int_0^t R(t-s)G(s, y(s))ds. \tag{25}$$

Starting from assumption (17) on the nonlinear term G, and from the relation (25), we want to investigate the asymptotic behaviour of the numerical solution to (24) when it is known that $\lim_{n \to \infty} x_n = 0$.

Theorem 6. *Consider Equation (24), and assume that (17) holds for the function G and that:*

1. $a(t)$ *is completely monotone and* $\lim_{t \to \infty} a(t) = 0$,
2. $p(t) \in L^1[0, +\infty)$, $p'(t) < 0$.

If $\lim_{t \to \infty} g(t) = 0$, *then the solution* $y(t)$ *and the numerical solution* y_n *obtained by the* (ρ, σ) *method* (6) *satisfy*

$$\lim_{t \to \infty} y(t) = 0, \text{ and } \lim_{n \to \infty} y_n = 0.$$

Proof. For assumption 1, the solution $x(t)$ of Equation (22) with a completely monotone kernel satisfies $\lim_{t \to \infty} x(t) = 0$. This also holds true for its numerical approximation (see, for example, [6]).

Considering Equation (25), it is

$$|y(t)| \leq |x(t)| + \int_0^t R(t-s)p(s)|y(s)|ds.$$

Since $R(t)$ is completely monotone and $p(t)$ is bounded, the solution of the equation

$$z(t) = |x(t)| + \int_0^t R(t-s)p(s)z(s)ds, \tag{26}$$

satisfies $\lim_{t \to \infty} z(t) = 0$. By using the comparison theorem (see, for example, [17]), $y(t)$ also tends to zero. Considering the numerical solution z_n of (26), we want to show, by means of Theorem 5, that $\lim_{n \to \infty} z_n = 0$. Then, y_n will also vanish.

This is true because all the assumptions of Theorem 5 are satisfied. Indeed:

(i) $|K(t,s)| = R(t,s)p(s)$, thus, $\frac{\partial}{\partial t}|K(t,s)| = R'(t-s)p(s) < 0$, and $\lim_{t \to \infty} K(t,s) = \lim_{t \to \infty} R(t-s)p(s) = 0$; as a matter of fact, as R is the resolvent of a completely monotone vanishing kernel, it is, in turn, a completely monotone vanishing kernel.

(ii) $|k(t,t)| = R(0)p(t) \in L^1[0, +\infty)$, since assumption 2. holds,

(iii) $\varphi'(t) = R(0)p'(t) \leq 0$, since assumption 2. holds,

(iv) $\lim_{t \to \infty} x(t) = 0$, as pointed out before, as it is the solution of the linear VIE (22), with a completely monotone kernel and vanishing forcing term.

□

5. Numerical Examples

In this section, we report some numerical experiments in order to experimentally prove the theoretical results illustrated in Section 4. For our experiments, we choose illustrative test equations and we use the (ρ, σ) method (6) with trapezoidal weights.

In our first example, we refer to Equation (5) with the kernel k given by

$$k(t,s) = 10se^{-s(t+1)}, \tag{27}$$

and the forcing term $f(t)$ such that the solution $y(t) = e^{-t}$. Since $f(t)$ tends to zero as t goes to infinity, all the assumptions of Theorem 3 are satisfied (for example, with $\bar{t} > 3.5$), and thus, both the numerical solution and the continuous one vanish. This is also clear in Figure 1.

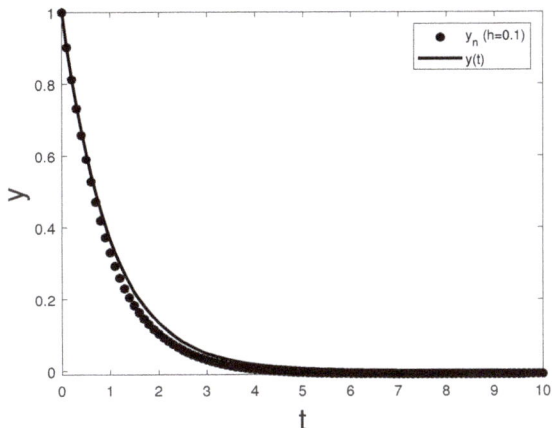

Figure 1. Numerical solution to problem (27) compared to the analytical one.

Now, consider Equation (16) with

$$k(t-s) = (t-s)e^{-2(t-s)}, \quad G(t,y) = 2y\frac{e^{-y^2 t}}{(1+t^2)(1+y^2)}, \text{ and } f(t) = e^{-t^2}. \quad (28)$$

In Figure 2, we draw the numerical solution obtained with step size $h = 0.1$, which clearly vanishes at infinity, according to Theorem 4, since all assumptions are accomplished with $p(t) = \frac{2}{1+t^2}$ and $\bar{t} > \frac{1}{2}$.

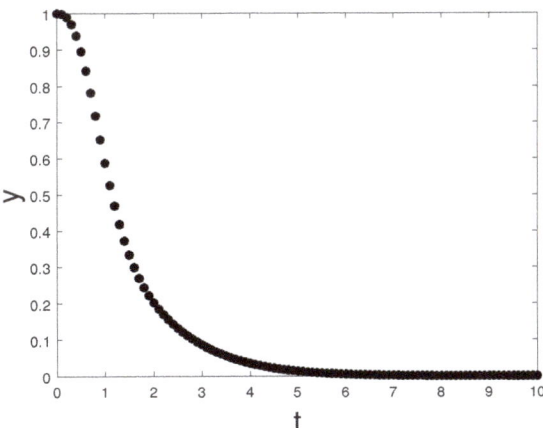

Figure 2. Numerical solution to problem (28) with $h = 0.1$.

Our third example consists in Equation (5) with

$$k(t,s) = \frac{1}{(1+2t-s)^2}, \quad (29)$$

and $f(t)$ such that the solution $y(t) = \frac{1}{t+1}$. According to Theorem 5, with $\varphi(t) = (1+t)^{-2}$, since $f(t)$ tends to zero, the numerical solution vanishes regardless of the step size h, thus replicating the asymptotic behaviour of the continuous one. This behaviour is shown in Figure 3.

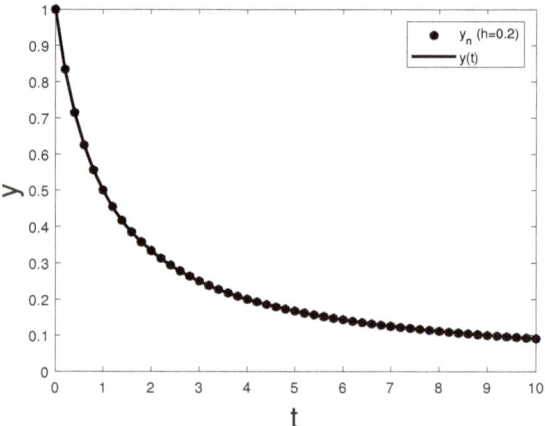

Figure 3. Numerical solution to problem (29) with $h = 0.2$, compared to the analytical one.

In all our experiments, we used sizes for the meshes that ensure reasonable accuracy in the numerical solution at finite times. Integration with larger discretisation steps naturally introduces greater errors on finite time intervals, but the numerical solution maintains the expected behaviour at infinity. Thus, this confirms the asymptotic-preserving characteristics of the numerical schemes without restrictions on h. This can be observed, for example, in Figure 4, again referring to example (29) with $h = 1$.

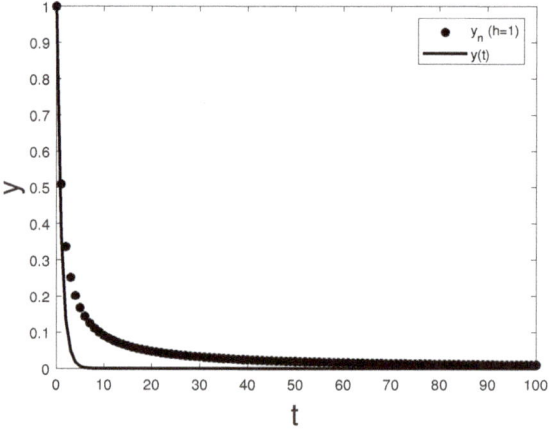

Figure 4. Numerical solution to problem (29) with $h = 1$, compared to the analytical one.

6. Conclusions

Starting from an idea developed in [3], here, we have introduced a technique for the analysis of the vanishing behaviour of the numerical solution to VIEs. This new approach, which is based on suitable splittings of the kernel function, allows one to preserve the character of the analytical solution even in the weighted sums that appear in the method, thus leading to unconditional stability results in many applications of interest.

Author Contributions: Both authors contributed equally to this work. Both authors have read and agreed to the published version of the manuscript.

Funding: This research was funded by the INdAM-GNCS 2020 project "Metodi numerici per problemi con operatori non locali".

Acknowledgments: The authors are grateful to the anonymous reviewers for their constructive comments, which helped to improve the manuscript.

Conflicts of Interest: The authors declare no conflict of interest.

References

1. Gripenberg, G.; Londen, S.O.; Staffans, O. *Volterra Integral and Functional Equations*; Cambridge University Press: Cambridge, UK, 1990.
2. Elaydi, S. *An Introduction to Difference Equations*; Springer: New York, NY, USA, 2005.
3. Messina, E.; Raffoul, Y.; Vecchio, A. Analysis of Perturbed Volterra Integral Equations on Time Scales. *Mathematics* **2020**, *8*, 1133. [CrossRef]
4. Győri, I.; Reynolds, D. On admissibility of the resolvent of discrete Volterra equations. *J. Differ. Equ. Appl.* **2010**, *16*, 1393–1412. [CrossRef]
5. Brunner, H.; van der Houwen, P. *The Numerical Solution of Volterra Equations*; North-Holland: Amsterdam, The Netherlands, 1986.
6. Lubich, C. On the stability of linear multistep methods for Volterra convolution equations. *IMA J. Numer. Anal.* **1983**, *3*, 439–465. [CrossRef]
7. Wolkenfelt, P.H.M. *Linear Multistep Methods and the Construction of Quadrature Formulae for Volterra Integral and Integro-Differential Equations*; Report NW 76/79; Mathematisch Centrum: Amsterdam, The Netherlands, 1979.
8. Messina, E.; Vecchio, A. Stability and Convergence of Solutions to Volterra Integral Equations on Time Scales. *Discret. Dyn. Nat. Soc.* **2015**, *2015*, 6. [CrossRef]
9. Messina, E.; Vecchio, A. A sufficient condition for the stability of direct quadrature methods for Volterra integral equations. *Numer. Algorithms* **2017**, *74*, 1223–1236. [CrossRef]
10. Agyingi, E.; Baker, C. Derivation of variation of parameters formulas for non-linear Volterra equations, using a method of embedding. *J. Integral Equ. Appl.* **2013**, *25*, 159–191. [CrossRef]
11. Miller, R. A nonlinear variation of constants formula for Volterra equations. *Math. Syst. Theory* **1972**, *6*, 226–234.
12. Miller, R. On the linearization of Volterra integral equations. *J. Math. Anal. Appl.* **1968**, *23*, 198–208. [CrossRef]
13. Burton, T. Kernel-Resolvents relations for an integral equation. *Tatra Mt. Math. Publ.* **2011**, *48*, 25–40. [CrossRef]
14. Brunner, H. *Collocation Methods for Volterra Integral and Related Functional Differential Equations*; Cambridge University Press: Cambridge, UK, 2004.
15. Vecchio, A. Volterra discrete equations: Summability of the fundamental matrix. *Numer. Math.* **2001**, *89*, 783–794. [CrossRef]
16. Gripenberg, G. On Positive, Nonincreasing Resolvents of Volterra Equations. *J. Differ. Equ.* **1978**, *30*, 380–390. [CrossRef]
17. Brunner, H. *Volterra Integral Equations: An Introduction to Theory and Applications*; Cambridge Monographs on Applied and Computational Mathematics; Cambridge University Press: Cambridge, UK, 2017. [CrossRef]

Article

A Three-Phase Fundamental Diagram from Three-Dimensional Traffic Data

Maria Laura Delle Monache [1], Karen Chi [2], Yong Chen [3], Paola Goatin [4], Ke Han [5], Jing-mei Qiu [6] and Benedetto Piccoli [7,*]

1. University Grenoble Alpes, Inria, CNRS, Grenoble INP, GIPSA-Lab, 38000 Grenoble, France; ml.dellemonache@inria.fr
2. Learning Lab, University of Pennsylvania, Philadelphia, PA 19104, USA; karenchi@sas.upenn.edu
3. Department of Biostatistics, Epidemiology and Informatics (DBEI), University of Pennsylvania, Philadelphia, PA 19104, USA; ychen123@mail.med.upenn.edu
4. Université Côte d'Azur, Inria, CNRS, LJAD, 06902 Sophia Antipolis, France; paola.goatin@inria.fr
5. Institute of System Science and Engineering, Southwest Jiaotong University, Chengdu 611756, China; kehan@swjtu.edu.cn
6. Department of Mathematics, University of Delaware, Newark, DE 19713, USA; jingqiu@udel.edu
7. Department of Mathematical Sciences, Rutgers University-Camden, Camden, NJ 08102, USA
* Correspondence: piccoli@camden.rutgers.edu

Abstract: This paper uses empirical traffic data collected from three locations in Europe and the US to reveal a three-phase fundamental diagram with two phases located in the uncongested regime. Model-based clustering, hypothesis testing and regression analyses are applied to the speed–flow–occupancy relationship represented in the three-dimensional space to rigorously validate the three phases and identify their gaps. The finding is consistent across the aforementioned different geographical locations. Accordingly, we propose a three-phase macroscopic traffic flow model and a characterization of solutions to the Riemann problems. This work identifies critical structures in the fundamental diagram that are typically ignored in first- and higher-order models and could significantly impact travel time estimation on highways.

Keywords: macroscopic models; traffic data; gap analysis; multi-phase models

MSC: 35L65; 76A30; 62H30

Citation: Delle Monache, M.L.; Chi, K.; Chen, Y.; Goatin, P.; Han, K.; Qiu, J.; Piccoli, B. A Three-Phase Fundamental Diagram from Three-Dimensional Traffic Data. *Axioms* **2021**, *10*, 17. https://doi.org/10.3390/axioms10010017

Academic Editor: Hari Mohan Srivastava

Received: 5 November 2020
Accepted: 28 January 2021
Published: 7 February 2021

Publisher's Note: MDPI stays neutral with regard to jurisdictional claims in published maps and institutional affiliations.

Copyright: © 2021 by the authors. Licensee MDPI, Basel, Switzerland. This article is an open access article distributed under the terms and conditions of the Creative Commons Attribution (CC BY) license (https://creativecommons.org/licenses/by/4.0/).

1. Introduction

In the last seventy years, many traffic flow models have been developed and researched. Two of the most commonly used macroscopic models are the celebrated first-order Lighthill–Whitham–Richards (LWR) model, [1,2] and the second-order Aw-Rascle–Zhang model [3,4]. In both cases, the so-called Fundamental Diagram (FD) provides a closure of the evolution equations, thus allowing a well-posed theory and well-grounded simulation tools (see [5]). The FD usually refers to the empirically observed flow-occupancy curve, which in mathematical terms refers to the functional relationship between flow and density (modeling counterpart of occupancy) or between average speed of vehicles and density. For macroscopic fluid-dynamic models, there is a rich discussion on FD (see, e.g., [5–10]).

In this article, we focus on the FD for single roads by proposing a new approach to study the fundamental relationship among flow, density and speed. We propose novel statistical methodologies to analyze traffic data from fixed sensors, focusing on the three-leg relationships among the flow, density and speed. In particular, rather than considering the FD as a two-quantity relationship (flow–density or speed–density), we analyze data in the three-dimensional space represented by flow, density and speed. This allows us to better exploit the statistical tools, in particular for the analysis of traffic regimes.

We recall that, in equilibrium regimes, the fundamental relationship $flow = density \times speed$ dictates that traffic measurement points should lie on a three-dimensional surface (see, e.g., [11] Figure 4.1). In reality, observed traffic largely deviates from equilibrium and usually exhibits *free* and *congested* phases, with the first corresponding to stable and regular traffic, while the second reflects delays and congestion. Moreover, in the early 2000s, Kerner [12] introduced a tree-phase traffic theory, based on the distinction among *free flow*, *synchronized flow* and *wide-moving jam*. The last two phases are associated with congested traffic.

In this paper, using clustering methodologies, we are able to identify three traffic regimes, which are distinct in a statistically significant fashion. Interestingly, two regimes appear in what is commonly referred to as the free flow traffic and the third corresponds to the congested phase. This analysis does not contradict Kerner's theory but rather points out that the static/stationary free-flow condition in the FD could exhibit two distinct phases, while the distinction of phases in congested traffic (e.g., Kerner's model) is mainly dynamic.

The second main empirical result of our paper is the clear evidence of the existence of a gap between the two phases of free flow and the congested one. While the appearance of such gap is best visualized in the 3D representation of the FD relationships, we use the classical flow–density relationship to statistically prove the existence of the gap. The main purpose is to prove the ubiquity (with respect to data collected at different geographical location and on different road types) of the gap in the classical setting and to enable a simpler analysis.

Building on the empirical evidence illustrated thus far, we propose a new three-phase macroscopic model. The LWR model is very popular in the traffic literature due to its simple mathematical representation. However, it has certain modeling limitations especially when it comes to describing complex wave structures such as stop and go waves, phantom jam and capacity drop. To overcome various limitations, Aw-Rascle [3] and independently Zhang [4] proposed a new model with conservation of a modified momentum. This so-called Aw-Rascle–Zhang (ARZ) model can be interpreted as part of a general family called General Second-Order Models (GSOM, see [13]). Such models consist of the usual conservation of mass and the advective transport of a Lagrangian (or single driver) variable, which can represent, for instance, the desired speed of drivers. A recently proposed model of this category is the Collapsed Generalized ARZ model (CGARZ [14]), where the driver speed depends on the Lagrangian variable only in the congested phase. Another line of research focuses on models showing two distinct phases, called the phase transition models [6,8,9].

Our proposed model is a combination of the features offered by the ARZ, CGARZ and phase transition models. Our three-phase model not only has the characteristics of a CGARZ model with a gap among phases when analyzed in the flow–density space, but also exhibits the newly discovered phase when analyzed in the speed–density space. After showing how our model performs in data fitting, we provide a complete characterization of the characteristic curves and the solutions of the Riemann problems. The latter are the building block for solutions to Cauchy problems (see [5]). To sum up, the main novelties and contributions of our paper are as follows:

- Unlike most studies that focus on traffic data from a single source, we use data from multiple geographic locations in Europe and the US and analyze the fundamental relationships among flow, density and speed in the 3D space instead of the commonly adopted two-variable representation of the FD. In addition, we use a set of statistical tools including model-based clustering, hypothesis testing and regression to analyze the traffic data.
- Following the above exercise, we discover three data clusters representing three traffic regimes, two of which are contained in the free-flow phase and the third corresponds to the congested phase. Moreover, we are able to detect a statistically significant gap between the first two regimes and the third one. These findings are validated using

multiple data sources, and the main features (regimes and gaps) are consistent across different geographical areas.
- Building on the first two, we propose a new three-phase macroscopic traffic flow model, which exhibits all the characteristics shown by our data analyses and combines the features of the ARZ, CGARZ and phase transition models. A complete characterization of solutions of the Riemann problems is provided.

The article is organized as follows. In Section 2, we introduce the datasets, their statistical analysis and the results obtained. Moreover, we describe the impacts of these results on traffic modeling. Lastly, in Section 3, we propose a new three-phase macroscopic model.

2. Data Analysis

In this section, we describe the data analyzed in the paper and then present the statistical analysis performed.

2.1. Experimental Data

We consider traffic data collected by static sensors (magnetic coils or radars) located on urban and extra-urban roads and highways. Sensors capture these traffic data regularly over a period of time. The sensor data provide the following aggregated quantities which are measured independently over a short time interval (3–10 min).

- Flux (denoted as f), also known as flow or volume, is the number of vehicles passing through a fixed location per unit of time.
- Velocity (denoted as v) is the average speed of vehicles per unit of time.
- Occupancy (denoted as o) is the percentage of time that a vehicle covers the sensor over the unit time of data collection.

Occupancy acts as a surrogate for the true density of traffic, as true density is practically difficult to capture, although there is some measurement error involved with its calculation. It is know that density and occupancy are correlated at lower densities, but this does not extend to higher densities. The data were collected from three different locations: Rome (Italy), Las Vegas (NV, USA) and Sophia Antipolis (France). The Rome data were provided by ATAC S.p.a. [15] (the municipal society for traffic monitoring and control of Rome) and refer to a road in the city of Rome, Viale del Muro Torto, which links the historical center with the northern area of the city. Data were collected over a period of a week on three sensors. Each collected quantity (occupancy, flow and speed) was aggregated on 1 min intervals. The data from Las Vegas were collected by the Regional Transportation Commission of Southern Nevada (RTC), Freeway and Arterial System of Transportation (FAST) [16]. The data were collected from 50 urban and freeway sensors over a period of five years and aggregated on 10 min intervals. The data from Sophia Antipolis were collected by the Département des Alpes-Maritimes [17] on two extra-urban sensors over a period of eight months and aggregated over 6 min intervals. For more details on the data, we refer the reader to Appendix A. Despite the fact that the data were aggregated at different intervals, the results, as shown below, are consistent. Since we primarily focused on the three traffic characteristics of flux, velocity and occupancy, we were conveniently positioned to analyze the data in three dimensions, a novel concept and approach that is described in the next section.

2.2. Statistical Tools
2.2.1. Cluster Analysis

Cluster analysis is the classification of data with a previously unknown structure and the partitioning of a dataset into meaningful subsets. Clustering sheds light on hidden or non-intuitive relationships between those data and their attributes. Each cluster contains a group of objects that are more closely related to each other than they would be as objects of other clusters. *The concept of distance is thus inherently crucial in the process of cluster analysis, as clusters are grouped based on the results of this measure.* Distance serves as a way to evaluate the closeness, as well as dissimilarity, of pairs of observations. There are at least two options to conduct cluster analysis for this traffic data: model-based clustering (e.g., mixture of normals) and non-parametric clustering (e.g., k-means). Although k-means is popular for complex and high-dimensional data, it is generally used for data involving variables of the same scale (hence, more suitable for data with spherical clusters, e.g., Euclidean distance in 3D), whereas our data consist of three variables of different scales. For this reason, model-based clustering has more flexibility in the shape of clusters; for instance, mixture models [18] can identify clusters in the traffic data that were ellipsoidal.

Empirical evaluations on the distributions of the three traffic variables through quantile–quantile (Q-Q) plot, Shapiro test and Box–Cox transformation have suggested that normal distributions are appropriate. Here, we propose the use of a finite mixture model with G multivariate normals [18]. Specifically, denote data $\mathbf{y}$ with independent trivariate observations (flux, velocity and occupancy) $\{\mathbf{y_1}, \mathbf{y_2}, \ldots, \mathbf{y_n}\}$ the likelihood for a mixture model with G components is

$$\ell(\theta_1, \theta_2, \ldots, \theta_G; \pi_1, \pi_2, \ldots, \pi_G | \mathbf{y}) = \prod_{i=1}^{n} \sum_{k=1}^{G} \pi_k f_k(\mathbf{y_i} | \theta_k),$$

where i stands for ith observation, $f_k(\cdot)$ and θ_k are the density function and model parameters of the kth cluster in the mixture and π_k is the probability that an observation belongs to the kth cluster, subject to the simplex constraint $\{\pi_k \geq 0; \sum_{k=1}^{G} \pi_k = 1\}$. Such a model can be fitted by the expectation-maximization (EM) algorithm and is implemented by the R package 'mclust'.

2.2.2. Three Phase Traffic

Figure 1 provides a 3D visualization and cluster analysis result on the Rome dataset, where observations in different clusters are marked by different colors. Previous knowledge assumed that traffic involves two clusters: free flow and congestion. Free flow corresponds to steady traffic flow at high speeds (and low densities), while congestion is characterized by low flux and reduced speeds. From this new 3D visualization of data, we can identify a third phase, which we call the **"free choice phase"**, which corresponds to the situation of a relatively empty road, whereby drivers choose their speed independently without influence from or interaction with other vehicles.

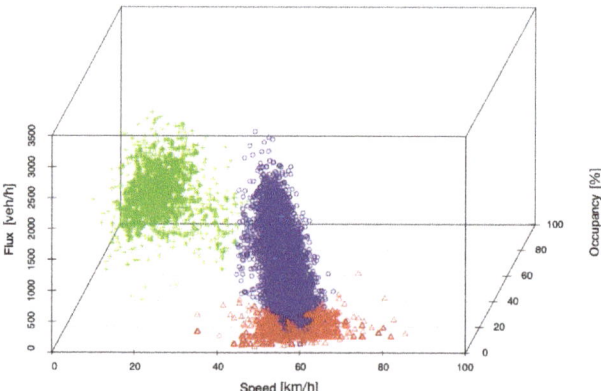

Figure 1. 3D visualization and cluster analysis results of Rome data suggest the existence of third phase (red) in addition to the free flow (blue) and congestion (green) phases.

In the free choice phase, the flow of cars is low while the speed is variable. Model selection procedures (e.g., Bayesian information criterion (BIC) or adjusted BIC) have been used to select the number of clusters, and the datasets from Rome, Nevada and Sophia

Antipolis have consistently suggested the existence of the third phase. Such additional phase is incorporated into the mathematical modeling.

Notice that our three phases are different from those indicated by Kerner [12]. Indeed, we have two sub-phases in the free phase cluster opposed to Kerner's model with two sub-phases in the congestion phase cluster.

2.2.3. Gap Analysis

We developed and applied a rigorous hypothesis testing procedure to the datasets to formally investigate the presence of phase transitions. Specifically, investigating the presence of phase transition can be formulated as testing the existence of a "gap region" at the upper portion of occupancy in the free phase and its proximity to the lower portion of occupancy in congestion. As shown in Figure A1 (left) in Appendix A, such gap region can be potentially masked by isolated points in the gap, which could be in fact due to measurement errors or random variations in flux and occupancy. To reduce the impact of these isolated points, we propose to take the upper quantile of the free phase (e.g., 95th percentile, denoted as ρ_{FP}) and the lower quantile of the congested phase (e.g., 5th percentile, denoted as ρ_C) and formally test for $H_0: \rho_{FP} \geq \rho_C$, i.e., there is no gap, against $H_a: \rho_{FP} < \rho_C$, i.e., there is a gap. Figure 2 illustrates these two scenarios of H_0 and H_a.

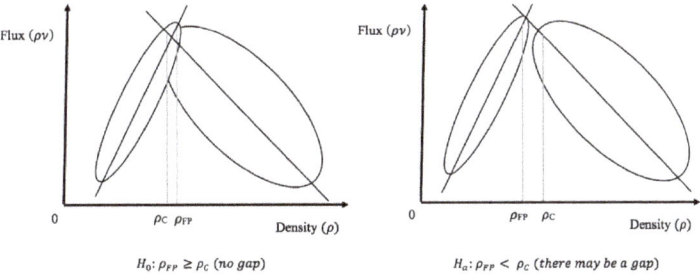

Figure 2. Illustration of hypothesis testing procedure for phase transition.

For given pre-specified percentiles q_{FP} and q_C (e.g., 95% and 5%), denote $\hat{\rho}_{FP}$ and $\hat{\rho}_C$ as the corresponding quantiles in the two clusters. The existence of phase transition can be formally tested by a *one-sided test* based on the Wald statistic

$$T = \min\{\hat{\rho}_{FP} - \hat{\rho}_C, 0\} / \sqrt{\text{var}(\hat{\rho}_{FP} - \hat{\rho}_C)}.$$

The variance $\text{var}(\hat{\rho}_{FP} - \hat{\rho}_C)$ can be approximated by

$$\text{var}(\hat{\rho}_{FP} - \hat{\rho}_C) = \text{var}(\hat{\rho}_{FP}) + \text{var}(\hat{\rho}_C) \approx \frac{q_{FP}(1 - q_{FP})}{n_{FP}\{f(F^{-1}(q_{FP}))\}^2} + \frac{q_C(1 - q_C)}{n_C\{f(F^{-1}(q_C))\}^2},$$

where $f(\cdot)$ and $F(\cdot)$ stand for the estimated distribution function and cumulative distribution function of ρ, respectively, and n_{FP} and n_C stand for the number of data points in the free and congestion phases, respectively. The first equality in the calculation of variance is due to the independence of $\hat{\rho}_{FP}$ and $\hat{\rho}_C$ as they are estimated from different sets of observations. The approximation is due to a standard result for asymptotic variance of percentile estimates (see [19]). The complete sets of results for the statistical analysis on the three datasets can be found in Appendix B.

3. A Macroscopic Second-Order Model Accounting for the 3 Phases

Following the approach of Colombo et al. [7], Fan et al. [14], we propose a new macroscopic model accounting for the three phases derived in the previous sections. In conservation form, the model can be expressed as

$$\partial_t \rho + \partial_x (\rho \, v(\rho, y/\rho)) = 0,$$
$$\partial_t y + \partial_x (y \, v(\rho, y/\rho)) = 0, \tag{1}$$

where the velocity function is chosen such that

$$v(\rho, y/\rho) = \begin{cases} v_{FC}(\rho, y/\rho), & \text{if } 0 < \rho \leq \rho_{FC}, \\ v_{FP}(\rho), & \text{if } \rho_{FC} < \rho \leq \rho_{FP}, \\ v_C(\rho, y/\rho), & \text{if } \rho_C \leq \rho \leq \rho_{max}, \end{cases} \tag{2}$$

for some $0 < \rho_{FC} < \rho_{FP} < \rho_C < \rho_{max}$, and it is continuous at ρ_{FC} and ρ_{FP}. In (1)–(2), the quantity $w = y/\rho \in [w_{min}, w_{max}]$ may represent various traffic characteristics, such as vehicles classes [20], aggressiveness [21], desired spacing [22] or perturbation from equilibrium [23], which are transported with the traffic stream. We refer to the variable $y = \rho w$ as a *total property* [14]. The function v defined in (2) must be:

1. Non-negative: $v(\rho, w) \geq 0$ for all $\rho \in [0, \rho_{max}], w \in [w_{min}, w_{max}]$;
2. Continuous: $v_{FC}(\rho_{FC}, w) = v_{FP}(\rho_{FC})$ and $v_C(\rho_{FP}, w) = v_{FP}(\rho_{FP})$ for all $w \in [w_{min}, w_{max}]$
3. Vanishing at maximal density: $v(\rho_{max}, w) = v_C(\rho_{max}, w) = 0$ for all $w \in [w_{min}, w_{max}]$
4. Non-decreasing with respect to w: $\dfrac{\partial v}{\partial w}(\rho, w) \geq 0$ for $\rho \in [0, \rho_{max}]$

With the above assumptions, the corresponding flux function $q(\rho, w) = \rho \, v(\rho, w)$ satisfies $q(0, w) = q(\rho_{max}, w) = 0$ for all w.

To take into account the possible presence of a gap, as suggested by our analysis, we fix the value $v_C^{max} \leq v_{FP}^{min} := v_{FP}(\rho_{FP})$ of the maximal speed in congestion, and let $\rho_C \in [\rho_{FP}, \rho_{max}[$ be the density value such that

$$v_C(\rho_C, w_{min}) = v_C^{max}.$$

Defining the velocity function (see Figure 3) as

$$v_g(\rho, w) = \begin{cases} v_{FC}(\rho, w), & \text{if } 0 < \rho \leq \rho_{FC}, \\ v_{FP}(\rho), & \text{if } \rho_{FC} < \rho \leq \rho_{FP}, \\ \min\{v_C^{max}, v_C(\rho, w)\}, & \text{if } \rho_C \leq \rho \leq \rho_{max}, \end{cases} \tag{3}$$

the corresponding flux function $q_g(\rho, w) := \rho v_g(\rho, w)$ displays the desired gap between the free-flow and congested phases (see Figure 4).

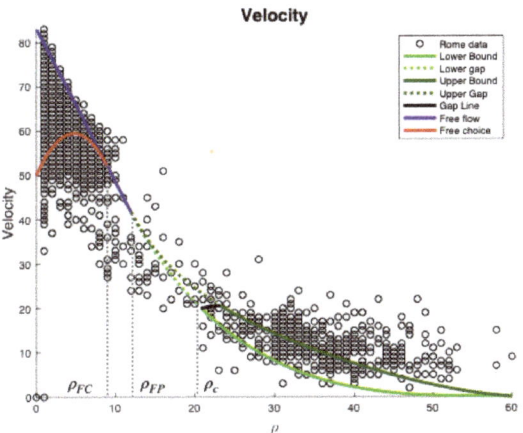

Figure 3. An example of speed function.

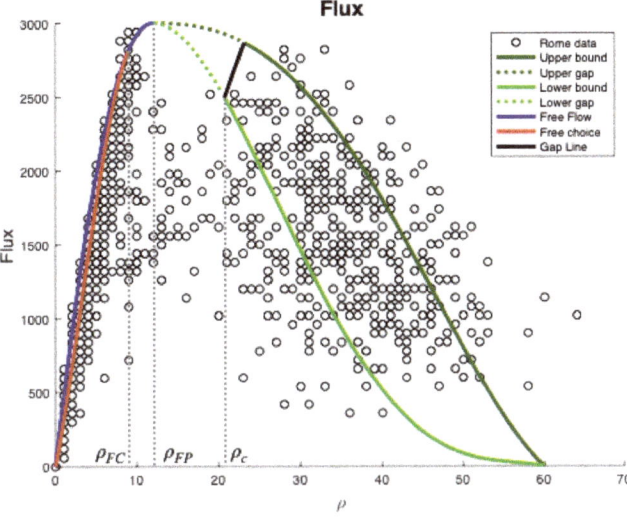

Figure 4. General (non-concave) fundamental diagram.

3.1. Riemann Solver

To simplify the construction, it is not restrictive to assume that the fundamental diagram is ρ-differentiable, i.e., we assume that

$$\frac{\partial v_{FC}}{\partial \rho}(\rho_{FC}, w) = v'_{FP}(\rho_{FC}) \quad \text{for all } w \in [w_{\min}, w_{\max}]$$

and

$$\frac{\partial v_C}{\partial \rho}(\rho_{FP}, w) = v'_{FP}(\rho_{FP}) \quad \text{for all } w \in [w_{\min}, w_{\max}].$$

System (1) is defined on the invariant domain

$$\Omega = \{(\rho, \rho w) \in [0, \rho_{\max}] \times [0, \rho_{\max} w_{\max}] : w \in [w_{\min}, w_{\max}]\}.$$

We note that, under the above assumptions on the velocity function v, $(\rho, y) \in \Omega$ if and only if $w \in [w_{\min}, w_{\max}]$ and $v(\rho, y/\rho) \in [0, v(0, w_{\max})]$. The eigenvalues are given by

$$\lambda_1(\rho, y/\rho) = v(\rho, y/\rho) + \rho \frac{\partial}{\partial \rho} v(\rho, y/\rho) \quad \text{and} \quad \lambda_2(\rho, y/\rho) = v(\rho, y/\rho), \qquad (4)$$

so the system is strictly hyperbolic for $\rho > 0$ as long as $\partial v(\rho, y/\rho)/\partial \rho \neq 0$. We note that the second characteristic field is linearly degenerate, giving origin to contact discontinuity waves, while the first characteristic field is genuinely non-linear if

$$\frac{\partial^2 q}{\partial \rho^2}(\rho, w) = 2 \frac{\partial v}{\partial \rho}(\rho, w) + \rho \frac{\partial^2 v}{\partial \rho^2}(\rho, w) < 0, \qquad \text{for } \rho \in [0, \rho_{\max}], \qquad (5)$$

holds. Moreover, the Riemann invariants of the systems are given by w and v. In particular, the iso-values $w = \text{const}$ correspond to waves of the first family (we recall that the system belongs to Temple class, i.e., shock and rarefaction curves coincide) and the contact discontinuities verify $v = \text{const}$. More precisely, in the strictly concave case (5) the elementary waves are constructed as follows.

- **1-rarefaction waves.** Two points $(\rho_l, \rho_l w_l)$ and $(\rho_r, \rho_r w_r)$ are connected by a 1-rarefaction wave if and only if

$$w_l = w_r \quad \text{and} \quad \lambda_1(\rho_l, w_l) < \lambda_1(\rho_r, w_r).$$

- **1-shock waves.** Two points $(\rho_l, \rho_l w_l)$ and $(\rho_r, \rho_r w_r)$ are connected by a 1-shock wave if and only if

$$w_l = w_r \quad \text{and} \quad \lambda_1(\rho_l, w_l) > \lambda_1(\rho_r, w_r).$$

In this case, the jump discontinuity moves with speed

$$\sigma = \frac{\rho_l v(\rho_l, w_l) - \rho_m v(\rho_m, w_m)}{\rho_l - \rho_m}.$$

- **2-contact discontinuity.** Two points $(\rho_l, \rho_l w_l)$ and $(\rho_r, \rho_r w_r)$ are connected by a 2-contact wave if and only if

$$v(\rho_l, w_l) = v(\rho_r, w_r).$$

In the general (non-concave) case (see Figure 4), the 1-waves consist of a concatenation of shocks and rarefactions (see ([24] Section 1)).

Based on the above elementary waves, the solution corresponding to general Riemann data $(\rho_l, \rho_l w_l)$, $(\rho_r, \rho_r w_r)$ can be constructed as follows. Let $(\rho_m, \rho_m w_m)$ be the intermediate point defined by

$$w_m = w_l, \qquad v(\rho_m, w_m) = v(\rho_r, w_r).$$

Setting $v_{w_l}(\rho) = v(\rho, w_l)$, ρ_m is given by

$$\rho_m = \begin{cases} v_{w_l}^{-1}(v(\rho_r, w_r)), & \text{if } v(\rho_r, w_r) < v(0, w_l), \\ 0, & \text{otherwise}. \end{cases}$$

In the latter case, a vacuum zone appears between the sector

$$v(0, w_l)\, t < x < v(\rho_r, w_r)\, t.$$

The complete solution is then given by a 1-wave connecting $(\rho_l, \rho_l w_l)$ and $(\rho_m, \rho_m w_m)$, followed by a 2-contact discontinuity between $(\tilde{\rho}_m, \tilde{\rho}_m w_m)$ and $(\rho_r, \rho_r w_r)$ (eventually separated by a vacuum zone if $v(\rho_r, w_r) > v(\rho_m, w_l)$ and $\rho_m = 0$).

The presence of the gap between $v_C^{\max}$ and $v_{FP}^{\min}$ does not modify the procedure, since the definition domain

$$\Omega_g = \left\{(\rho, \rho w) \in \Omega : v(\rho, w) \in [0, v_C^{\max}] \cup [v_{FP}^{\min}, v(0, w_{\max})]\right\}$$

is still invariant. We set $\Omega_g = \Omega_{FP} \cup \Omega_C$ with

$$\Omega_{FP} = \left\{(\rho, \rho w) \in \Omega : v(\rho, w) \in [v_{FP}^{\min}, v(0, w_{\max})]\right\},$$
$$\Omega_C = \{(\rho, \rho w) \in \Omega : v(\rho, w) \in [0, v_C^{\max}]\}.$$

We can distinguish the following cases:
- If $(\rho_l, \rho_l w_l)$ and $(\rho_r, \rho_r w_r)$ belongs both to Ω_{FP} or Ω_C, the Riemann solver is defined as above.
- If $(\rho_l, \rho_l w_l) \in \Omega_C$ and $(\rho_r, \rho_r w_r) \in \Omega_{FP}$, the intermediate point $(\rho_m, \rho_m w_m)$ belongs to Ω_{FP}. Let $(\rho_c, \rho_c w_c) \in \partial \Omega_C$ the point defined by

$$w_c = w_l, \qquad v(\rho_c, w_c) = v_C^{\max}.$$

The solution is composed by 1-waves connecting $(\rho_l, \rho_l w_l)$ and $(\rho_c, \rho_c w_l)$, a phase-transition jump between $(\rho_c, \rho_c w_l)$ and $(\rho_{FP}, \rho_{FP} w_l)$ moving with speed

$$\sigma = \frac{\rho_c v_C^{\max} - \rho_{FP} v_{FP}^{\min}}{\rho_c - \rho_{FP}},$$

followed by 1-waves connecting $(\rho_{FP}, \rho_{FP} w_l)$ and $(\rho_m, \rho_m w_m)$ and eventually a 2-contact from $(\rho_m, \rho_m w_m)$ to $(\rho_r, \rho_r w_r)$.
- If $(\rho_l, \rho_l w_l) \in \Omega_{FP}$ and $(\rho_r, \rho_r w_r) \in \Omega_C$, the intermediate point $(\rho_m, \rho_m w_m)$ belongs to Ω_C. Therefore, the solution always contains a 1-wave (shock phase-transition) from $(\rho_l, \rho_l w_l)$ to $(\rho_m, \rho_m w_m)$, followed by a 2-contact discontinuity. Notice that the solution may also contain an intermediate 1-wave in the congested phase.

4. Numerical Scheme and Simulations

4.1. Numerical Scheme

For simplicity, let us rewrite problem (1) in compact form:

$$\partial_t \mathbf{u} + \partial_x \mathbf{f}(\mathbf{u}) = 0 \qquad \mathbf{u} \in \Omega_{FC} \cup \Omega_{FP} \cup \Omega_C \qquad (6)$$

where $\mathbf{u} = (\rho, y)$ and

$$\mathbf{f}(\mathbf{u}) = \begin{cases} (\rho v_{FC}, y v_{FC}), & \text{if } 0 < \rho \le \rho_{FC}, \\ (\rho v_{FP}, y v_{FP}), & \text{if } \rho_{FP} < \rho \le \rho_{FP}, \\ (\rho v_C, y v_C), & \text{if } \rho_C < \rho \le \rho_{\max}. \end{cases}$$

Let us fix constant space step Δx and time step Δt and $\nu = \Delta t / \Delta x$. Let us define the mesh interfaces $x_{j+1/2} = j \Delta x$ for $j \in \mathbb{Z}$ and the intermediate times $t^n = n \Delta t$ for $n \in \mathbb{N}$. A piecewise constant approximated solution $\mathbf{u}_\nu(x, t^n)$ of $\mathbf{u}$ is given by

$$\mathbf{u}_\nu(x, t^n) = \mathbf{u}_j^n \text{ for all } x \in C_j = [x_{j-1/2}; x_{j+1/2}[, \ j \in \mathbb{Z}, \ n \in \mathbb{N}.$$

In this paper, we use the numerical scheme introduced in [25]. This scheme is a modified Godunov scheme composed of two steps. The first step looks at the evolution in time of the Cauchy problem, while the second step projects it onto piecewise constant functions:

Step 1: Evolution in time.
This step consists in solving the Riemann problem at each cell interface $x_{j+1/2}$ with initial

data $(\mathbf{u}_j^n, \mathbf{u}_{j+1}^n)$, obtaining an exact solution $\overline{\mathbf{u}}_\nu(x, t^{n+1}-)$.

Step 2: Projection to time t^{n+1}

Once all Riemann problems at interfaces are solved, Chalons and Goatin [25] proposed a new averaging procedure. The idea is that, since the solution can contain states in different phases, the average is not done on the regular mesh cells but on modified non-uniform cells that contain only values belonging to the same phase. We denote this modified cells by $\overline{C}_j^n = [\overline{x}_{j-1/2}^n, \overline{x}_{j+1/2}^n[$. Afterwards, a sampling strategy allows us to recover a piecewise constant solution on the initial mesh cells C_j.

We define the new interface $\overline{x}_{j+1/2}^n$ at time as t^{n+1}

$$\overline{x}_{j+1/2}^n = x_{j+1/2} + \sigma_{j+1/2}^n \Delta t, \quad j \in \mathbb{Z} \tag{7}$$

and the new space intervals

$$\overline{\Delta x_j^n} = \overline{x}_{j+1/2}^n - \overline{x}_{j-1/2}^n, \quad j \in \mathbb{Z},$$

with $\sigma_{j+1/2}^n = \sigma(\mathbf{u}_j^n, \mathbf{u}_{j+1}^n)_{j \in \mathbb{Z}}$ the characteristic speeds of propagation of phase transitions at interfaces. Then, we average the solution of Step 1 on the cells $\overline{C}_j^n$, obtaining a piecewise constant approximate solution $\overline{\mathbf{u}}_j^{n+1}$ on a non-uniform mesh with

$$\overline{\mathbf{u}}_j^{n+1} = \frac{1}{\overline{\Delta x_j^n}} \int_{\overline{x}_{j-1/2}^n}^{\overline{x}_{j+1/2}^n} \overline{\mathbf{u}}_\nu(x, t^{n+1}-) \, dt, \quad j \in \mathbb{Z}.$$

The modified Godunov scheme then reads:

$$\overline{\mathbf{u}}_j^{n+1} = \frac{\Delta x}{\overline{\Delta x_j^n}} \mathbf{u}_j^n - \frac{\Delta t}{\overline{\Delta x_j^n}} (\mathbf{f}^{n,-}(\mathbf{u}_j^n, \mathbf{u}_{j+1}^n) - \mathbf{f}^{n,+}(\mathbf{u}_j^n, \mathbf{u}_{j-1}^n)) \text{ for all } j \in \mathbb{Z}, \tag{8}$$

with

$$\overline{\mathbf{f}}^{n,\pm}(\mathbf{u}_j^n, \mathbf{u}_{j+1}^n) = \mathbf{f}(\mathcal{RS}(\sigma_{j+1/2}^{n,\pm}; \mathbf{u}_j^n, \mathbf{u}_{j+1}^n)) - \sigma_{j+1/2}^n \mathcal{RS}(\sigma_{j+1/2}^{n,\pm}; \mathbf{u}_j^n, \mathbf{u}_{j+1}^n), \tag{9}$$

and $\mathcal{RS}$ the solution to the Riemann problem as given in Section 3.1.

We then project the solution onto the original mesh C_j using a well distributed random sequence $(a_n) \in]0, 1[$ as follows:

$$\mathbf{u}_j^{n+1} = \begin{cases} \overline{\mathbf{u}}_{j-1}^{n+1} & \text{if } a_{n+1} \in \left]0, \frac{\Delta t}{\Delta x} \max\{\sigma_{j-1/2}^n, 0\}\right[, \\ \overline{\mathbf{u}}_j^{n+1} & \text{if } a_{n+1} \in \left[\frac{\Delta t}{\Delta x} \max\{\sigma_{j-1/2}^n, 0\}, 1 + \frac{\Delta t}{\Delta x} \min\{\sigma_{j+1/2}^n, 0\}\right[, \\ \overline{\mathbf{u}}_{j+1}^{n+1} & \text{if } a_{n+1} \in \left[1 + \frac{\Delta t}{\Delta x} \min\{\sigma_{j+1/2}^n, 0\}, 1\right[. \end{cases} \tag{10}$$

Following Chalons and Goatin [25], we consider the van der Corput random sequence defined by

$$a_n = \sum_{k=0}^m i_k 2^{-(k+1)},$$

where $n = \sum_{k=0}^m i_k 2^k$, $i_k = 0, 1$, is the binary expansion of $n \in \mathbb{N}$.

4.2. Numerical Simulations

We compared our model (1) to the model in [25] in a simulation on a single road on length 1 with $x \in [-0.5, 0.5]$ with an initial data

$$\rho(x, 0) = \begin{cases} 0.1 & \text{if } x \leq 0 \\ 0.6 & \text{if } x > 0 \end{cases} \tag{11}$$

We point out that our model is profoundly different from the phase transition models, even with gap between phases. The main reason is that the second order model (1) admits phase transitions, i.e., shock waves connecting phases, but as classical first family waves, and allows different speeds also in free choice for different values of the variable w. This is well captured by the simulation in Figure 5. An initial condition with one backward moving shock is perturbed by a boundary datum presenting oscillations in the w variable. As a result, for large times, small oscillations are visible on the left (free flow) and large oscillations are propagated through the shock (congested flow). On the other side, the phase transition model of Chalons and Goatin [25] is insensitive to such oscillations in the w variable, as shown in Figure 6.

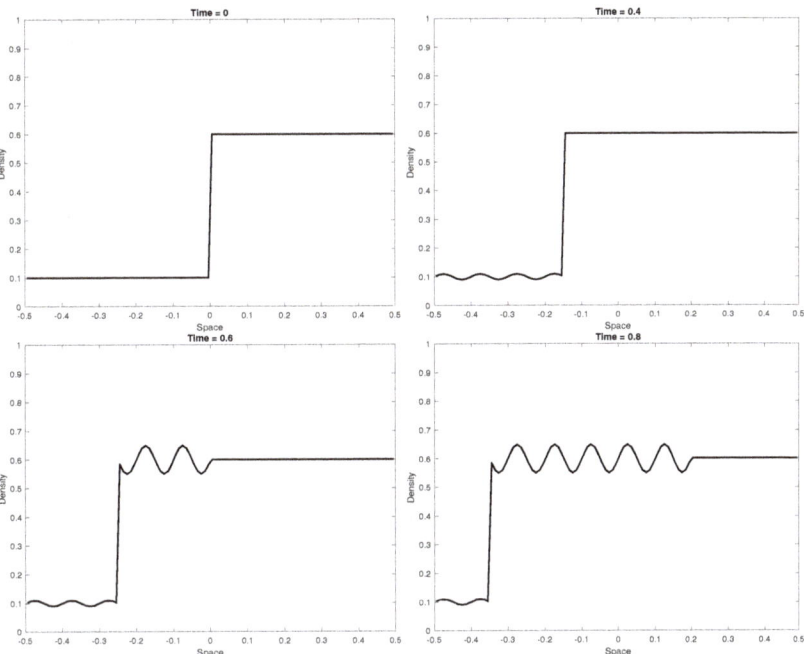

Figure 5. Evolution of the density for our model on a single road.

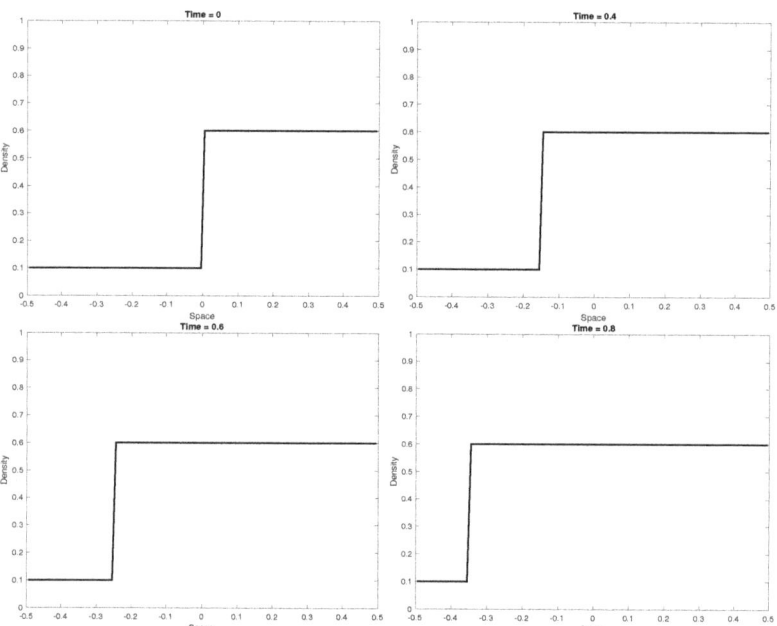

Figure 6. Evolution of the density for a classical phase transition model on a single road.

5. Conclusions

We analyzed three different datasets collected from different locations in Europe and the US from fixed sensors. Representing data via a three-dimensional fundamental diagram, we showed the presence of three traffic phases, two in free flow regime and one in congested flow regime, and of a statistically significant gap between free and congested flow. Based on these results, we designed a new second-order macroscopic model that is capable describing analytically the gap and the three different phases. Moreover, a characterization of Riemann problem solutions and a numerical example are provided, illustrating the difference with phase transition models.

Author Contributions: Software, M.L.D.M. and K.C.; Validation, M.L.D.M.; Visualization, M.L.D.M. and K.H.; Data curation, Y.C. and K.H.; Supervision, Y.C., P.G. and B.P.; Formal analysis, P.G. and B.P.; Methodology, P.G., J.-m.Q. and B.P.; Investigation, J.-m.Q. All authors have read and agreed to the published version of the manuscript.

Funding: Supported by the French National Research Agency under the "Investissements d'avenir" program (ANR-15-IDEX-02). Supported in part by the National Institute of Health grants 1R01LM012607 and 1R01AI130460. Supported by the NSF project Grant CNS No. 1446715, the NSF project KI-Net Grant DMS No. 1107444 and the Lopez chair endowment.

Acknowledgments: M.L.D.M. acknowledges that this article was developed in the framework of the Grenoble Alpes Data Institute, supported by the French National Research Agency under the "Investissements d'avenir" program (ANR-15-IDEX-02). Y.C.'s research is supported in part by the National Institute of Health grants 1R01LM012607 and 1R01AI130460. B.P. acknowledges the support of the NSF project Grant CNS No. 1446715, the NSF project KI-Net Grant DMS No. 1107444 and the Lopez chair endowment. The authors thank ATAC S.p.a. for providing the traffic data from the city of Rome and the Départment des Alpes Maritimes for providing data from Sophia Antipolis.

Conflicts of Interest: The authors declare no conflict of interest.

Appendix A. Data Description

Our dataset sources are:

1. *Rome*: Three sensors over one week (June 2006) aggregated every 1 min
2. *Las Vegas*: Fifty sensors over five years (2010–2015) aggregated over every 10 min
3. *Sophia Antipolis*: Four sensors for eight months (January–August 2014) every 6 min

We use the information collected in Rome as the primary example to illustrate the data structure. The Rome dataset contains data of each minute of the entire day for one week; thus, 10,080 observations were collected. Since our primary dataset from Rome consists of one week of observations, we only analyzed data for one week in the other locations as well. The datasets from Las Vegas and Sophia Antipolis were used to validate the results from the Rome data.

Figure A1 illustrates the pairwise plots among the three measured variables based on the dynamic data collected from a sensor located in the road Viale del Muro Torto in the city of Rome on a Monday. These plots can provide useful insight on the functional relationship between these variables in two-dimensional space. For instance, the plot of flux against occupancy suggests a linear relationship with small variation when occupancy is less than a threshold (known as the **free phase**) and much larger variation when occupancy is larger than the threshold (known as the **congestion phase**). Furthermore, *both flux vs. speed and flux vs. occupancy plots suggest a possible "gap" between free and congestion phases, which corresponds to phase transition. These are important features that need to be taken into consideration in the mathematical modeling.* Such pairwise plots are useful to generate data-driven hypotheses that need to be formally tested statistically and validated across different datasets.

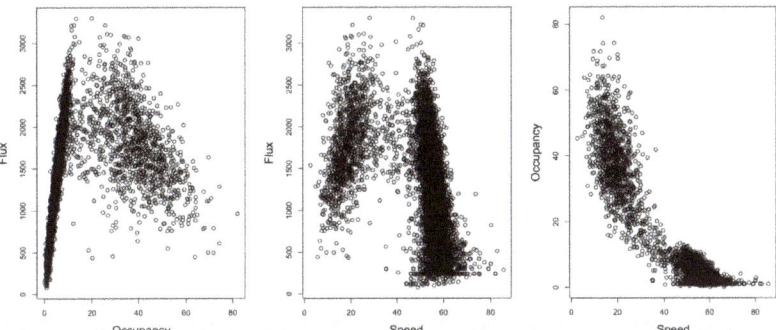

Figure A1. Pairwise scatterplots of Rome Data in the road Viale del Muro Torto: flux vs. occupancy (**left**); flux vs. speed (**middle**); and occupancy vs. speed (**right**).

Appendix B. Results of Cluster Analysis

We conducted model-based cluster analyses on three datasets. We found statistically significant gaps between free and congested phases in all three datasets. We also used regression analysis to demonstrate the existence of free choice phase in explaining the variability in observed flux.

Table A1 presents the results from the described gap analysis of the Rome, Nevada and Sophia datasets. After "trimming" a very small percent of data (e.g., 3%) and considering (97%, 3%) quantiles of free and congestion phases, the test statistics suggested strong evidence of a gap (indicating phase transition) in the three datasets.

Table A1. Results of model-based cluster analysis using Rome, Las Vegas and Sophia Antipolis data. Phase: estimated phase using cluster analysis. FP, free phase; C, congestion phase. Density, estimated density value at the percentile.

Dataset	Percentile (%)	Phase	Density	Test Statistic	p-Value
Rome	97.5	FP	10	0	1
	2.50	C	10		
	97	FP	9	−1.79	0.037
	3	C	10		
	95	FP	9	−3.15	<0.001
	5	C	11		
Las Vegas	97.5	FP	12	−0.91	0.18
	2.50	C	13		
	97	FP	12	−2.17	0.015
	3	C	14		
	95	FP	11	−5.75	<0.001
	5	C	16		
Sophia	97.5	FP	5	−2.82	0.002
	2.50	C	10		
	97	FP	5	−6.44	<0.001
	3	C	12		
	95	FP	4	−5.79	<0.001
	5	C	12		

Graphical representations of the results are illustrated below. To clarify the color code for the graphs, the region colored in red corresponds to the Free Choice phase, blue to the Free Flow phase and green to Congestion. The Free Flow phase in this three cluster model corresponds to the remainder of the original conception of Free Flow without Free Choice. The Free Choice and Free Flow phases from the three cluster model are collectively referred to as the Free Phase from here on.

Figure A2 illustrates the clustering performed by 'mclust' of the Rome data on a two-dimensional level. Pairs plots present the data according to each pair of variables: velocity and flux, occupancy and flux and velocity and occupancy. This type of plot provides insight on the shape and characteristics of the data in 2D. For instance, we observed some sort of gap between Free Phase and Congestion, which can be most easily viewed in the pairs plot of the variables occupancy and flux. Figure 1 is the three-dimensional representation of the Rome data, a novel approach to visualizing traffic data. Through this 3D plot of data, the proposed gap between Free Phase and Congestion is even more noticeable, reinforcing our observations from the 2D case. Specifying 'mclust' to filter through the data for two or three clusters indicated there is some margin of difference between the original Free Flow phase in the two cluster model and the Free Phase in the three cluster model; the latter model is generally neater than the former. The disparity is minimal and perhaps insignificant, although it is worth noting.

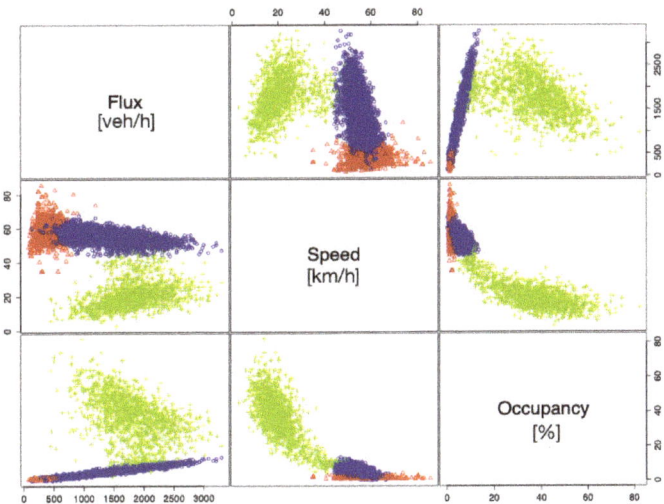

Figure A2. Pairs plot of clustered Rome data.

Cluster analyses using the Las Vegas and Sophia Antipolis data corroborate the results obtained with the Rome data. Las Vegas data collected from highway sensors 25 and 99, which we refer to as Nevada 25 and Nevada 99, respectively, clusters into nine phases through 'mclust'. However, we forced R to choose only two and three clusters to match the original two-phase model of traffic flow proposed by the field and the three-phase model discovered in this study. The results are reported in Figures A3 and A4. The data and analyses suggest that the three-cluster model was preferred using Bayesian information criterion (BIC) for model selection [18].

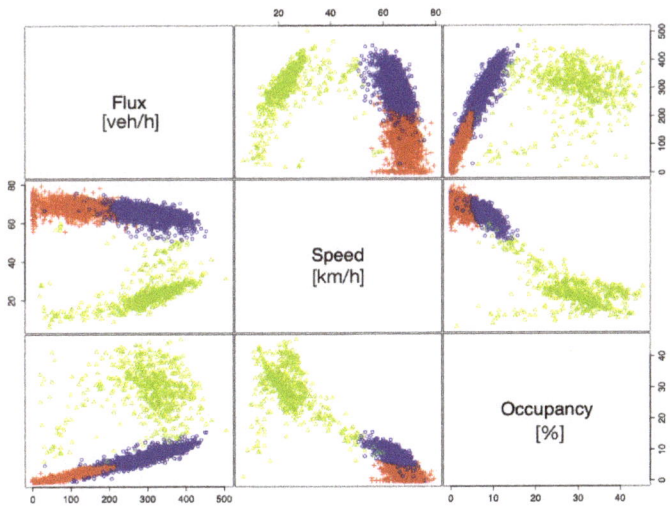

Figure A3. Pairs plot of clustered Las Vegas data.

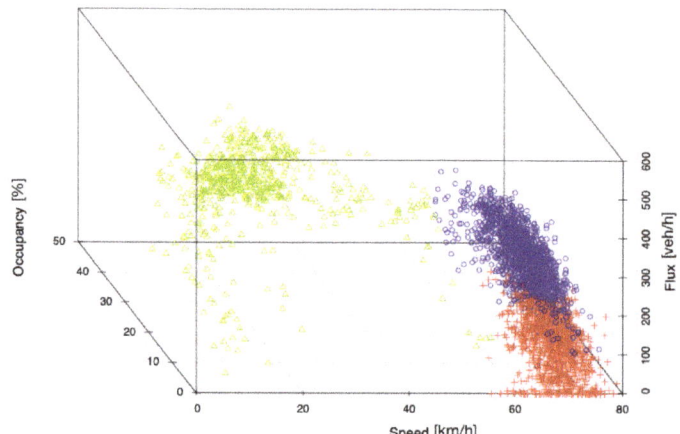

Figure A4. 3D plot of clustered Las Vegas data.

The data from Sophia Antipolis also demonstrate the existence of the Free Choice phase. These two datasets both have a wide range of observed speeds at low levels of occupancy, as shown in the pairs plots and 3D plots in Figures A5 and A6.

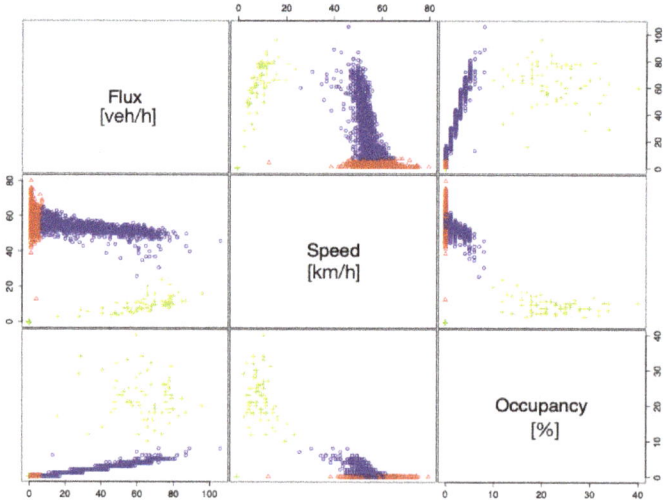

Figure A5. Pairs plot of clustered Sophia Antipolis data.

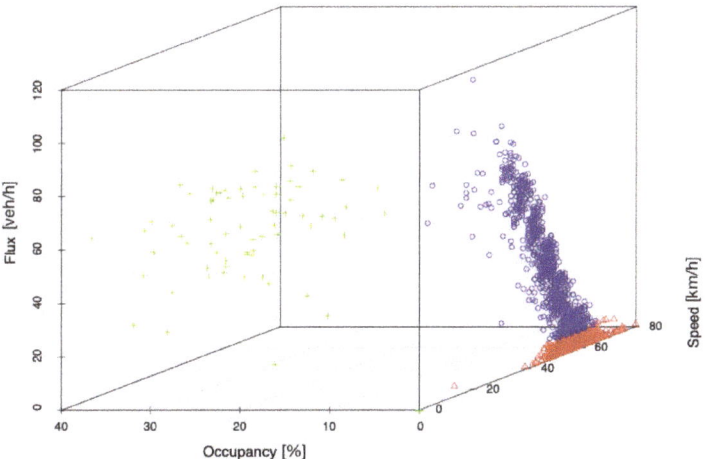

Figure A6. 3D plot of clustered Sophia Antipolis data.

Quantifying the Improved Goodness of Fit Through RSS Comparisons

We conducted further analyses by using residual sum of squares (RSS) values to compare various models considered in this paper. The RSS value was calculated as the sum of differences between the observed and fitted values of flux, where the fitted values of flux were obtained from two or three cluster models with or without speed as an additional predictor in addition to the occupancy. The RSS is an objective measure of the remaining variability of the flux that has not been explained by a particular model.

With Rome data, we considered a baseline model with two-phase and occupancy as the only predictor. In other words, this is a 2D and two-phase model as in the existing literature. We calculated the RSS value for this model and compared with the RSS values from more complex models. Specifically, we found that adding speed into the model (i.e., a 3D and two-phase model) explained additionally 1.2% of the variability in observed flux, on top of the baseline 2D and two-phase model. Furthermore, adding the Free Choice phase alone (i.e., a 2D and three-phase model) reduced 13.6% of the remaining variability. Adding both speed and the free choice phase to the model (i.e., a 3D and three-phase model) further reduced 16.0% of the variability in observed flux, compared to the baseline model.

The results of these comparisons suggest that the three-cluster model is indeed superior to the two-cluster model and that 3D rendering of the data is appropriate. The percent change for adjustment of number of clusters as well as dimensions both increases indicates a clear improvement; the three-cluster model is better than the two-cluster model, 3D analysis of data is more informative than that in 2D and the 3D three-cluster model provides the more favorable RSS value overall. These results are consistent across our datasets (see Figures A7–A10 and Tables A2–A6). These results have important implications in understanding traffic flow. They confirm the utility of analyzing this type of data in three dimensions and reveal the presence of a third phase.

Table A2. RSS analysis with two clusters with flux and occupancy. β_0 is the value of the intercept and β_1 the occupancy.

	β_0	β_1	R^2	Adj. R^2	RSS
Free Phase	77.71	249.18	0.952	0.952	157,765,974
Congestion	2184.09	−11.00	0.09746	0.09676	283,255,197

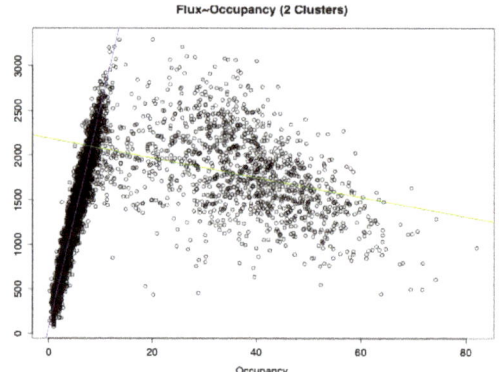

Figure A7. Flux vs. occupancy: RSS analysis.

Table A3. RSS analysis with two clusters with flux, occupancy and speed. β_0 is the value of the intercept, β_1 the occupancy, and β_2 the speed.

	β_0	β_1	β_2	R^2	Adj. R^2	RSS
Free Phase	−248.7	253.88	5.48	0.953	0.953	154,308,365
Congestion	1941.54	−8.11	6.64	0.1032	0.1018	281,448,225

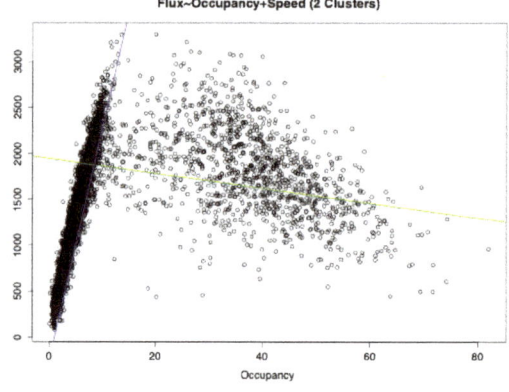

Figure A8. Flux vs. occupancy + speed: RSS analysis.

Table A4. RSS analysis with three clusters with flux and occupancy. β_0 is the value of the intercept and β_1 the occupancy.

	β_0	β_1	R^2	Adj. R^2	RSS
Free Choice	79.56	201.3	0.563	0.5628	14,269,933
Free Flow	187.68	231.36	0.9175	0.9175	126,714,084
Congestion	2302.55	−13.87	0.1618	0.1611	240,324,480

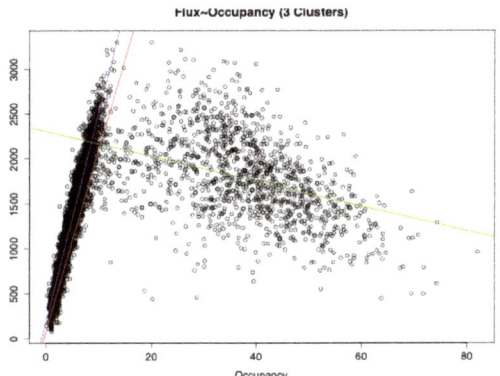

Figure A9. Flux vs. occupancy: RSS analysis.

Table A5. Residual sum squared analysis with three clusters with flux, occupancy and speed. β_0 is the value of the intercept and β_1 the occupancy, β_2 the speed.

	β_0	β_1	β_2	R^2	Adj. R^2	RSS
Free Choice	−153.88	200.36	4.03	0.6033	0.603	12,951,197
Free Flow	−316.91	239.01	8.43	0.9193	0.9192	123,984,911
Congestion	2065.61	−11.05	6.5	0.1676	0.1663	283,641,319

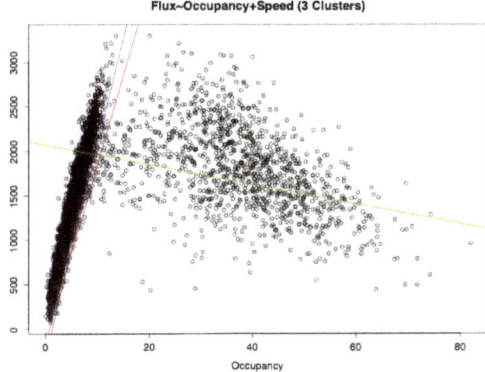

Figure A10. Flux vs. occupancy + speed: RSS analysis.

Table A6. RSS improvement.

	% RSS Improvement	
	2 Clusters	3 Clusters
2D	FP: 13.2%	FC: 9.2%
		FF: 2.2%
	C: 15.8%	C: 0.7%
3D	FP: 11.3%	FC: -
		FF: -
	C: 15.2%	C: -

References

1. Lighthill, M.J.; Whitham, G.B. On kinematic waves. II. A theory of traffic flow on long crowded roads. *Proc. R. Soc. A* **1955**, *229*, 317–346.
2. Richards, P.I. Shock waves on the highway. *Oper. Res.* **1956**, *4*, 42–51. [CrossRef]
3. Aw, A.; Rascle, M. Resurrection of "second order" models of traffic flow. *SIAM J. Appl. Math.* **2000**, *60*, 916–938. [CrossRef]
4. Zhang, H.M. A non-equilibrium traffic model devoid of gas-like behavior. *Transp. Res. Part B Methodol.* **2002**, *36*, 275–290. [CrossRef]
5. Garavello, M.; Han, K.; Piccoli, B. *Models for Vehicular Traffic on Networs*; Series on Applied Mathematics; American Institute of Mathematical Sciences: Springfield, MO, USA, 2016; Volume 9.
6. Colombo, R.M. Hyperbolic phase transitions in traffic flow. *SIAM J. Appl. Math.* **2003**, *63*, 708–721. [CrossRef]
7. Colombo, R.M.; Marcellini, F.; Rascle, M. A 2-phase traffic model based on a speed bound. *SIAM J. Appl. Math.* **2010**, *70*, 2652–2666. [CrossRef]
8. Colombo, R.M.; Goatin, P.; Piccoli, B. Road networks with phase transitions. *J. Hyperbolic Differ. Eq.* **2010**, *7*, 85–106. [CrossRef]
9. Goatin, P. The Aw–Rascle vehicular traffic flow model with phase transitions. *Math. Comput. Model.* **2006**, *44*, 287–303. [CrossRef]
10. Coclite, G.; Garavello, M.; Piccoli, B. Traffic flow on a road network. *SIAM J. Math. Anal.* **2005**, *36*, 1862–1886. [CrossRef]
11. Gerlough, D.L.; Huber, M.J. *Traffic Flow Theory*; No. HS-006 783; Transportation Research Board: Washington, DC, USA, 1976.
12. Kerner, B.S. Three-phase traffic theory and highway capacity. *Phys. A* **2004**, *333*, 379–440. [CrossRef]
13. Lebacque, J.P.; Khoshyaran, M.M. A variational formulation for higher order macroscopic traffic flow models of the GSOM family. *Transp. Res. Part Methodol.* **2013**, *57*, 245–265. [CrossRef]
14. Fan, S.; Sun, Y.; Piccoli, B.; Seibold, B.; Work, D.B. A Collapsed Generalized Aw-Rascle-Zhang Model and Its Model Accuracy. *arXiv* **2017**, arXiv:physics.soc-ph/1702.03624.
15. ATAC S.p.A. Available online: http://www.atac.roma.it/ (accessed on 3 April 2017).
16. Regional Transportation Commission of Southern Nevada (RTC), Freeway and Arterial System of Transportation (FAST) Division. Available online: http://bugatti.nvfast.org/ (accessed on 3 April 2017).
17. Département des Alpes Maritimes. Available online: https://www.departement06.fr (accessed on 3 April 2017).
18. Fraley, C.; Raftery, A.E. Model-based clustering, discriminant analysis, and density estimation. *J. Am. Stat. Assoc.* **2002**, *97*, 611–631. [CrossRef]
19. Gross, A.J.; Clark, V. *Survival Distributions: Reliability Applications in the Biomedical Sciences*; John Wiley & Sons: Hoboken, NJ, USA, 1975.
20. Fan, S.; Work, D.B. A heterogeneous multiclass traffic flow model with creeping. *SIAM J. Appl. Math.* **2015**, *75*, 813–835. [CrossRef]
21. Fan, S.; Herty, M.; Seibold, B. Comparative model accuracy of a data-fitted generalized Aw-Rascle-Zhang model. *Netw. Heterog. Media* **2014**, *9*, 239–268. [CrossRef]
22. Zhang, P.; Wong, S.; Dai, S. A conserved higher-order anisotropic traffic flow model: Description of equilibrium and non-equilibrium flows. *Transp. Res. Part B Methodol.* **2009**, *43*, 562–574. [CrossRef]
23. Blandin, S.; Work, D.; Goatin, P.; Piccoli, B.; Bayen, A. A general phase transition model for vehicular traffic. *SIAM J. Appl. Math.* **2011**, *71*, 107–127. [CrossRef]
24. Bianchini, S. The semigroup generated by a Temple class system with non-convex flux function. *Differ. Integral Eq.* **2000**, *13*, 1529–1550.
25. Chalons, C.; Goatin, P. Godunov scheme and sampling technique for computing phase transitions in traffic flow modeling. *Interfaces Free Boundaries* **2008**, *10*, 197–221. [CrossRef]

Article

Input-to-State Stability of a Scalar Conservation Law with Nonlocal Velocity

Simone Göttlich [1,*], Michael Herty [2] and Gediyon Weldegiyorgis [3]

[1] Department of Mathematics, University of Mannheim, 68131 Mannheim, Germany
[2] IGPM, RWTH Aachen University, Templergraben 55, 52056 Aachen, Germany; herty@igpm.rwth-aachen.de
[3] Department of Mathematics and Applied Mathematics, University of Pretoria, Pretoria 0002, South Africa; gediyon@aims.ac.za
* Correspondence: goettlich@uni-mannheim.de

Abstract: In this paper, we study input-to-state stability (ISS) of an equilibrium for a scalar conservation law with nonlocal velocity and measurement error arising in a highly re-entrant manufacturing system. By using a suitable Lyapunov function, we prove sufficient and necessary conditions on ISS. We propose a numerical discretization of the scalar conservation law with nonlocal velocity and measurement error. A suitable discrete Lyapunov function is analyzed to provide ISS of a discrete equilibrium for the proposed numerical approximation. Finally, we show computational results to validate the theoretical findings.

Keywords: conservation laws; feedback stabilization; input-to-state stability; numerical approximations; nonlocal velocity

MSC: 35L65; 93D15; 65N08

Citation: Göttlich, S.; Herty, M.; Weldegiyorgis, G. Input-to-State Stability of a Scalar Conservation Law with Nonlocal Velocity. *Axioms* **2021**, *10*, 12. https://doi.org/10.3390/axioms10010012

Received: 6 November 2020
Accepted: 14 January 2021
Published: 21 January 2021

Publisher's Note: MDPI stays neutral with regard to jurisdictional claims in published maps and institutional affiliations.

Copyright: © 2020 by the authors. Licensee MDPI, Basel, Switzerland. This article is an open access article distributed under the terms and conditions of the Creative Commons Attribution (CC BY) license (https://creativecommons.org/licenses/by/4.0/).

1. Introduction

The nature of modern high-volume production is characterized by a large number of items passing through many production steps. This type of production system has fluid-like properties and has been modeled successfully by continuum models [1–5]. In these models, the product at different production stages and the speed of production are the quantities of interest.

Specifically, in the manufacturing system of a factory that involves a highly re-entrant system where products visit machines multiple times, such as the production of semiconductor devices, a continuum model has been introduced in [3] that is inspired by the Lighthill–Whitham traffic model [6]. The dynamics of this model is mathematically given by hyperbolic partial differential equation of the form

$$\partial_t \rho(t,x) + \lambda(W(t))\partial_x \rho(t,x) = 0, \quad t \in [0,+\infty), \; x \in [0,1], \qquad (1)$$

where $\rho(t,x)$ is the product density which describes the total mass $W(t)$ at the time t and the production stage x,

$$W(t) = \int_0^1 \rho(t,x)dx, \quad t \in (0,+\infty). \qquad (2)$$

Contrary to classical traffic flow models, the differential equation depends on the **nonlocal** quantity (2). The function $\lambda(W(t))$ is a velocity. In production systems, it is natural to assume that the velocity function is positive and decreasing as the total mass is increasing. In the manufacturing system, the initial density of products at production stage x is taken as the initial data

$$\rho(0,x) = \rho_0(x), \quad x \in [0,1], \qquad (3)$$

and the influx is used to control the system or stabilize the system at an equilibrium. Since the velocity is positive, we only require boundary conditions at $x = 0$, i.e., the influx

$$\rho(t,0)\lambda(W(t)) = U(t), \quad t \in [0,+\infty). \tag{4}$$

Under suitable assumptions on λ, ρ_0 and U, the existence and uniqueness of a classical solution of the Cauchy problem for the scalar conservation law Equation (1) with Equations (3) and (4) is proven in [7–10].

General stabilization problems with boundary controls have been studied in the past years in [11–22] for hyperbolic systems and recently in [7,10] for scalar conservation laws with nonlocal velocity. The focus is to derive an asymptotic stability around a given equilibrium such that solutions to the conservation laws reach the equilibrium state as time tends to infinity. Such a property is attained by an exponential stability result and presented for example in ([21], Theorem 2.3) for quasi-linear hyperbolic systems. Further references also on hyperbolic balance laws other hyperbolic systems may be found in the recent book [15].

However, when boundary controls are subjected to unknown disturbances, solutions reaching the given equilibrium point are influenced by the disturbances and a notion of asymptotic stability is required. The concept of input-to-state stability (ISS) [11,20,23] has been used to describe asymptotic stability. Concerning an asymptotic behavior of classical solutions, the Lyapunov method is used to investigate sufficient conditions to achieve an exponential stability in [16,17] for hyperbolic systems and in [7,10] for scalar conservation laws with nonlocal velocity. The Lyapunov method is also used for ISS of (local) hyperbolic systems in [11,20]. For the numerical analysis of asymptotic behavior of numerical solutions discretized by a first-order finite volume scheme, a discrete Lyapunov function is used to prove exponential stability results for hyperbolic systems in [24–28] and for scalar conservation laws with nonlocal velocity in [10], and ISS results for (local) hyperbolic systems could be established recently in [29,30]. Please note that the previous given references refer to ISS for hyperbolic systems. However, the theory of ISS has also been developed for other systems as for example, linear systems, time-delay equations or parabolic differential equations. A detailed review of those results is beyond the scope of this presentation and we refer the interested reader to the recent review article [31] for additional references and a review of the state-of-the-art in this field.

The previously given references refer all to ISS theory for hyperbolic problems. However, it is worth mentioning that there exists a huge amount of literature on ISS stability for problems related to other differential equations. We can not review those at this point but would like to point to some references on ISS theory for infinite-dimensional problems [32,33] and for linear [34], semi-linear [35] and nonlinear [36] parabolic system with boundary inputs. A systematic treatment of ISS using (linear) operator theory has been presented for example in [37] and non-coercive Lyapunov theory for ISS in [38,39].

Our focus in this work is hyperbolic problems. In connection with (hyperbolic) scalar conservation law with nonlocal velocity, in [10], the authors have studied global feedback stabilization of the closed-loop system in Equation (1) under the feedback law

$$U(t) - \rho^*\lambda(\rho^*) = k\Big(\rho(t,1)\lambda(W(t)) - \rho^*\lambda(\rho^*)\Big), \quad t \in (0,+\infty), \tag{5}$$

where $k \in [0,1)$ is the feedback parameter and $\rho^* \in \mathbb{R}$ is a given equilibrium. They generalize the stabilization results of [7] by using a Lyapunov function. In particular, for a given equilibrium $\rho^* = 0$ and a general velocity function $\lambda \in C^1([0,+\infty);[0,+\infty))$, the global stabilization result in L^2 for the closed-loop system of Equations (1), (3) and (5) is generalized to L^p ($p \geq 1$). Then, the global stabilization result in L^2 for the closed-loop system of Equations (1), (3) and (5) with a family of velocity functions

$$\lambda(s) = \frac{A}{B+s}, \quad s \in [0,+\infty) \quad \text{with} \quad A > 0, \ B > 0, \tag{6}$$

is obtained for a given equilibrium $\rho^* > 0$. By using a discrete Lyapunov function, they also established stabilization results for a discrete scalar conservation law with nonlocal velocity and using a first-order finite volume scheme.

In this paper, we study ISS for the closed-loop system of Equations (1) and (3) under the feedback law defined by

$$U(t) - \rho^*\lambda(\rho^*) = k\Big((\rho(t,1) + d(t))\lambda(W(t)) - \rho^*\lambda(\rho^*)\Big), \; t \in (0, +\infty), \qquad (7)$$

where $d(t) \in \mathbb{R}$ is a bounded perturbation in the measurement. In particular, we use an ISS-Lyapunov function to investigate sufficient and necessary conditions for ISS in L^2 for an equilibrium $\rho^* \geq 0$ and the velocity function defined by Equation (6). The numerical analysis of sufficient and necessary conditions for ISS is performed by using a discrete ISS-Lyapunov function for numerical solution obtained by a first-order finite volume scheme. Moreover, we provide numerical simulations to illustrate theoretical results for some velocity functions of type Equation (6).

The paper is organized as follows: In Section 2, we present stabilization results of ISS for a scalar conservation law with nonlocal velocity and measurement error. The numerical discretization of stabilization results of ISS for the scalar conservation law with nonlocal velocity and measurement error is presented in Section 3. Finally, in Section 4, we show numerical simulations for the scalar conservation law with nonlocal velocity and measurement error to illustrate the theoretical results.

2. Asymptotic Stability of a Scalar Conservation Law with Nonlocal Velocity and Measurement Error

We study ISS of a closed-loop system of scalar conservation laws with nonlocal velocity and measurement error of the form:

$$\begin{cases} \partial_t \rho(t,x) + \lambda(W(t))\partial_x \rho(t,x) = 0, & t \in (0, +\infty), \; x \in (0,1), \\ \rho(0,x) = \rho_0(x), & x \in (0,1), \\ U(t) - \rho^*\lambda(\rho^*) = k((\rho(t,1) + d(t))\lambda(W(t)) - \rho^*\lambda(\rho^*)), & t \in (0, +\infty), \\ \rho(t,0)\lambda(W(t)) = U(t), & t \in [0, +\infty), \\ W(t) = \int_0^1 \rho(t,x)dx, & t \in (0, +\infty), \end{cases} \qquad (8)$$

where $\rho(t,x)$ is the product density, $\lambda(\cdot) \in C^1([0, +\infty), (0, +\infty))$ is the velocity function, $W(t)$ is total mass, $U(t)$ is the controller and $k \in [0,1)$ is a non-negative feedback parameter, $\rho^* \geq 0$ is an equilibrium solution and $d(t) \in \mathbb{R}$ is a bounded (known) perturbation in the measurement. A weak solution of the closed-loop system in Equation (8) is defined below.

Definition 1 (Weak solution). *Fix $T > 0$. A function $\rho \in C^0([0,T]; L^1(0,1))$ is called a weak solution to Equation (8) if for every $s \in (0,T]$ and every $\varphi \in C^1([0,s] \times [0,1])$ satisfying*

$$\varphi(s,x) = 0, \; \forall x \in [0,1] \quad \text{and} \quad \varphi(t,1) = \kappa\varphi(t,0), \; \forall t \in [0,s],$$

the following equation holds:

$$\int_0^s \int_0^1 \rho(t,x)(\partial_t \varphi(t,x) + \lambda(W(t))\partial_x \varphi(t,x))dxdt$$
$$+ \int_0^s ((1-k)\rho^*\lambda(\rho^*) + d(t))\varphi(t,0)dt + \int_0^1 \rho(0,x)\varphi(0,x)dx = 0.$$

Let $d \equiv 0$, $\rho^* \geq 0$, $p \in [1, +\infty)$ and $k \in [0,1]$ be given. Then, the existence and uniqueness of the non-negative weak solution $\rho \in C^0([0, +\infty); L^p(0,1))$ and the non-negative classical solution $\rho \in C^1([0, +\infty) \times [0,1])$ of the closed-loop system in Equation (8) are available in [7,10].

We now analyze ISS for the system Equation (8) with $\rho^* \geq 0$ in the sense of the following definitions. This is also known as global ISS. Note that ISS Lyapunov functions can be defined within a very general setting and we refer to ([31], Definition 2.11) for such a definition. In Definition (3) below, we introduce ISS-Lyapunov functions tailored to system Equation (8).

Definition 2 (Input-to-state stability (ISS))**.** *Let $D > 0$. An equilibrium $\rho^* \geq 0$ of the closed-loop system in Equation (8) is exponential ISS in L^2-norm with respect to any disturbance function $d(\cdot) \in L^\infty(0,\infty)$ such that $\|d\|_{L^\infty(0,\infty)} \leq D$ if there exist positive constants $\gamma_1, \gamma_2, \gamma_3$ independent of d such that, for every initial condition $\rho_0(x) \in L^2(0,1)$, the L^2-solution to the closed-loop system in Equation (8) satisfies*

$$\|\rho(t,\cdot) - \rho^*\|_{L^2} \leq \gamma_2 e^{-\gamma_1 t}\|\rho_0 - \rho^*\|_{L^2} + \gamma_3 \|d(s)\|_{L^\infty(0,t)}, \quad t \in [0,+\infty). \tag{9}$$

Hence, the equilibrium ρ^* is ISS with respect to disturbances $d \in \mathcal{D} := \{d(\cdot) \in L^\infty(0,\infty) : \|d\|_{L^\infty} \leq D\}$.

Definition 3 (ISS-Lyapunov function)**.** *The function $\mathbf{L} : L^2(0,1) \to \mathbb{R}_+$ is said to be an ISS-Lyapunov function for the closed-loop system in Equation (8) if*

(i) *there exist positive constants $\alpha_1 > 0$ and $\alpha_2 > 0$ such that for all solutions $\rho \in C^0([0,\infty); L^2(0,1))$ and $t \in [0,+\infty)$*

$$\alpha_1 \|\rho(t,\cdot) - \rho^*\|_{L^2}^2 \leq \mathbf{L}(\rho(t,\cdot)) \leq \alpha_2 \|\rho(t,\cdot) - \rho^*\|_{L^2}^2, \tag{10}$$

(ii) *there exist positive constants $\eta > 0$ and $\nu > 0$ such that for all solutions $\rho \in C^0([0,\infty); L^2(0,1))$ and $t \in [0,+\infty)$*

$$\frac{d}{dt}\mathbf{L}(\rho(t,\cdot)) \leq -\eta \mathbf{L}(\rho(t,\cdot)) + \nu d^2(t).$$

For a notion of differentiability of $\mathbf{L}$, we also refer for example to ([31], Section 2.2). To simplify the notation we also introduce the function

$$\mathcal{L}(t) := \mathbf{L}(\rho(t,\cdot)), \tag{11}$$

where $\rho \in C^0([0,\infty); L^2(0,1))$ is the solution to Equation (8).

Theorem 1 (ISS for $\rho^* \geq 0$)**.** *Fix any $\rho^* \geq 0$, $k \in [0,1)$, $R > 0$, $D > 0$ and any $\rho_0 \in L^2(0,1)$ satisfying $\rho_0 \geq 0$ a.e. in $(0,1)$. Assume further*

$$\|\rho_0(\cdot) - \rho^*\|_{L^2(0,1)} \leq R. \tag{12}$$

Assume there exists a non-negative almost everywhere weak solution $\rho \in C^0([0,+\infty); L^2(0,1))$ to the Cauchy problem in Equation (8) where λ is given by Equation (6).

Then, the steady-state ρ^* of the system in Equation (8) is exponential ISS in L^2-norm with respect to any disturbance function $d \in \{d(\cdot) \in L^\infty(0,\infty) : \|d\|_{L^\infty} \leq D\}$.

Before we begin the proof of Theorem 1, we consider the following transformation at the equilibrium ρ^*,

$$\tilde{\rho}(t,x) := \rho(t,x) - \rho^*, \quad \widetilde{W}(t) := W(t) - \rho^*, \quad \tilde{\rho}_0(x) := \rho_0(x) - \rho^*,$$
$$\tilde{\lambda}_{\widetilde{W}}(t) := \lambda(\rho^* + \widetilde{W}(t)), \quad \widetilde{U}(t) := \tilde{\lambda}_{\widetilde{W}}(t)\tilde{\rho}(t,0).$$

Then, the system in Equation (8) with Equation (6) can be rewritten as follows for $t \in (0, +\infty)$:

$$\begin{cases} \partial_t \tilde{\rho}(t,x) + \tilde{\lambda}_{\widetilde{W}}(t) \partial_x \tilde{\rho}(t,x) = 0, \ x \in (0,1), \\ \tilde{\rho}(0,x) = \tilde{\rho}_0(x), \ x \in (0,1), \\ \widetilde{U}(t) = k\tilde{\lambda}_{\widetilde{W}}(t)(\tilde{\rho}(t,1) + d(t)) + (1-k)\rho^*(\lambda(\rho^*) - \tilde{\lambda}_{\widetilde{W}}(t)), \\ \tilde{\lambda}_{\widetilde{W}}(t) := \lambda(\rho^* + \widetilde{W}(t)), \\ \widetilde{W}(t) = \int_0^1 \tilde{\rho}(t,x) dx \geq -\rho^*, \\ \lambda(s) = \frac{A}{B+s}, \ \text{with} \ A > 0, \ B > 0, \ s \in [0, +\infty). \end{cases} \quad (13)$$

By using the velocity function Equation (6) in Equation (13), we have

$$\rho^*(\lambda(\rho^*) - \tilde{\lambda}_{\widetilde{W}}(t)) = \theta \tilde{\lambda}_{\widetilde{W}}(t) \widetilde{W}(t), \quad t \in [0, +\infty), \quad (14)$$

where

$$\theta := \frac{\rho^*}{B + \rho^*} < 1.$$

For convenience, until the end of this proof, we omit the symbol "~". Then, the system in Equation (13) with Equation (14) can be rewritten in the following form for $t \in (0, +\infty)$:

$$\begin{cases} \partial_t \rho(t,x) + \lambda_W(t) \partial_x \rho(t,x) = 0, \ x \in (0,1), \\ \rho(0,x) = \rho_0(x), \ x \in (0,1), \\ U(t) = k\lambda_W(t)(\rho(t,1) + d(t)) + (1-k)\theta \lambda_W(t) W(t) \text{ with } \theta = \frac{\rho^*}{B+\rho^*}, \\ \lambda_W(t) := \lambda(\rho^* + W(t)), \\ \rho(t,0)\lambda_W(t) = U(t), \\ W(t) = \int_0^1 \rho(t,x) dx \geq -\rho^*, \\ \lambda(s) = \frac{A}{B+s}, \ \text{with} \ A > 0, \ B > 0, \ s \in [0, +\infty). \end{cases} \quad (15)$$

With the above notation, the assumption in Equation (12) of Theorem 1 reads

$$\|\rho_0\|_{L^2(0,1)} \leq R. \quad (16)$$

Proof. The following proof of Theorem 1 is an extension of the proof of Theorem 3.2 in [10]. Since C^1-functions are dense in $L^2(0,1)$, we can analyze ISS for the system Equation (15) with non-negative weak solution $\rho \in C^0([0, +\infty); L^2(0,1))$ as follows: For $\phi \in L^2(0,1)$, we first define a candidate ISS-Lyapunov function by

$$\mathbf{L}(\phi) = \int_0^1 \phi^2(x) e^{-\beta x} dx + a \left(\int_0^1 \phi(x) dx \right)^2.$$

and then we have according to (11)

$$\mathcal{L}(t) := \mathbf{L}(\rho(t, \cdot)) = \int_0^1 \rho^2(t,x) e^{-\beta x} dx + aW^2(t), \quad \forall t \in [0, +\infty), \quad (17)$$

where $\beta > 0$ and $a \in \mathbb{R}$ are constants. By definition of W and Hölder inequality, we have

$$W(t)^2 = \left(\int_0^1 \rho(t,x) e^{-\frac{1}{2}\beta x} e^{\frac{1}{2}\beta x} dx \right)^2 \leq \int_0^1 e^{\beta x} dx \int_0^1 e^{-\beta x} \rho^2(t,x) dx. \quad (18)$$

Hence, if

$$a > -\frac{\beta}{e^\beta - 1}, \quad (19)$$

then $\mathcal{L}(t) > 0$ for all $t \geq 0$. We will further assume from now on that $a \leq 0$. Furthermore, for $0 < C_1 := C_1(\beta) = \frac{e^\beta - 1}{\beta}$ we obtain

$$W^2(t) \leq C_1 \int_0^1 e^{-\beta x} \rho^2(t,x) dx \tag{20}$$

and for $C_2 := C_2(a,\beta) = C_1 + \max\{a,1\} > 0$ we have

$$\mathcal{L}(t) \leq C_2 \int_0^1 e^{-\beta x} \rho^2(t,x) dx. \tag{21}$$

Since $a < 0$ we also obtain

$$(1 + aC_1) \int_0^1 e^{-\beta x} \rho^2(t,x) dx \leq \mathcal{L}(t). \tag{22}$$

Summarizing, there exist positive constants $C_i = C_i(a,\beta), i \in \{3,4\}$ such that for all $t \geq 0$

$$W^2(t) \leq C_3 \int_0^1 \rho^2(t,x) e^{-\beta x} dx \leq \mathcal{L}(t) \leq C_4 \int_0^1 \rho^2(t,x) e^{-\beta x} dx \tag{23}$$

and therefore $\mathcal{L}$ is equivalent to the L^2-norm of ρ. Note that for $\rho^* = 0$ we may set $a = 0$ in Equation (17). The time derivative of the candidate ISS-Lyapunov function in Equation (17) is given by:

$$\begin{aligned} \frac{d\mathcal{L}}{dt}(t) &= \int_0^1 2\rho(t,x)\rho_t(t,x) e^{-\beta x} dx + 2aW(t) \frac{dW}{dt}(t) \\ &= -\beta \lambda_W(t) \int_0^1 \rho^2(t,x) e^{-\beta x} dx + \frac{1}{\lambda_W(t)} \left(\lambda_W^2(t) \rho^2(t,0) - \lambda_W^2(t) \rho^2(t,1) e^{-\beta} \right) \\ &\quad + 2aW(t)(\lambda_W(t)\rho(t,0) - \lambda_W(t)\rho(t,1)) \\ &= -\beta \lambda_W(t) \int_0^1 \rho^2(t,x) e^{-\beta x} dx + A_1(t), \end{aligned}$$

where $A_1(t)$ contains all contributions due to the boundary conditions. In the following we will analyze and estimate A_1. Note that $\lambda_W(t)\rho(t,0) = U(t)$ and U is given by Equation (15). More precisely, we will use the following estimate for any $\epsilon > 0$

$$\begin{aligned} 2aW(t)U(t) &= 2aW(t)(k\lambda_W(t)(\rho(t,1) + d(t)) + (1-k)\theta\lambda_W(t)W(t)) \\ &\leq k^2 d^2(t) \lambda_W(t) \frac{1}{\epsilon} + \epsilon a^2 W^2 \lambda_W(t) + 2aW(t)(k\lambda_W(t)\rho(t,1) + (1-k)\theta\lambda_W(t)W(t)), \\ U^2(t) &= (k\lambda_W(t)(\rho(t,1) + d(t)) + (1-k)\theta\lambda_W(t)W(t))^2 \\ &\leq (1+\epsilon)(k\lambda_W(t)\rho(t,1) + (1-k)\theta\lambda_W(t)W(t))^2 + k^2 d^2(t) \lambda_W^2(t)(1 + \frac{1}{\epsilon}). \end{aligned}$$

In order to simplify the notation of the following computations, we neglect the time dependence and we define

$$y := y(t) := \lambda_W(t)\rho(t,1), \quad b_1 := b_1(t) := (1 + \frac{2}{\epsilon}) k^2 \lambda_W(t) d^2(t) + \epsilon a^2 W^2(t) \lambda_W(t).$$

Then, we have

$$A_1(t) \le b_1 + \frac{1}{\lambda_W}\left((1+\epsilon)(ky + (1-k)\theta\lambda_W W)^2 - y^2 e^{-\beta}\right) + 2aW((k-1)y + (1-k)\theta\lambda_W W)$$

$$= b_1 + \frac{(1+\epsilon)k^2 - e^{-\beta}}{\lambda_W}\left(y + \lambda_W W\left(\frac{a(k-1)}{(1+\epsilon)k^2 - e^{-\beta}} - \frac{(k-1)k(1+\epsilon)\theta}{(1+\epsilon)k^2 - e^{-\beta}}\right)\right)^2$$

$$+ \lambda_W W^2\left(\theta^2(k-1)^2(1+\epsilon) - 2a\theta(k-1) - \frac{(k-1)^2}{(1+\epsilon)k^2 - e^{-\beta}}(a - k(1+\epsilon)\theta)^2\right)$$

$$= b_1 + \frac{(1+\epsilon)k^2 - e^{-\beta}}{\lambda_W}(\ldots)^2$$

$$- \lambda_W W^2 \frac{(k-1)^2}{(1+\epsilon)k^2 - e^{-\beta}}\left(a^2 - 2ak(1+\epsilon)\theta + k^2(1+\epsilon)^2\theta^2 + \frac{2a\theta((1+\epsilon)k^2 - e^{-\beta})}{(k-1)} - \theta^2(1+\epsilon)((1+\epsilon)k^2 - e^{-\beta})\right)$$

$$= b_1 + \frac{(1+\epsilon)k^2 - e^{-\beta}}{\lambda_W}(\ldots)^2$$

$$- \lambda_W W^2 \frac{(k-1)^2}{(1+\epsilon)k^2 - e^{-\beta}}\left(a^2 - 2a\theta\frac{k(1+\epsilon) - e^{-\beta}}{1-k} + \theta^2(1+\epsilon)e^{-\beta}\right)$$

$$= b_1 + \frac{(1+\epsilon)k^2 - e^{-\beta}}{\lambda_W}(\ldots)^2$$

$$- \lambda_W W^2 \frac{(k-1)^2}{(1+\epsilon)k^2 - e^{-\beta}}\left(a - \theta\frac{k(1+\epsilon) - e^{-\beta}}{1-k}\right)^2$$

$$- \lambda_W W^2 \frac{(k-1)^2}{(1+\epsilon)k^2 - e^{-\beta}}\left(\theta^2(1+\epsilon)e^{-\beta} - \theta^2\left(\frac{k(1+\epsilon) - e^{-\beta}}{1-k}\right)^2\right).$$

Even so, it is not necessary that the proof simplifies if ϵ is chosen depending on β. We set for

$$\epsilon := \epsilon(\beta) = \beta^2. \tag{24}$$

For any fixed $0 \le k < 1$ and all $0 < \beta^2 < \epsilon^*$ with $\epsilon^* := \min\{1, \frac{1}{2}\frac{1-k}{k}\}$, we have

$$(1+\beta^2)k^2 < (1+\beta^2)k < 1 \tag{25}$$

and hence for all $\beta < \min\{\sqrt{\epsilon^*}, \beta^*\}$ with $\beta^* := -\ln((1+\epsilon^*)k)$, we have

$$e^{-\beta} > (1+\beta^2)k > (1+\beta^2)k^2. \tag{26}$$

Furthermore, consider

$$a(\beta) = \theta\frac{k(1+\beta^2) - e^{-\beta}}{1-k} < 0. \tag{27}$$

For $\beta \to 0$, we have $\lim_{\beta \to 0} a(\beta) = -\theta > -1$ and we have $\lim_{\beta \to 0} \frac{\beta}{e^\beta - 1} = 1$. Hence, there exists a $\beta^{**} > 0$ such that for all $\beta \le \min\{\beta^*, \beta^{**}\}$ and for $a(\beta)$ as given by Equation (27), the inequalities (26), (27) and (19) hold true. Using the inequality (26) and the particular choice for $a(\beta)$ and $\epsilon(\beta)$, we obtain for all β sufficiently small

$$A_1(t) \le b_1(t) + \theta^2 \lambda_W W^2 \frac{(k-1)^2(1+\beta^2)e^{-\beta} - (k(1+\beta^2) - e^{-\beta})^2}{e^{-\beta} - (1+\beta^2)k^2} \tag{28}$$

$$= \left(1 + \frac{2}{\beta^2}\right)k^2 \lambda_W d^2 + \beta^2 a^2(\beta)W^2\lambda_W + \theta^2 \lambda_W W^2 b_2(\beta, k), \tag{29}$$

$$b_2(\beta, k) := \frac{(k-1)^2(1+\beta^2)e^{-\beta} - (k(1+\beta^2) - e^{-\beta})^2}{e^{-\beta} - (1+\beta^2)k^2}. \tag{30}$$

Using the estimate (20) to bound W^2 and using $k < 1$, we obtain

$$A_1(t) \leq \left(1 + \frac{2}{\beta^2}\right) k^2 \lambda_W d^2 + \lambda_W \theta^2 \frac{e^\beta - 1}{\beta} \left(\beta^2 \frac{a^2(\beta)}{\theta^2} + b_2(\beta, k)\right) \int_0^1 e^{-\beta x} \rho^2(t, x) dx$$

$$\leq \lambda_W \left(1 + \frac{2}{\beta^2}\right) d^2 + \lambda_W \theta^2 b_3(\beta, k) \int_0^1 e^{-\beta x} \rho^2(t, x) dx,$$

$$b_3(\theta, k) := \frac{e^\beta - 1}{\beta} \left(\beta^2 \frac{a^2(\beta)}{\theta^2} + b_2(\beta, k)\right).$$

An elementary computation shows that $f(\beta, k)$ has the following properties

$$b_3(0, k) = 0 \text{ and } \partial_\beta b_3(0, k) = 1.$$

Replacing b_3 by a second-order Taylor expansion in β at $\beta = 0$ therefore yields the estimate

$$A_1(t) \leq \left(1 + \frac{2}{\beta^2}\right) \lambda_W d^2 + \theta^2 \lambda_W \left(\beta + O(\beta^2)\right) \int_0^1 e^{-\beta x} \rho^2(t, x) dx. \tag{31}$$

Now, we proceed with the estimate of $\frac{d}{dt}\mathcal{L}(t)$ as

$$\frac{d}{dt}\mathcal{L}(t) \leq -\beta \lambda_W(t) \int_0^1 e^{-\beta x} \rho^2(t, x) dx \left(1 - \theta^2 + O(\beta)\right) + \left(1 + \frac{2}{\beta^2}\right) \lambda_W(t) d^2(t). \tag{32}$$

Since $\theta < 1$ there exists $0 < \bar{\beta} < \min\{\beta^*, \beta^{**}\}$ sufficiently small, such that

$$0 < 1 - \theta^2 + O(\beta). \tag{33}$$

Using the estimate (22) there is a constant $0 < \eta := \eta(k, \rho^*)$, we obtain

$$\frac{d}{dt}\mathcal{L}(t) \leq -\eta \lambda_W(t) \mathcal{L}(t) + \left(1 + \frac{2}{\beta^2}\right) \lambda_W(t) d^2(t). \tag{34}$$

By definition, we have that $0 \leq \lambda_W(t)$. Next, we show that $\lambda_W(t)$ is bounded from below by a positive constant. This requires to obtain an upper bound on $W(t)$. The previous inequality (34) yields the following bound on $W^2(t)$ for $C_5 := C_5(\bar{\beta}) = \frac{C_1}{(1+a)C_1}$ and for $C_6 := C_6(\bar{\beta}) = \left(1 + \frac{2}{\beta^2}\right)$:

$$\frac{1}{C_5} W^2(t) \leq \mathcal{L}(t) \leq e^{-\eta \int_0^t \lambda_W(s) ds} \mathcal{L}(0) + \int_0^t C_6 \lambda_W(s) d^2(s) e^{-\eta \int_s^t \lambda_W(r) dr} \tag{35}$$

$$\leq \mathcal{L}(0) + \frac{C_6}{\eta} \|d(t)\|_{L^\infty(0,t)} \left(1 - e^{-\eta \int_0^t \lambda_W(s) ds}\right). \tag{36}$$

By assumption $\|\rho_0\|^2 \leq R^2$. By definition we have $-\rho^* \leq W(t)$ and therefore

$$-\rho^* \leq W(t) \leq \sqrt{C_5 R^2 + \frac{C_5 C_6}{\eta} \|d(\cdot)\|_{L^\infty(0,t)}}. \tag{37}$$

Due to the definition of λ_W, it is uniformly bounded from above by $\sigma_1 := \frac{A}{B}$. Furthermore, we have that $W(t)$ is bounded from above due to Equation (37) and since d is bounded. Hence, λ_W is bounded from below by $\sigma_2 = \sigma_2(\|d\|_{L^\infty(0,\infty)}, R, \bar{\beta})$. Note that the L^∞ norm of the disturbances are uniformly bounded by the constant D. This yields that for all $t \geq 0$

$$\sigma_1 \geq \lambda_W(t) \geq \sigma_2. \tag{38}$$

Using the previous estimate for λ_W in Equation (34) yields the assertion. The decay rate η^* of the Lyapunov function is $\eta^* = \eta \sigma_2$ and $\nu = \sigma_1 C_6$. □

Some remarks are in order.

Remark 1. *Note that the rate $\eta = \eta(\rho^*, k)$ as a function of k tends to zero as k tends to one, this can be seen for example in Equation (26) defining the upper bound for $\bar{\beta}$. Similarly, if $\theta \to 1$, i.e., $\rho^* \to \infty$, we observe that $\eta \to 0$ due to Equation (33).*

The bound on $W(t)$ is required to obtain the exponential decay. Therefore, the final rate depends on the constant R and we refer to Equation (37) and following for its detailed dependence. Note that in the case $\rho^* = 0$ we may set $a = 0$ and therefore no bound on W is necessary.

Further, the result holds true for any solution $\rho \in C^0([0, \infty); L^2(0, 1))$ and hence uniqueness of solutions is not required. Regarding existence of solutions, it might be possible to extend recent results [40–42]. However, so far existence results in the case $d \equiv 0$ exist [10].

Note that the decay rate η will be dependent on the bound of the disturbance as well as on R, but will be uniform with respect to ρ_0 provided that ρ_0 fulfills (12).

In ([7], Lemma 3.5) it has been shown that in the case $d \equiv 0$ and $\rho^* = 0$ exponential stability does not hold if $k > 1$.

For $\rho^* = 0$, Theorem 1 holds true for any velocity function $\lambda(\cdot) \in C^1([0, +\infty), (0, +\infty))$. This case is similar to a problem studied in [10]. Therein, a detailed discussion of the case $d \equiv 0$ has been presented and we refer in particular to ([10], Theorem 3.1).

3. Numerical Study of Asymptotic Stability of a Scalar Conservation Law with Nonlocal Velocity and Measurement Error

In the following section, we extend the result to a proper discretization of the continuous dynamics. The following results are based on similar estimates as in the previous section and it is a minor extension of the proof presented in ([10], Section 4.2). In order to not repeat the estimates obtained in [10], we will use a similar notation and mostly report on the changes in estimates due to the additional disturbance d. As seen in the previous proof in Equation (24), it is possible to chose $\epsilon = \beta^2$ and we will do so in the following proof directly. This simplifies the notation and reduces the technicality of the computations.

As in ([10], Section 4.2) we introduce a first-order Upwind discretization of the closed-loop system in Equation (8). To this end we divide the spatial domain $[0, 1]$ using an equidistant grid with cell width Δx and $J \in \mathbb{N}$ cells such that $\Delta x J = 1$. The cell centers are denoted by $x_j = (j - \frac{1}{2}) \Delta x$, $j \in \{1, \ldots, J\}$ and, the boundary of the domain are x_0 and x_J, respectively. Moreover, we discretize $W(t)$ by

$$W^n = \Delta x \sum_{j=1}^{J} \rho_j^n, \quad n \in \{1, 2, \ldots\}, \tag{39}$$

with the point wise values of the solution $\rho_j^n = \rho(t^n, x_j)$. Further, we define the discrete values λ^n by

$$\lambda^n := \lambda(W^n) = \frac{A}{B + W^n}, \quad A > 0, B > 0, \tag{40}$$

where $t^n = n \Delta t$, $n \in \{0, 1, \ldots\}$ denotes the discrete time such that the time step size Δt satisfies a stability condition due to Courant–Friedrichs–Lewy condition (CFL). This condition states that Δt is chosen such that

$$0 < r^n := \frac{\lambda^n \Delta t}{\Delta x} \leq 1, \quad \forall n \in \{0, 1, \ldots\}. \tag{41}$$

Since $\lambda^n \leq \frac{A}{B}$ for all $n \geq 0$, we can choose a possibly small but fixed Δt such that the previous condition (41) holds true for all n with fixed Δt and Δx. This choice allows to take a uniform grid in time. As in the continuous case we have $\rho^* > 0$. For the given initial values $\bar{\rho}^0 = (\rho_0^0, \rho_1^0, \ldots, \rho_J^0)^\top$ with $\rho_j^0 \geq 0$, $j \in \{0, \ldots, J\}$, we employ a first-order finite volume scheme, given by the explicit Upwind method, to discretize the system in Equation (8).

$$\begin{cases} \rho_j^{n+1} = (1-r^n)\rho_j^n + r^n \rho_{j-1}^n, & j \in \{1,\ldots,J\}, \ n \in \{0,1,\ldots\}, \\ \rho_0^{n+1} = k\rho_J^{n+1} + (1-k)\frac{\rho^*\lambda(\rho^*)}{\lambda^{n+1}} + kd^{n+1}, & n \in \{0,1,\ldots\}. \end{cases} \quad (42)$$

We now define discrete version of ISS and ISS-Lyapunov function as follows:

Definition 4 (Discrete ISS). *Let $D > 0$. An equilibrium $\rho^* \geq 0$ of the discrete closed-loop system in Equation (42) is ISS in L^2-norm with respect to discrete disturbances $d^n \leq D$, $n \in \{1,2,\ldots\}$ if there exist positive real constants $\gamma_1 > 0$, $\gamma_2 > 0$ and $\gamma_3 > 0$ such that, for every initial condition $\rho_j^0, j \in \{1,\ldots,J\}$, the solution $\rho_j^n, j \in \{1,\ldots,J\}, n \in \{0,1,\ldots\}$ to the discrete closed-loop system in Equation (42) satisfies*

$$\|\overrightarrow{\rho}^n - \rho^*\|_{L^2_{\Delta x}} \leq \gamma_2 e^{-\gamma_1 t^n} \|\overrightarrow{\rho}^0\|_{L^2_{\Delta x}} + \gamma_3 \max_{0 \leq s < n}(|d^s|), \ n \in \{1,2,\ldots\}, \quad (43)$$

where $\overrightarrow{\rho}^n = (\rho_j^n)_{j=1}^J$ and

$$\|\overrightarrow{\rho}^n\|_{\ell^2}^2 := \Delta x \sum_{j=1}^J \left(\rho_j^n\right)^2, \quad n \in \{0,1,\ldots\}.$$

Definition 5 (Discrete ISS-Lyapunov function). *A function $\mathbf{L} : \mathbb{R}^J \to \mathbb{R}_0^+$ is said to be a discrete ISS-Lyapunov function for the discrete closed-loop system in Equation (42) if*

(i) *there exist positive constants $\alpha_1 > 0$ and $\alpha_2 > 0$ such that for all $n \in \{0,1,\ldots\}$*

$$\alpha_1 \|\overrightarrow{\rho}^n - \rho^*\|_{\ell^2}^2 \leq \mathbf{L}(\overrightarrow{\rho}^n) \leq \alpha_2 \|\overrightarrow{\rho}^n - \rho^*\|_{\ell^2}^2, \quad (44)$$

(ii) *there exist positive constants $\eta > 0$ and $\nu > 0$ such that for all $n \in \{0,1,\ldots\}$*

$$\frac{\mathbf{L}(\overrightarrow{\rho}^{n+1}) - \mathbf{L}(\overrightarrow{\rho}^n)}{\Delta t} \leq -\eta \mathbf{L}(\overrightarrow{\rho}^n) + \nu(d^n)^2.$$

To simplify the notation later on we will define the sequence of discrete values $\mathcal{L}^n$ by

$$\mathcal{L}^n := \mathbf{L}(\overrightarrow{\rho}^n), \ n \in \{0,1,\ldots\} \quad (45)$$

and where $\overrightarrow{\rho}^n$ are given as solution to the system in (42).

Theorem 2. *(Discrete ISS for $\rho^* \geq 0$) Assume that the CFL condition in Equation (41) holds. Let $D > 0$. For every $\rho^* \geq 0$, every $k \in [0,1)$, every $R > 0$ and for every initial data $\vec{\rho}^0 = (\rho_0^0, \rho_1^0, \ldots, \rho_J^0)^\top$ with $\rho_j^0 \geq 0, j \in \{1,\ldots,J\}$ and*

$$\|\vec{\rho}^0 - \rho^*\vec{e}\|_{\ell^2} \leq R, \quad (46)$$

where $\vec{e} = \overbrace{(1,\ldots,1)^\top}^{J+1}$, the solution $\vec{\rho}^n = (\rho_0^n, \rho_1^n, \ldots, \rho_J^n)^\top$ to the system in Equation (42) satisfies $\rho_j^n \geq 0, j \in \{0,\ldots,J\}, n \in \{0,1,\ldots\}$ and the steady-state ρ^* of the discrete system in Equation (42) is ISS in L^2-norm with respect any discrete disturbance function $d^n, n \in \{1,2,\ldots\}$ such that $d^n \leq D$.

In order to analyze the ISS of the discrete system in Equation (42) by the discrete Lyapunov method, we use the following transformation

$$\tilde{\rho}_j^n = \rho_j^n - \rho^*, \ \widetilde{W}^n = \Delta x \sum_{j=1}^J \tilde{\rho}_j^n, \ \tilde{\lambda}_{\widetilde{W}}^n = \lambda\left(\rho^* + \widetilde{W}^n\right), \ \tilde{r}^n = \frac{\Delta t}{\Delta x}\tilde{\lambda}_{\widetilde{W}}^n, \ n \in \{0,1,\ldots\}. \quad (47)$$

For simplicity, we omit the symbol "~" in Equation (47) and discretize the system in Equation (15) as follows

$$\begin{cases} \rho_j^{n+1} = (1-r^n)\rho_j^n + r^n \rho_{j-1}^n, \ j \in \{1,\ldots,J\}, \ n \in \{0,1,\ldots\}, \\ \rho_0^{n+1} = k\rho_j^{n+1} + (1-k)\theta W^{n+1} + kd^{n+1} \text{ with } \theta = \frac{\rho^*}{B+\rho^*}, \ n \in \{0,1,\ldots\}, \\ r^n = \frac{\Delta t}{\Delta x} \lambda_W^n, \ n \in \{0,1,\ldots\}, \\ \lambda_W^n = \lambda(\rho^* + W^n), \ n \in \{0,1,\ldots\}, \\ W^n = \Delta x \sum_{j=1}^J \rho_j^n \geq -\rho^*, \ n \in \{0,1,\ldots\}, \\ \lambda(s) = \frac{A}{B+s}, \ s \geq 0. \end{cases} \quad (48)$$

Thus, the assumption in Equation (46) in Theorem 2 is now expressed as

$$\|\vec{\rho}^0\|_{\ell^2} \leq R. \quad (49)$$

Note that the proof of Theorem 2 is an extension of the proof of Theorem 4.2 in [10]. Thus, some details of the proof can be found in [10] and we will point to the corresponding estimates in order to reduce the technicality of the proof.

Proof. As in the continuous case the proof simplifies if $\rho^* = 0$. Therefore, we consider in the forthcoming proof only the more interesting case

$$\rho^* > 0. \quad (50)$$

Since the initial data $\rho_j^0 \geq 0, j \in \{0,\ldots,J\}$, by the discrete system in Equation (48) and the CFL condition in Equation (41), we have $\rho_j^n \geq 0, j \in \{0,\ldots,J\}, n \in \{0,1,\ldots\}$.

Consider the following candidate Lyapunov function Equation (17) for any $\vec{\phi} \in \mathbb{R}^J$

$$\mathbf{L}(\vec{\phi}) = \Delta x \sum_{j=1}^J (\phi_j)^2 e^{-\beta x_j} + a\left(\Delta x \sum_{j=1}^J \phi_j\right)^2.$$

where $\beta > 0$. In particular, we set a

$$a = \theta \frac{k - e^{-\beta}}{1-k} < 0 \quad (51)$$

and since $\theta < 1$ there exists β^* sufficiently small such that $0 > a > -\frac{\beta}{e^\beta - 1}$, see ([10], (3.25), (3.26)).

According to (45), the values of $\mathbf{L}$ at the solution $\vec{\rho}^n$ at time t^n for $n \geq 0$ are given by

$$\mathcal{L}^n = \|\vec{\rho}^n\|_\beta^2 + a(W^n)^2, \quad (52)$$

$$\|\vec{\rho}^n\|_\beta^2 := \Delta x \sum_{j=1}^J (\rho_j^n)^2 e^{-\beta x_j}. \quad (53)$$

For fixed $k \in [0,1)$, we assume as in [10] there exists a β^{**} such that for $0 < \beta < \beta^{**}$

$$\exp(-\beta) > k > k^2 \text{ and } \beta < 1-k, \quad (54)$$

holds true and that

$$0 < \Delta x < 1. \tag{55}$$

As a first step, we prove that $\mathcal{L}^n$ is equivalent to $\|\vec{\rho}^n\|_\beta^2$. This part does not dependent on the boundary condition ρ_0^{n+1} and is therefore analogous to [10]. In particular, due to estimate ([10], (4.32), (4.34)) we have for all $n \geq 1$

$$(W^n)^2 \leq \Delta x^2 \sum_{j=1}^{J}(\rho_j^n)^2 e^{-\beta x_j}\sum_{j=1}^{J} e^{\beta x_j} \leq \frac{\Delta x(e^\beta - 1)}{1 - e^{-\beta \Delta x}}\|\vec{\rho}^n\|_\beta^2 \tag{56}$$

$$\leq (1+\beta)^2 \|\vec{\rho}^n\|_\beta^2 \leq (1+3\beta)\|\vec{\rho}^n\|_\beta^2. \tag{57}$$

Due to the bounds on a, we obtain the estimate ([10], (4.38)) for all $n \geq 0$

$$\|\vec{\rho}^n\|_\beta^2 \geq \mathcal{L}^n \geq \|\vec{\rho}^n\|_\beta^2\left(1 + \theta\frac{k - e^{-\beta}}{1-k}\frac{\Delta x(e^\beta - 1)}{1 - e^{-\beta\Delta x}}\right) \tag{58}$$

$$\geq (1 - \theta(1+3\beta))\|\vec{\rho}^n\|_\beta^2 \geq \frac{1-\theta}{2}\|\vec{\rho}^n\|_\beta^2, \tag{59}$$

where the last inequality is true provided that

$$0 < \beta \leq \min\{1, \beta^*, \beta^{**}, \frac{1-\theta}{6\theta}\}. \tag{60}$$

Furthermore, the discrete weighted norm is equivalent to the ℓ^2-norm as in ([10], (4.39)) for all $n \geq 0$

$$e^{-\beta}\|\vec{\rho}^n\|_{\ell^2}^2 \leq \|\vec{\rho}^n\|_\beta^2 \leq \|\vec{\rho}^n\|_{\ell^2}. \tag{61}$$

As a second step, we estimate a finite difference approximation to the temporal derivative of $\mathcal{L}$.

$$\frac{\mathcal{L}^{n+1} - \mathcal{L}^n}{\Delta t} = \frac{\Delta x}{\Delta t}\sum_{j=1}^{J}\left[\left(\rho_j^{n+1}\right)^2 - \left(\rho_j^n\right)^2\right]e^{-\beta x_j} \tag{62}$$

$$+ \frac{a(\Delta x)^2}{\Delta t}\left[\left(\sum_{j=1}^{J}\rho_j^{n+1}\right)^2 - \left(\sum_{j=1}^{J}\rho_j^n\right)^2\right]. \tag{63}$$

Precisely, as in [10], we use the discrete scheme (48), the CFL condition (41) that ensures $0 < r^j \leq 1$ and the convexity $z \to z^2$ to estimate for all $i = 1, \ldots, J$ and $n \geq 0$

$$(\rho_i^{n+1})^2 = [(1-r^n)\rho_i^n + r^n\rho_{i-1}^n]^2 \leq (1-r^n)(\rho_i^n)^2 + r^n(\rho_j^n)^2. \tag{64}$$

Then, we obtain the discrete counterpart to the integration by parts formula

$$\frac{\mathcal{L}^{n+1} - \mathcal{L}^n}{\Delta t} \leq \lambda_W^n \left(\sum_{j=1}^{J} (\rho_{j-1}^n)^2 e^{-\beta x_{j-1}} e^{-\beta \Delta x} - \sum_{j=1}^{J} (\rho_j^n)^2 e^{-\beta x_j} \right) \tag{65}$$

$$+ \frac{a(\Delta x)^2}{\Delta t} \left(\left(\sum_{j=1}^{J} \rho_j^n - r^n \rho_J^n + r^n \rho_0^n \right)^2 - \left(\sum_{j=1}^{J} \rho_j^n \right)^2 \right) \tag{66}$$

$$= \lambda_W^n e^{-\beta \Delta x} \left(\frac{1}{\Delta x} \|\vec{\rho}^n\|_\beta^2 - e^{-\beta} (\rho_J^n)^2 + (\rho_0^n)^2 \right) - \frac{\lambda_W^n}{\Delta x} \|\vec{\rho}^n\|_\beta^2 \tag{67}$$

$$+ \frac{a}{\Delta t} \left(\left(W^n - r^n \Delta x \rho_J^n + r^n \Delta x \rho_0^n \right)^2 - (W^n)^2 \right) \tag{68}$$

$$= \frac{e^{-\beta \Delta x} - 1}{\Delta x} \lambda_W^n \|\vec{\rho}\|_\beta^2 + A_1^n. \tag{69}$$

Here, the last line is as in ([10], (4.29)) except that the boundary term ρ_0^n that is part of A_1^n includes now the disturbance d^n. We split the boundary condition at $x = 0$ as

$$\rho_0^n = \bar{\rho}_0^n + k d^n, \quad \bar{\rho}_0^n := k \rho_J^n + (1-k) \theta W^n \tag{70}$$

and obtain

$$A_1^n = \lambda_W^n e^{-\beta \Delta x} \left((\bar{\rho}_0^n + k d^n)^2 - e^{-\beta} (\rho_J^n)^2 \right) \tag{71}$$

$$+ a \lambda_W^n \left(r^n \Delta x \left(\bar{\rho}_0^n + k d^n - \rho_J^n \right)^2 + 2 \left(\bar{\rho}_0^n + k d^n - \rho_J^n \right) W^n \right). \tag{72}$$

As in the continuous case, we estimate

$$(\bar{\rho}_0^n + k d^n)^2 \leq (1 + \beta^2)(\bar{\rho}_0^n)^2 + (1 + \frac{1}{\beta^2})(k d^n)^2 \tag{73}$$

and similarly for the term $2 k d^n W^n$ and $\left(\bar{\rho}_0^n + k d^n - \rho_J^n \right)^2$, respectively. Hence, we obtain

$$A_1^n \leq A_2^n + A_3^n + A_4^n,$$

$$A_2^n := \lambda_W^n e^{-\beta \Delta x} \left((\bar{\rho}_0^n)^2 - e^{-\beta} (\rho_J^n)^2 \right) + a \lambda_W^n \left(r^n \Delta x \left(\bar{\rho}_0^n - \rho_J^n \right)^2 + 2 \left(\bar{\rho}_0^n - \rho_J^n \right) W^n \right),$$

$$A_3^n := \beta^2 \lambda_W^n e^{-\beta \Delta x} (\bar{\rho}_0^n)^2 + \beta^2 \|a\| \lambda_W^n r^n \Delta x \left(\bar{\rho}_0^n - \rho_J^n \right)^2 + \beta^2 \lambda_W^n (W^n)^2,$$

$$A_4^n := \lambda_W^n e^{-\beta \Delta x} (1 + \frac{1}{\beta^2})(k d^n)^2 + \|a\| \lambda_W^n r^n \Delta x (1 + \frac{1}{\beta^2})(k d^n)^2 + \frac{1}{\beta^2} \|a\| \lambda_W^n (k d^n)^2.$$

Next, we estimate A_3^n and A_4^n. Here, we use that a defined by (51), λ_W^n, are bounded by

$$\|a\| \leq \frac{\beta}{e^\beta - 1} \leq 1, \quad \lambda_W^n \leq \frac{A}{B}, \quad \text{and } r^n \leq 1,$$

respectively, and that r^n, Δx and θ are all bounded by one. Additionally, we have a bound on $(W^n)^2$ due to (56) and $\beta \leq 1$ by (60) such that

$$(\bar{\rho}_0^n)^2 \leq 2 \left(\rho_J^n \right)^2 + 2(W^n)^2 \leq (2 + 2(1+3)) \|\vec{\rho}^n\|_\beta^2, \text{ and } \left(\bar{\rho}_0^n - \rho_J^n \right)^2 \leq 22 \|\vec{\rho}^n\|_\beta^2.$$

Hence, there exists a constant $C > 0$ such that A_3^n and A_4^n are estimated by

$$A_3^n \leq C \beta^2 \lambda_W^n \|\vec{\rho}^n\|_\beta^2 \text{ and } A_4^n \leq (1 + \frac{3}{\beta^2}) \lambda_W^n (d^n)^2. \tag{74}$$

A crucial estimate is now performed on A_2^n. Due to the previous estimates as well as due to Equation (70) we have that A_2^n coincides with ([10], A_2) and hence we may use the same estimates ([10], (4.31), (4.34)) to obtain

$$A_2^n \leq \lambda_W^n \theta^2 (W^n)^2 \Big((k - e^{-\beta})(2 - e^{\beta \Delta x}) + e^{-\beta \Delta x}(1-k) \Big)$$

$$\leq \lambda_W^n \theta^2 (1 + 3\beta) \|\vec{\rho}^n\|_\beta^2 \Big((k - e^{-\beta})(2 - e^{\beta \Delta x}) + e^{-\beta \Delta x}(1-k) \Big)$$

$$\leq \lambda_W^n \theta^2 (1 + 3\beta) \|\vec{\rho}^n\|_\beta^2 \, \beta.$$

The previous estimates allow to estimate the discrete temporal derivative of $\mathcal{L}$ in Equation (65) for $n \geq 0$:

$$\frac{\mathcal{L}^{n+1} - \mathcal{L}^n}{\Delta t} \leq \frac{e^{-\beta \Delta x} - 1}{\Delta x} \lambda_W^n \|\vec{\rho}\|_\beta^2 + A_2^n + A_3^n + A_4^n$$

$$\leq \left((-\beta + \frac{\Delta x}{2}\beta^2) + \theta^2(\beta + 3\beta^2) + C\beta^2 \right) \lambda_W^n \|\vec{\rho}\|_\beta^2 + (1 + \frac{3}{\beta^2})\lambda_W^n (d^n)^2,$$

$$\leq -\beta \left(1 - \frac{\beta}{2} - \theta^2 (1 + 3\beta) - C\beta \right) \lambda_W^n \|\vec{\rho}\|_\beta^2 + (1 + \frac{3}{\beta^2})\lambda_W^n (d^n)^2,$$

$$\leq -\beta \frac{1 - \theta^2}{2} \lambda_W^n \|\vec{\rho}^n\|_\beta^2 + (1 + \frac{3}{\beta^2})\lambda_W^n (d^n)^2.$$

The last inequality holds true provided that $0 < \beta$ is sufficiently small such that (60) and

$$\beta \leq \frac{1 - \theta^2}{7 + 2C}. \tag{75}$$

hold true.

Finally, it remains to show that λ_W^n is bounded from below by a strictly positive number. This is equivalent to show that W^n is bounded from above and similar to the continuous analysis. Note that due to $\|\vec{\rho}^n\|_\beta^2 \geq \mathcal{L}^n$ and therefore

$$\frac{\mathcal{L}^{n+1} - \mathcal{L}^n}{\Delta t} \leq -b_1 \lambda_W^n \mathcal{L}^n + b_2 (d^n)^2, \tag{76}$$

$$b_1 := b_1(\beta) = \beta \frac{1 - \theta^2}{2}, \quad b_2 := b_2(\beta) := 1 + \frac{3}{\beta^2}. \tag{77}$$

Solving recursively (76), we obtain with $\prod_{r=n+1}^n (\cdot) = 1$

$$\mathcal{L}^{n+1} \leq \prod_{m=0}^n (1 - \Delta t b_1 \lambda_W^m) \mathcal{L}^0 + b_2 \Delta t \sum_{m=0}^n \lambda_W^m (d^m)^2 \prod_{r=m+1}^n (1 - b_1 \Delta t \lambda_W^r) \tag{78}$$

$$\leq \exp\left(-b_1 \Delta t \sum_{m=0}^n \lambda_W^m \right) \mathcal{L}^0 + \max_{0 \leq s \leq n} (d^s)^2 b_2 \Delta t \sum_{m=0}^n \lambda_W^m \prod_{r=m+1}^n (1 - b_1 \Delta t \lambda_W^r). \tag{79}$$

The following equalities show that the last term of the previous sum can be bounded independent of λ_W^n:

$$-\frac{1}{b_1 \Delta t} \sum_{m=0}^{n} -b_1 \Delta t \lambda_W^m \prod_{r=m+1}^{n} (1 - b_1 \Delta t \lambda_W^r) \tag{80}$$

$$= -\frac{1}{b_1 \Delta t} \sum_{m=0}^{n} (1 - b_1 \Delta t \lambda_W^m - 1) \prod_{r=m+1}^{n} (1 - b_1 \Delta t \lambda_W^r) \tag{81}$$

$$= -\frac{1}{b_1 \Delta t} \sum_{m=0}^{n} \left(\prod_{r=m}^{n}(1 - b_1 \Delta t \lambda_W^r) - \prod_{r=m+1}^{n} (1 - b_1 \Delta t \lambda_W^r) \right) \tag{82}$$

$$= -\frac{1}{b_1 \Delta t} \prod_{r=0}^{n}(1 - b_1 \Delta t \lambda_W^r) - 1 \tag{83}$$

$$= \frac{1}{b_1 \Delta t} \left(1 - \prod_{r=0}^{n}(1 - b_1 \Delta t \lambda_W^r) \right). \tag{84}$$

Note that since $b_1 < 1$ and Δt fulfills the CFL condition (41) we have that for all $n \geq 0$,

$$b_1 \Delta t \lambda_W^n \leq b_1 \Delta x \leq 1$$

and therefore $1 - b_1 \Delta t \lambda_W^r$ is non–negative. In addition, by definition $-\rho^* \leq W^n$ and due to (59) and (60), we have

$$(W^n)^2 \leq 4\|\bar{\rho}^n\|_\beta^2 \leq \frac{8}{1-\theta} \mathcal{L}^n. \tag{85}$$

Combing the previous estimate, (79) and (84), we obtain

$$\frac{1-\theta}{8}(W^n)^2 \leq \mathcal{L}^n \leq \exp\left(-b_1 \Delta t \sum_{m=0}^{n} \lambda_W^m \right) \mathcal{L}^0 + \max_{0 \leq s \leq n} (d^s)^2 \frac{b_2}{b_1} \left(1 - \prod_{r=0}^{n}(1 - b_1 \Delta t \lambda_W^r)\right)$$

$$\leq \mathcal{L}^0 + \max_{0 \leq s \leq n} (d^s)^2 \frac{b_2}{b_1} \leq \|\bar{\rho}^0\|_\beta^2 + \max_{0 \leq s \leq n} (d^s)^2 \frac{2(\beta^2+3)}{\beta^3(1-\theta^2)}.$$

Since the norm of $\|\rho^0\|_{l^2}$ is bounded according to assumption (49), this shows that W^n is bounded from above by constant $c = c(R, \theta, \beta, \|d\|_{\ell^\infty})$. This implies that there exists a constant $0 > \sigma_2 = \sigma_2(R, \rho^*, \theta, \beta, \|d\|_{\ell^\infty})$ such that

$$\sigma_2 \leq \lambda_W^n \leq \frac{A}{B}, \ \forall n \geq 0. \tag{86}$$

Note that the norm $\|d\|_{\ell^\infty}$ can be bounded by D by assumption.

In the last step we now use the bound on λ_W^n to obtain the exponential decay of $\mathcal{L}^n$. Using (86) in estimate (76), we obtain for all $n \geq 0$

$$\frac{\mathcal{L}^{n+1} - \mathcal{L}^n}{\Delta t} \leq -\beta \frac{1-\theta^2}{2} \sigma_2 \mathcal{L}^n + (1 + \frac{3}{\beta^2}) \frac{A}{B}(d^n)^2 = -\eta \mathcal{L}^n + \nu(d^n)^2, \tag{87}$$

and $\eta := \beta \frac{1-\theta^2}{2} \sigma_2 > 0$ and $\nu := (1 + \frac{3}{\beta^2}) \frac{A}{B}$. This concludes the proof in the discrete case. □

4. Numerical Simulations

In this section, we illustrate the theoretical results in Sections 2 and 3 by providing numerical computations of ISS of a scalar conservation law with nonlocal velocity and boundary measurement error. We apply the discretization introduced in the previous section and we chose $A = B = 1$ which leads to the velocity function

$$\lambda(W(t)) = \frac{1}{1+W(t)}, \quad \text{with} \quad W(t) = \int_0^1 \rho(t,x)dx. \tag{88}$$

As measurement error, we consider

$$d(t) = 2.4 \times 10^{-3} \sin(t), \quad t \in (0, \infty). \tag{89}$$

4.1. Example 1

In this example, we consider the equilibrium solution $\rho^* = 0$ and an initial condition $\rho_0(x) = 1 + \sin(2\pi x)$ for $x \in [0,1]$. In the figures following, we show the decay of the discrete L^2-error $\|\bar{\rho}^n - \rho^*\|_{\ell^2}$ of the system Equation (8) for two given CFL conditions 0.5 and 0.9 in Table 1, respectively. Here, CFL=$a \leq 1$ is a stronger condition than (41) and it implies that Δt is such that

$$\lambda_W^n \frac{\Delta t}{\Delta x} \leq a < 1, \, n \geq 0. \tag{90}$$

A value CFL≤ 1 improves the stability of the scheme at the expense of additional artificial diffusion of the scheme. Due to the artificial diffusion and the disturbance we observe only approximately the excepted first-order convergence with respect to Δx of the Upwind scheme. In Figure 1, the convergence of the solution of the system in Equation (8) to the equilibrium for different values of k is shown. As expected we observe that as k increases the rate of decay of the Lyapunov function decreases. Furthermore, we observe that below the mesh accuracy of $\Delta x = 10^{-3}$ no further decay is observed.

Table 1. Comparison of $\|\bar{\rho}^n - \rho^*\|_{\ell^2}^2$ for different number of grid points J with $\rho^* = 0$, $k = 0.3$ and $T = 10$.

	(a) CFL = 0.5.	
J	$\|\bar{\rho}^n - \rho^*\|_{\ell^2}$	order
100	1.9171 e-05	–
200	1.1899 e-05	0.6881 e+00
400	6.9631 e-06	0.7730 e+00
800	3.7638 e-06	0.8875 e+00
1600	1.5902 e-06	1.2430 e+00
	(b) CFL = 0.9.	
J	$\|\bar{\rho}^n - \rho^*\|_{\ell^2}$	order
100	1.3831 e-05	–
200	8.1304 e-06	0.7665 e+00
400	4.8604 e-06	0.7423 e+00
800	2.8262 e-06	0.7822 e+00
1600	1.1624 e-06	1.2818 e+00

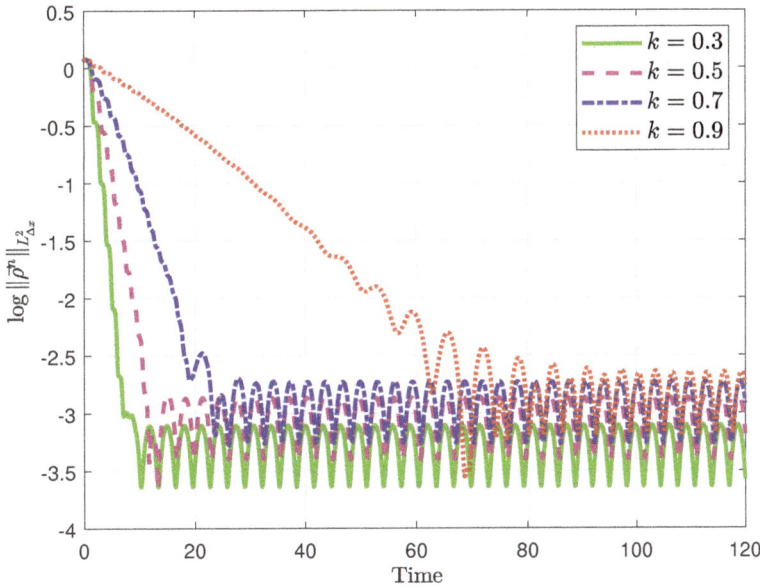

Figure 1. Comparison of log-scale of $\|\vec{\rho}^n - \rho^*\vec{e}\|_{L^2_{\Delta x}}$ with Courant–Friedrichs–Lewy condition (CFL) = 0.75 and $\rho^* = 0$.

4.2. Example 2

We repeat the previous experiment for a non-zero steady state, i.e., we choose $\rho^* = 1$ and as initial condition $\rho_0(x) = 2 + 2\sin(2\pi x)$ $x \in [0,1]$. We show similar results as above for the system in Equation (8) which are presented in Table 2 and Figure 2.

Table 2. Comparison of $\|\vec{\rho}^n - \rho^*\|_{\ell^2}$ of the solution for number of grids J with $\rho^* = 1$, $k = 0.3$ and $T = 20$.

(a) CFL = 0.5.		
J	$\|\vec{\rho}^n - \rho^*\|_{\ell^2}$	order
100	3.0916 e-04	–
200	1.5261 e-04	1.0185 e+00
400	7.2438 e-05	1.0750 e+00
800	3.1425 e-05	1.2048 e+00
1600	1.0567 e-05	1.5723 e+00
(b) CFL = 0.9.		
J	$\|\vec{\rho}^n - \rho^*\|_{\ell^2}$	order
100	2.8645 e-04	–
200	1.4299 e-04	1.0024 e+00
400	6.9982 e-05	1.0309 e+00
800	3.0215 e-05	1.2117 e+00
1600	1.0128 e-05	1.5769 e+00

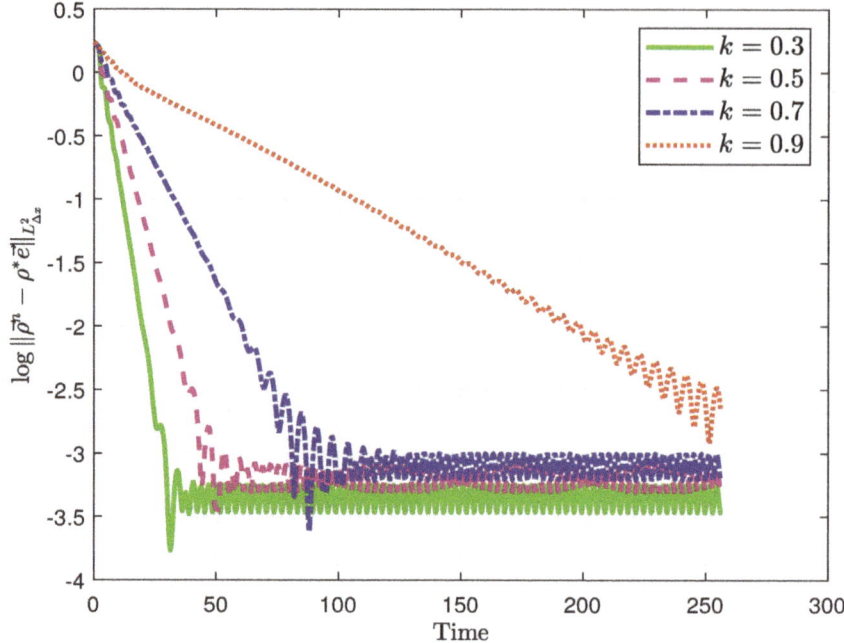

Figure 2. Comparison of log-scale of $\|\vec{\rho}^n - \rho^* \vec{e}\|_{\ell^2}$ with CFL = 0.75 and $\rho^* = 1$.

5. Conclusions and Outlook

This paper considered input-to-state stability (ISS) for a scalar conservation law with nonlocal velocity and boundary measurement error. An ISS-Lyapunov function is employed to investigate conditions for ISS of an equilibrium for the scalar conservation law with nonlocal velocity and measurement error. Numerical study of a decay of ISS-Lyapunov function is analyzed. Finally, numerical simulations illustrate the theoretical results.

Possible extensions might be to consider also ISS with respect to the L^2-norm in time in the continuous and discrete case.

A drawback of Theorem 1 is the fact that the system might not have a solution a priori. As stated in Remark 1, it might be possible to extend results [40–42] to obtain a continuous in time and L^2-space solution for the presented problem. This is subject of future work.

Author Contributions: S.G., M.H. and G.W. contributed equally to the derivation, formal analysis, writing of draft and revision as well editing. M.H. and S.G. acquired funding for this project through the German Research Foundation (DFG). All authors have read and agreed to the published version of the manuscript.

Funding: This research was funded by DFG under grant number HE5386/18-1, HE5386/19-1 and GO1920/10-1.

Institutional Review Board Statement: Not applicable.

Informed Consent Statement: Not applicable.

Conflicts of Interest: The authors declare no conflict of interest.

References

1. Armbruster, D.; Degond, P.; Ringhofer, C. A model for the dynamics of large queuing networks and supply chains. *SIAM J. Appl. Math.* **2006**, *66*, 896–920. [CrossRef]
2. Herty, M.; Klar, A.; Piccoli, B. Existence of solutions for supply chain models based on partial differential equations. *SIAM J. Math. Anal.* **2007**, *39*, 160–173. [CrossRef]

3. He, F.; Armbruster, D.; Herty, M.; Dong, M. Feedback control for priority rules in re-entrant semiconductor manufacturing. *Appl. Math. Model.* **2015**, *39*, 4655–4664. [CrossRef]
4. D'Apice, C.; Göttlich, S.; Herty, M.; Piccoli, B. *Modeling, Simulation, and Optimization of Supply Chains*; Society for Industrial and Applied Mathematics (SIAM): Philadelphia, PA, USA, 2010; p. x+206. A continuous approach.
5. Chen, H.; Harrison, J.M.; Mandelbaum, A.; Van Ackere, A.; Wein, L.M. Empirical Evaluation of a Queueing Network Model for Semiconductor Wafer Fabrication. *Oper. Res.* **1988**, *36*, 202–215. [CrossRef]
6. Helbing, D. Traffic and related self-driven many-particle systems. *Rev. Mod. Phys.* **2001**, *73*, 1067–1141. [CrossRef]
7. Coron, J.M.; Wang, Z. Output Feedback Stabilization for a Scalar Conservation Law with a Nonlocal Velocity. *SIAM J. Math. Anal.* **2013**, *45*, 2646–2665. [CrossRef]
8. Coron, J.M.; Kawski, M.; Wang, Z. Analysis of a conservation law modeling a highly re-entrant manufacturing system. *Discret. Contin. Dyn. Syst. Ser. B* **2010**, *14*, 1337–1359. [CrossRef]
9. Shang, P.; Wang, Z. Analysis and control of a scalar conservation law modeling a highly re-entrant manufacturing system. *J. Differ. Equ.* **2011**, *250*, 949–982. [CrossRef]
10. Chen, W.; Liu, C.; Wang, Z. Global Feedback Stabilization for a Class of Nonlocal Transport Equations: The Continuous and Discrete Case. *SIAM J. Control Optim.* **2017**, *55*, 760–784. [CrossRef]
11. Tanwani, A.; Prieur, C.; Tarbouriech, S. Stabilization of linear hyperbolic systems of balance laws with measurement errors. In *Control Subject to Computational and Communication Constraints*; Springer: Cham, Switzerland, 2018; Volume 475, pp. 357–374. Lect. Notes Control Inf. Sci.
12. Dos Santos, V.; Bastin, G.; Coron, J.M.; d'Andréa Novel, B. Boundary control with integral action for hyperbolic systems of conservation laws: stability and experiments. *Automat. J. IFAC* **2008**, *44*, 1310–1318. [CrossRef]
13. Bastin, G.; Coron, J.M.; d'Andréa Novel, B. On Lyapunov stability of linearised Saint-Venant equations for a sloping channel. *Netw. Heterog. Media* **2009**, *4*, 177–187. [CrossRef]
14. Bastin, G.; Coron, J.M. Exponential stability of semi-linear one-dimensional balance laws. In *Feedback Stabilization of Controlled Dynamical Systems*; Springer: Cham, Switzerland, 2017; Volume 473, pp. 265–278. Lect. Notes Control Inf. Sci.
15. Bastin, G.; Coron, J.M. *Stability and Boundary Stabilization of 1-D Hyperbolic Systems*; Springer: Berlin/Heidelberg, Germany, 2016. [CrossRef]
16. Coron, J.M.; d'Andréa Novel, B.; Bastin, G. A strict Lyapunov function for boundary control of hyperbolic systems of conservation laws. *IEEE Trans. Automat. Control* **2007**, *52*, 2–11. [CrossRef]
17. Diagne, A.; Bastin, G.; Coron, J.M. Lyapunov exponential stability of 1-D linear hyperbolic systems of balance laws. *Autom. J. IFAC* **2012**, *48*, 109–114. [CrossRef]
18. Gugat, M.; Perrollaz, V.; Rosier, L. Boundary stabilization of quasilinear hyperbolic systems of balance laws: Exponential decay for small source terms. *J. Evol. Equ.* **2018**, *18*, 1471–1500. [CrossRef]
19. Cen, L.H.; Xi, Y.G. Stability of boundary feedback control based on weighted Lyapunov function in networks of open channels. *Acta Automat. Sin.* **2009**, *35*, 97–102. [CrossRef]
20. Prieur, C.; Mazenc, F. ISS-Lyapunov functions for time-varying hyperbolic systems of balance laws. *Math. Control Signals Syst.* **2012**, *24*, 111–134. [CrossRef]
21. Coron, J.M.; Bastin, G.; d'Andréa Novel, B. Dissipative boundary conditions for one-dimensional nonlinear hyperbolic systems. *SIAM J. Control Optim.* **2008**, *47*, 1460–1498. [CrossRef]
22. Zhang, L.; Prieur, C. Necessary and Sufficient Conditions on the Exponential Stability of Positive Hyperbolic Systems. *IEEE Trans. Automat. Control* **2017**, *62*, 3610–3617. [CrossRef]
23. Lamare, P.; Auriol, J.; Di Meglio, F.; Aarsnes, U.J.F. Robust output regulation of 2×2 hyperbolic systems: Control law and Input-to-State Stability. In Proceedings of the 2018 Annual American Control Conference (ACC), Milwaukee, WI, USA, 27–29 June 2018; pp. 1732–1739. [CrossRef]
24. Banda, M.K.; Weldegiyorgis, G.Y. Numerical boundary feedback stabilisation of non-uniform hyperbolic systems of balance laws. *Int. J. Control* **2020**, *93*, 1428–1441. [CrossRef]
25. Banda, M.K.; Herty, M. Numerical discretization of stabilization problems with boundary controls for systems of hyperbolic conservation laws. *Math. Control Relat. Fields* **2013**, *3*, 121–142. [CrossRef]
26. Göttlich, S.; Schillen, P. Numerical discretization of boundary control problems for systems of balance laws: Feedback stabilization. *Eur. J. Control* **2017**, *35*, 11–18. [CrossRef]
27. Gerster, S.; Herty, M. Discretized feedback control for systems of linearized hyperbolic balance laws. *Math. Control Relat. Fields* **2019**, *9*, 517–539. [CrossRef]
28. Göttlich, S.; Schillen, P. Numerical feedback stabilization with applications to networks. *Discret. Dyn. Nat. Soc.* **2017**, 6896153. [CrossRef]
29. Weldegiyorgis, G.Y.; Banda, M.K. Input-to-State Stability of Non-uniform Linear Hyperbolic Systems of Balance Laws via Boundary Feedback Control. *Appl. Math. Optim.* **2020**. [CrossRef]
30. Bastin, G.; Coron, J.M.; Hayat, A. Input-to-State Stability in sup norms for hyperbolic systems with boundary disturbances. *arXiv* **2020**, arXiv:2004.12026.
31. Mironchenko, A.; Prieur, C. Input-to-State Stability of Infinite-Dimensional Systems: Recent Results and Open Questions. *SIAM Rev.* **2020**, *62*, 529–614. [CrossRef]

32. Dashkovskiy, S.; Mironchenko, A. Input-to-state stability of infinite-dimensional control systems. *Math. Control Signals Syst.* **2013**, *25*, 1–35. [CrossRef]
33. Karafyllis, I.; Krstic, M. *Input-to-State Stability for PDEs*; Communications and Control Engineering Series; Springer: Cham, Switzerland, 2019; p. xvi+287. [CrossRef]
34. Karafyllis, I.; Krstic, M. ISS with respect to boundary disturbances for 1-D parabolic PDEs. *IEEE Trans. Automat. Control* **2016**, *61*, 3712–3724. [CrossRef]
35. Zheng, J.; Zhu, G. Input-to-state stability with respect to boundary disturbances for a class of semi-linear parabolic equations. *Autom. J. IFAC* **2018**, *97*, 271–277. [CrossRef]
36. Zheng, J.; Zhu, G. Input-to-state stability for a class of one-dimensional nonlinear parabolic PDEs with nonlinear boundary conditions. *SIAM J. Control Optim.* **2020**, *58*, 2567–2587. [CrossRef]
37. Jacob, B.; Nabiullin, R.; Partington, J.R.; Schwenninger, F.L. Infinite-dimensional input-to-state stability and Orlicz spaces. *SIAM J. Control Optim.* **2018**, *56*, 868–889. [CrossRef]
38. Jacob, B.; Mironchenko, A.; Partington, J.R.; Wirth, F. Noncoercive Lyapunov functions for input-to-state stability of infinite-dimensional systems. *SIAM J. Control Optim.* **2020**, *58*, 2952–2978. [CrossRef]
39. Mironchenko, A.; Wirth, F. Non-coercive Lyapunov functions for infinite-dimensional systems. *J. Differ. Equ.* **2019**, *266*, 7038–7072. [CrossRef]
40. Ferrante, F.; Prieur, C. Boundary control design for conservation laws in the presence of measurement disturbances. *Math. Control Signals Syst.* **2021**. [CrossRef]
41. Dus, M.; Ferrante, F.; Prieur, C. On L^∞ stabilization of diagonal semilinear hyperbolic systems by saturated boundary control. *ESAIM Control Optim. Calc. Var.* **2020**, *26*, 23. [CrossRef]
42. Hayat, A. Global exponential stability and Input-to-State Stability of semilinear hyperbolic systems for the L^2 norm. *arXiv* **2020**, arXiv:2011.12682.

MDPI
St. Alban-Anlage 66
4052 Basel
Switzerland
Tel. +41 61 683 77 34
Fax +41 61 302 89 18
www.mdpi.com

Axioms Editorial Office
E-mail: axioms@mdpi.com
www.mdpi.com/journal/axioms

www.ingramcontent.com/pod-product-compliance
Lightning Source LLC
LaVergne TN
LVHW070429100526
838202LV00014B/1560